FARM ANIMAL GENETIC RESOU

Cover photo
A rare breed: the Hebridean, two-horned, ram lamb (Calanais McCritchie).
©Simon Tupper.

Farm Animal Genetic Resources

Edited by

G. Simm, B. Villanueva, K.D. Sinclair, S. Townsend

 NOTTINGHAM University Press

Nottingham University Press
Manor Farm, Main Street, Thrumpton
Nottingham, NG11 0AX, United Kingdom

NOTTINGHAM

First published 2004
© British Society of Animal Science

British Library Cataloguing in Publication Data
Farm Animal Genetic Resources
I. Simm, G., II. Villanueva, B., III. Sinclair, K.D., IV. Townsend, S.

ISBN 1-897676-15-8

Page layout and design by Nottingham University Press, Nottingham
Printed and bound by Hobbs the Printers, Hampshire, England

CONTENTS

Section 1: Policy issues

Section 2: Quantitative and molecular genetic basis for conservation

Section 3: Reproductive techniques to support conservation

Section 4: Conservation in action

Posters

DEVELOPMENTS IN THE CONSERVATION OF FARM ANIMAL GENETIC RESOURCES

The papers in this book are based on presentations given at a conference held in Edinburgh in November 2002, which was co-organised by the British Society of Animal Science (BSAS), the Department for Environment, Food and Rural Affairs (Defra), the Rare Breeds Survival Trust (RBST) and the Sheep Trust.

A number of factors led us to believe that such a conference was timely:

1. The growing international effort, led by The Food and Agriculture Organisation of the United Nations (FAO), to co-ordinate conservation of Farm Animal Genetic Resources. The conference coincided with the publication by Defra of the UK Country Report on Farm Animal Genetic Resources, as part of this international effort.
2. The Foot and Mouth Disease outbreak in the UK in 2001 and 2002 brought into sharp focus the real threat to rare and geographically isolated livestock breeds, and the need for greater co-ordination of conservation efforts.
3. The growth in scientific knowledge in areas relevant to conservation – especially in quantitative genetics, helping to improve the management of populations at risk, in molecular genetics, improving our capability to characterise and prioritise populations for conservation, and in reproductive biology, providing more effective techniques for conservation.

The chapters in this book have been organised into four sections: Policy issues, Quantitative and molecular genetic basis for conservation, Reproductive techniques to support conservation, and Conservation in action – a series of case studies illustrating some of the techniques covered earlier in the book.

We hope that the book will provide a useful update for those interested in conservation and sustainable utilisation of farm animal genetic resources. The chapters have been edited, but responsibility for the content remains with the authors.

G. Simm, B. Villanueva, K.D. Sinclair, S. Townsend
Editors

1

Conservation of farm animal genetic resources – a global view

R.A. Cardellino
Animal Production and Health Division, FAO, 00100 Rome, Italy

Abstract

Farm animal genetic resources face a double challenge. On the one hand the demand for animal products is increasing in developing countries. The Food and Agriculture Organisation of the United Nations (FAO) has estimated that demand for meat will double by 2030 (2000 basis) and demand for milk will more than double in this 30-year period. On the other hand, animal genetic resources are disappearing rapidly worldwide. Over the past 15 years, 300 out of 6000 breeds identified by FAO have become extinct, and 1 to 2 breeds disappear every week. FAO has been requested by its member countries to develop and implement a global strategy for the management of farm animal genetic resources. It is important to conserve local breeds because many of them utilise lower quality feed, are more resilient to climatic stress, are more resistant to local parasites and diseases, and represent a unique source of genes for improving health and performance traits of industrial breeds. It is important also to develop and utilise local breeds that are genetically adapted to their environments. Genotype x environment interactions are important especially where extreme environments are involved. Most of these production environments are harsh, with very limited natural and managerial inputs, and they are not limited to developing countries. Animals genetically adapted to these conditions will be more productive at lower costs. They will support food, agriculture and cultural diversity, and will be effective in achieving local food security objectives. In many countries local communities depend on these adapted genetic resources. Their disappearance or drastic modification, for example by crossbreeding, absorption or replacement by exotic breeds, will have tremendous impacts on these human populations. Most breeds at risk are not supported by any established conservation activity or related policy, and breed extinction rates are increasing.

Importance of livestock and of local breeds

The FAO is an intergovernmental organisation founded in 1945 to fight hunger and poverty in the world (www.fao.org). The poorest regions of the world, where malnutrition is a serious problem, are located primarily in Africa, Asia, Latin America and Eastern Europe. All of these regions depend on livestock for the nourishment of their people, especially in rural communities.

Livestock is an important component of food security and, in general, of human livelihood in most developing countries, accounting for more than 40% of overall agricultural output, and serving as a source of food (milk, meat, eggs), shelter and protection (fibre, hides), energy (animal draught, transport), fuel and fertiliser (manure), savings (cash value of animals) and cultural values. Local communities manage and utilise local breeds for their survival. The disappearance or reduction of these locally-adapted animal populations will mean the migration of rural populations to the already overcrowded urban areas, food insecurity and social disintegration of rural communities. Given the interdependence between the livestock and the crop components of production systems in low-input conditions, the erosion of local breeds will also have negative effects on the yield of local crops.

Domestic animal biological diversity at risk

By 2020 global consumption is expected go up by 120M tons of meat and 240M tons of milk as compared with 1993. On the other hand, animal genetic resources are eroding rapidly. Over the past 15 years, 300 out of 6000 breeds identified by FAO have become extinct. The Domestic Animal Diversity Information System (www.fao.org/DAD-IS) database provides the elements for an early warning system for animal genetic resources, and for the production of the World Watch List for Domestic Animal Diversity (WWL-DAD). Over 6300 breed populations of the 30 mammalian and avian species recorded by 180 countries, together with their wild relatives, newly domesticated species and related feral populations, are listed and described. Data quantity and quality increased since 1990 after FAO encouraged country reporting of animal genetic resources information.

The global database requires further development but the trend in erosion of animal genetic resources is confirmed. Overall figures available show a 10% increase in the number of recorded breeds at risk since 1995, and a 13% increase since 1993. Figure 1 shows the number of breeds at risk in 9 of the most important species used for food and agriculture.

It should be remarked that only 15 species of domestic animals account for 90% of livestock production. One can consider the case of chickens. This is one species that is undergoing rapid industrialization globally and the genetic base is eroding rapidly. Many poultry breeding companies have got rid of their old genetic stocks and much of the remaining genetic diversity in chickens is in backyard poultry and in fancy breeds. Pigs are following the same trend. The mechanization of agriculture has endangered many breeds of horses and donkeys.

Figure 1
Breeds at risk (from World Watch List for Domestic Animal Diversity at www.fao.org/ DAD-IS)

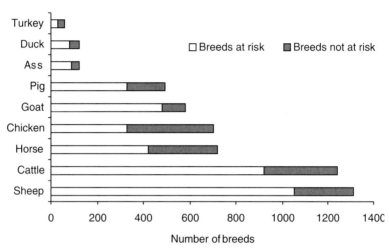

The third edition of the WWL-DAD compiled by FAO tells us that 1350 breeds of farm animals currently face extinction and that in the world, 1 to 2 breeds of farm animals are lost every week. There are several factors which place breeds at risk of loss and threaten domestic animal diversity. By far the greatest cause for genetic erosion is the growing trend to global reliance on a very limited number of modern breeds suited for the high input/output needs of industrial agriculture. This trend is of paramount concern as about 50% of the total variation at the quantitative level is between breeds, the remainder being common to all breeds. Hence moving to a few breeds would eliminate a considerable amount of variation in the species, in addition to jeopardising readily available gene combinations in other remaining unique gene resources.

In developed countries progress in advanced breeding and reproductive technology have lead to substantial increases in agricultural production in some production systems. The basis for this success was the possibility to develop and apply these technologies, and to access many diverse breeding populations with desired genes or gene combinations. This was further amplified by the possibility to access germplasm worldwide

3

and the development and easy movement of highly selected breeds. What was successful on one hand was deleterious on the other, since improvement programmes this century have only concentrated on a few breeds in each species, using high level inputs, and also upon just one or two traits with the improvement activity being carried out in comparatively benign environments. Proliferation was amplified through application of reproductive technologies, mainly through artificial insemination. Other modern biotechnologies, such as embryo transfer and cloning, once it becomes more efficient, may further aggravate the problem if adequate precautions are not taken. The result to date is that a large number of breeds and strains which were highly adapted to very specific environmental and feeding conditions are now threatened, or extinct.

For the developing world, there are several primary factors responsible for diminishing animal genetic diversity:

1. The introduction of exotic germplasm whereby exotic or other, often non-adapted, breeds have been introduced followed by rapid spread through indiscriminate crossbreeding. This has frequently arisen through wrong advice, often given in externally funded projects, and this has been exacerbated in many cases by flawed and misleading comparisons being made between the indigenous breed and the exotic germplasm. The net result has been that some indigenous breeds or landraces have been lost or displaced.

2. Changes in breeders' preferences to other breeds have occurred because of short-term socio-economic influences. These influences may arise from poor agricultural policies which promote fast solutions that are not sustainable in the long term, or from changing (possibly transient) market requirements for the products.

3. The complete ecosystem in which the breed was developed may be under threat and the decline is a symptom of some wider forces at work.

4. Natural disasters such as drought and diseases, as well as wars and other forms of political unrest and instability.

The locally adapted indigenous breeds in developing countries often have low absolute production figures while productivity itself is often remarkably high, when the production environment and the level of input are taken into consideration. Indigenous breeds produce and reproduce despite the sometimes very harsh environmental conditions, and are considered an important asset since, over time, they have developed valuable adaptive traits. This productivity in harsh environments is critically important since the vast majority of the world cannot sustain high input/output systems.

Ex-situ conservation

There is very little information on current *ex-situ* conservation in the world, but few countries are investing in it. FAO member countries have not yet considered a global approach to *ex-situ* conservation. It is widely recognised that there are major financial, technical and legal constraints for the setting up and maintenance of national and regional gene repositories. A global animal gene repository is still a remote possibility. Costs of collecting, processing and storing genetic material are beyond the available means of most developing countries. Efficient techniques for cryopreservation are not available for all farm animal species and many technical challenges still exist, with important differences in biological responses even between breeds of the same species. Legal aspects related to property rights, access, benefits and transfer of stored animal genetic material have not yet been considered by FAO member countries. The Commission on Genetic Resources for Food and Agriculture of FAO may consider in or after 2004 a treaty on animal genetic resources similar in scope to the treaty on plant genetic resources recently adopted.

The issues involved in ownership and use of biological material from a cryobank are basically legal issues and as such are subject to legislation regulating and protecting genetic resources. There are at least three levels of legal instruments that are relevant to ownership and use of genetic resources: (a) international treaties and conventions which the country or countries involved have signed; (b) national legislation covering genetic resources; (c) material transfer agreements or commercial transactions between providers and users of genetic resources, regulated by common law. All these legal instruments must be in mutual agreement in order to achieve harmonious and practical procedures for regulating and protecting genetic resources and the rights of parties involved, and prevent the proliferation of bilateral litigation. It is desirable that the most fundamental issues and basic principles for genetic resources are covered by multilateral agreements so that bilateral negotiation is greatly simplified and all countries have equal opportunity. Most international legal instruments deal with plant genetic resources.

For animal genetic resources, the Convention on Biological Diversity (CBD) and the Bonn Guidelines of the CBD apply. There is no equivalent of the International Treaty on Plant Genetic Resources for Food and Agriculture (ITPGRFA) for animal genetic resources. There is also no equivalent for animal genetic resources of the global system of gene banks that exists for plant genetic resources. Gene flow and commercialization issues for plant genetic resources as opposed to animal genetic resources have many and important differences.

The legal problems related to ownership and use of genetic material from a gene repository, in this case a cryobank for animal genetic resources, must be discussed at the international level, both in the public and in the private sector, and at the national level, also both in the private and in the public sector. With the exception of the CBD, most international agreements deal with plant genetic resources. In particular the ITPGRFA was a response to the need to reach international multilateral agreement on how to handle the global collections of germplasm under the Consultative Group on International Agricultural Research (CGIAR) system. There is no equivalent of this global genetic repository system for animal genetic resources and there is no indication that such an *ex situ* conservation global effort will be implemented in the near future. Therefore, we are dealing at the moment with national gene banks that can either be in the public sector or in the private sector.

National legislation regulates the transfer of animal genetic resources within the country and internationally between the country and another country (bilaterally). There may be barriers to export of genetic material due to protectionist measures, such as the ban on Merino exports that Australia held for several decades. There may be barriers to import of genetic material due to a variety of reasons, among them sanitary, or commercial, such as the historical ban on zebu imports to Brazil. Ownership of genetic material in a cryobank rests with the originator of the material unless otherwise specified. Use of genetic material from a cryobank must be regulated by specific Materials Transfer Agreements (MTAs) designed for that purpose. Such MTAs must be in harmony with national legislation and with international agreements (notably the CBD). They also have to contain clauses for the protection of intellectual property rights according to national legislation and to international agreements signed by the country. They should also consider the rights of the originators, as recommended by the Bonn Guidelines of the CBD.

In-situ conservation

A practical way of achieving *in-situ* conservation of animal genetic resources is through sustainable utilisation of a breed or specific animal population. This is perhaps the avenue that offers most promise at the moment, in spite of a relative poor track record of success primarily in developing countries but also in developed countries. Breeding strategies applied to animal populations have often failed due to a number of reasons, among them technically inadequate proposals, less than necessary participation from stakeholders, lack of infrastructure for implementation, poorly or wrongly defined selection objectives and goals, lack of continuity of the breeding programmes and low economic incentives for producers.

The engagement of communities in developing these resources seems essential, whether these are formal breeders' societies, organised livestock owners or people whose living depends on the breed. Indigenous knowledge and traditional good practices should not be ignored. It is important to ascertain how communities manage local animal genetic resources, how they cope with threats to these resources, and how their actions are related to environments and ecosystems. Subsidies to endangered breeds are in place in many developed countries. A way to further secure conservation of a breed is by developing niche markets for specialty products. FAO will encourage countries to implement sustainable breeding strategies for domestic animal populations, and will make efforts to provide and disseminate technical tools to cover a range of situations and production systems.

Management of small populations at risk

In the Secondary Guidelines for the Development of National Farm Animal Genetic Resources Management Plans (www.fao.org/DAD-IS) the basic principles regarding management of small populations at risk are outlined. These guidelines state the importance of the objectives for conservation of animal genetic resources. The objectives are not to conserve for the sake of conservation but for the economic, environmental, scientific, social and cultural benefits arising from the breeds. Further, the conservation activity has an important risk-reducing element in avoiding reliance on just a few breeds.

A step-by-step manual has been provided to assist countries and their stakeholders plan and act to conserve those breeds considered to have the potential for future contributions:

1. The first step is to evaluate the present situation of the breed by population censuses and surveys. These reveal the population dynamics of breeds within each agro-ecosystem and aids decisions on which breeds are at risk.

2. Information obtained in the first step is essential for the second step, which aids decisions upon which breeds should be conserved and

by which strategy. It is highly recommended that in-situ conservation should be the preferred conservation method and that cryoconservation should be used only after having exhausted all other conservation alternatives. *In-situ* conservation has the advantages that simple technologies are required to implement it, and that it allows the animals to adapt to changing environmental conditions and diseases, and the animals are readily available for expanded use. A component of the second step which needs to be addressed as part of the decision-making process, is to make a technical design for conservation strategies; addressing how many animals, how should they be maintained and (in cryoconservation) how many samples are needed. The main difference is between the *in vivo* and cryoconservation strategies which have been examined separately.

3. The third step, which must be conducted in parallel with the second step, is to construct a thorough organisation, communication and training plan for the activity. This addresses the ownership and responsibilities of stakeholders: private farmers, the state, NGOs and private companies.

4. The fourth step is to turn the plan into action.

It may be that organised conservation of a particular breed may not go beyond the second step, since it is unlikely that funding and economics will allow for conservation of all breeds, and choices between breeds will have to be made. It is hoped that the process described here will maximise the chances of successfully conserving any breed and help in making the right choices. The logical step-by-step nature of these guidelines should also help to write well-founded and coherent project proposals that maximise the chances of attracting funding bodies.

The Convention on Biological Diversity

International awareness of the essential role of animal genetic resources in food and agriculture is gradually increasing. Agricultural biological diversity has been discussed by the Conferences of the Parties (COP) to the Convention on Biological Diversity (CBD). At the second Conference in 1995, by decision II/15 parties recognised the special nature of agricultural biodiversity, its distinctive features and problems, needing distinctive solutions. The major discussion on agrobiodiversity took place at the COP in Buenos Aires in 1996, where by decision iii/11 parties decided to develop a programme of work on agricultural biodiversity. Moreover, parties strongly endorsed the further development of the global strategy for the management of farm animal genetic resources by FAO and supported development of inventories to better understand the status of farm animal genetic resources and measures necessary for their conservation and sustainable utilisation. The fifth COP to the CBD in 2000, by decision V/5 endorsed a multi-year work programme on agricultural biodiversity which includes four programme elements: I. Assessments; II. Adaptive management; III. Capacity building; IV. Mainstreaming.

The contribution of animal genetic resources to food security, poverty alleviation and rural development has been recognised by the World Food Summit 1996. While developing Agenda 21, the Commission on Sustainable Development of the United Nations strongly emphasised the importance of promoting sustainable agriculture and rural development (SARD) and underlined the essential need to ensure the conservation and sustainable use of genetic resources in achieving sustainable agriculture. At its eighth session in 2000, the Commission on Sustainable Development adopted a decision on SARD that urges governments to implement and actively contribute to the further development of the global strategy. Sustainable agriculture has also been an important agenda item at the World summit on sustainable development (Rio + 10) in 2002.

The global strategy for the management of farm animal genetic resources

FAO officials in charge of programmes of the organisation do not enjoy academic freedom and therefore guidance comes from the governing bodies where FAO member countries are represented. FAO Conference meets every two years and includes all FAO member countries and many other intergovernmental and international non-governmental organisations as observers. Council meets annually and is composed of some member countries elected by the Conference.

In the specific case of genetic resources, the intergovernmental forum is the Commission on Genetic Resources for Food and Agriculture (CGRFA) where all countries are represented. This commission deals with all aspects of genetic resources (plant, animal, forest and fisheries) and holds regular sessions every two years. Technical guidance is provided by a working group of this commission, the Intergovernmental Technical Working Group on Animal Genetic Resources (ITWG-AnGR).

FAO has been requested by its member countries, through the CGRFA and its ITWG-AnGR to develop and implement a global strategy for the management of farm animal genetic resources. This is intended to serve as a strategic framework to guide international efforts in the animal genetic resources sector. The global strategy is necessary to enhance awareness of the multiple roles and values of animal genetic resources. It provides guidelines for establishing national, regional and global policies, strategies and actions, and can serve to facilitate and coordinate the activities of many independent organisations that have an interest in animal genetic resources within the broader context of SARD.

The most important role of the global strategy is to assist countries in developing their capacity to manage their animal genetic resources for food and agriculture. In this respect they need to plan, design and implement sound livestock production systems that are sustainable and cost-effective over time. The global strategy is also necessary to promote the establishment of cost-effective approaches to conserving animal genetic resources which are not at present of interest to farmers. The key component of the Global Strategy is the Country-based *Planning and Implementation Infrastructure*, which includes five structural elements:

- The Global Focal Point at FAO Headquarters leads the planning, development and implementation of the overall strategy; develops and maintains the information and communication systems; oversees preparation of guidelines; co-ordinates the activity amongst the regions; prepares reports and meeting documents; facilitates policy discussions; identifies training, education and technology transfer needs; develops programme and project proposals; and mobilises donor resources.

- Regional Focal Points facilitate regional communications; provide technical assistance and leadership; co-ordinate training, research and planning activities amongst countries; initiate development of regional policies; assist in identifying project priorities and proposals, and interact with government agencies, donors, research institutions and non-governmental organisations.

- National Focal Points lead, facilitate and co-ordinate country activities, identify capacity-building needs; develop project proposals; assist with the development and implementation of country policy, and interface with the range of country stakeholders, including the country focus for biological diversity, and with the Regional Focal Point and the Global Focal Point.

- The Donor and Stakeholder Involvement Mechanism is meant to mobilise the range of stakeholders, providing broad-based support for the Global Strategy. The Global Focal Point seeks to ensure stakeholder involvement in all major aspects of the Global Strategy, using a variety of communication means. The Stakeholder mechanism provides additional opportunity for non-governmental contribution.

- The <u>Domestic Animal Diversity Information System (DAD-IS)</u> (http://www.fao.org/dad-is/) functions as the Clearing House Mechanism for the Global Strategy. It is a widely available and easily accessible global data and information system. Development and use of such global facility makes it possible to effectively share data and information among countries. DAD-IS is an advanced communication and information tool that allows a rapid and cost-effective distribution of guidelines, reports and meeting documents; and provides a mechanism to exchange views and address specific information requests, by linking breeders, scientists and policy makers. A key feature is the DAD-IS breeds database, which provides the basis of the Early Warning System for Animal Genetic Resources, and makes it possible to produce the World Watch List for Domestic Animal Diversity, the third edition which was released in 2001.

First report on the state of the world's animal genetic resources

As part of the global strategy for the management of farm animal genetic resources, FAO has invited 188 countries to participate in the First report on the state of the world's animal genetic resources, to be completed before 2006. To date 140 countries have agreed to submit country reports. Among items of the reports are the actual measures, if any, carried out for conservation of animal genetic resources. In Animal Genetic Resources Information Bulletin (FAO) number 30, available at www.fao.org/DAD-IS, the Guidelines for preparation of country reports can be found.

These Guidelines are for use in assisting the development of Country Reports as strategic policy documentation covering the state of animal genetic resources, of the art and capacity to manage these resources, and of country needs and priorities. The Guidelines serve to help support conduct of the country-driven State of the World Process for Animal Genetic Resources, preparation for which is being coordinated globally by FAO. The Country Reports will serve as the formative documentation in this Process and the involvement of all stakeholders in the development of these Reports is strongly encouraged.

Preparation of the first country-driven Report on the State of the World's Animal Genetic Resources has been initiated as an essential element of the Global Strategy. The first critical step in the process for developing the first Report on the State of the World's Animal Genetic Resources will be the preparation of Country Reports. The objective of the country and global assessments is to provide a comprehensive analysis of the status and trends of the world's animal biodiversity and of their underlying causes, as well as of local knowledge of its management.

The task is to go beyond description of the resources: to analyse and report on the state of these resources and capacities to manage them, to draw lessons from past experiences and identify problems and priorities. It also provides an important opportunity to look ahead and identify potential and likely needs, demands, trends, and national capacity building requirements in all aspects of the management of AnGR. While it is essential to understand the state of the resources and management capacities, Country Reports must also assess the underlying policies that affect both the resources and the existing capacity to manage them. The strategic priority actions report and the global report will be based on: Country Reports, thematic studies and reports from international non-governmental organisations.

The Country Report for the SoW-AnGR will be a strategic policy document covering the three strategic questions: Where are we? Where do we need to be? How do we get to where we need to be? The Country Reports will be used in planning and implementing priority country action. In addition, the Country Report will serve as documentation for development of the regional and global Strategic Priority Actions Reports and, subsequently, the first global Report itself on the SoW-AnGR. Country Reports should follow the agreed Guidelines for the Development of Country Reports and should consider the state of all important farm animal genetic resources in the country, the state of the art and the national capacity to manage these resources, as well as country priorities and needs for action.

Country Reports provide an assessment in three major areas:

- the **State of Diversity**: an assessment of the state of conservation, erosion and utilisation of farm animal agricultural biodiversity, and an analysis of the underlying processes;

- the **State of Country Capacity** to manage AnGR including existing AnGR policies, management plans, institutional infrastructures, human resources and equipment; and

- the **State of the Art** and the available methodologies and technologies to assist farmers, breeders, scientists to better understand, use, develop, and conserve AnGR, and thereby contribute to global food security and rural development.

In each country, the preparation of the Country Report will also facilitate the development of a comprehensive national databank for use in planning and implementing follow-up action, and in training and further capacity building.

Countries were asked to nominate a National Focal Point designating their National Co-ordinator. The National Co-ordinator co-ordinates the development of the country network and overall management of AnGR and is the official contact for communication with the Global Focal Point. Keeping in mind that the process involves both scientific and policy matters, the establishment of a National Consultative Committee is recommended to identify the primary areas and issues that need to be addressed in the preparation of the Country Report, frame out the report and oversee its preparation. It is essential that the National Consultative Committee has diverse representation, and also develops a broader network to ensure opportunities for the full range of stakeholders to contribute to the Country Report. International Organizations are also being invited to contribute to the SoW-AnGR preparatory process in the form of reports. The long-term aim of the SoW-AnGR process is for countries and regions to build on the analyses contained in the Country Reports to plan and implement appropriate AnGR management.

The response of countries to the invitation of the Director General of FAO, to participate in the First Report on the State of the World's Animal Genetic Resources and submit a Country Report, has been very good. During part of 2001 and 2002 FAO has trained almost 400 professionals in 178 countries, in the preparation of their Country Reports. To-date, 24 Country Reports have been officially received by FAO, and 68 draft country reports have been received by the Regional facilitators. At the moment FAO has a team of nearly 15 consultants working directly in over 14 country groupings. Many countries have undertaken the organisation of national stakeholder workshops to elaborate their animal genetic resources policies leading to the country reports.

FAO has organised to-date 14 sub-regional workshops to present draft (or a few final) country reports. This promotes regional cooperation since it is a way for countries that may be experiencing delays, to catch-up with those in a more advanced state of country report preparation, and learn from their experiences. These sessions are coordinated by the regional facilitators (FAO consultants).

FAO has implemented technical and/or financial support to 115 countries. This is the right moment to recognise the financial support to FAO from the Governments of The Netherlands and Finland, and from the Nordic Gene Bank. FAO thanks especially the collaboration of the World Association for Animal Production (WAAP) with whom it entered into an important agreement to provide technical and operational support for the State of the World reporting process, including training and country follow-up. FAO considers this cooperation a prime example of the effective collaboration between our organization and an International Non-Governmental Organization.

All cited bibliography and related publications can be found in www.fao.org/DAD-IS.

2

The conservation of animal genetic resources – a European perspective

E. Martyniuk
Department of Animal Genetics and Breeding, Warsaw Agricultural University, 02-786 Warszawa, ul. Ciszewskiego 8, Poland

Abstract

The paper provides an historical overview of conservation activities undertaken in Europe to maintain native livestock breeds, including the motivation and methods applied in conservation programmes and the contribution of various stakeholders. The current state of conservation activities is presented, based on reports provided by the National Coordinators on animal genetic resources (AnGR) during annual Workshops. These Workshops have been convened jointly by the Food and Agriculture Organisation of the United Nations (FAO) and the European Association for Animal Production since 1995, and are conducted within the framework of the FAO Global Strategy for Management of Farm Animal Genetic Resources. Analysis includes policy and legislation development, state and mode of financial support, conservation approaches and public awareness and education initiatives. The paper describes the establishment of the European Regional Focal Point for AnGR, its terms of reference, and ongoing and future activities. Questions regarding a vision of future needs and developments in AnGR are raised in this paper, both from a technical and policy context.

Where we came from - a historical perspective

National activities

The European Region has a long history of using, developing and exchanging animal genetic resources (AnGR). The region has benefited significantly from these resources, and over time, has developed expertise in their management. Since the beginning of the 20th century, the major challenge for the European livestock sector has been to increase production to supply a growing human population with animal products. Production increases were based on the enhancement of individual performance, and have been achieved through both the development of new technologies in animal husbandry and rapid progress in genetic improvement programmes. These latter efforts gradually resulted in the genetic uniformity of commercially used breeds. Possible negative consequences of the application of new reproduction methods were foreseen as early as 1954, when Nordic cattle breeders expressed their concerns about the possible loss of genetic variation due to artificial insemination (Hanson, 1954, cited by Maijala et al., 1992b).

The erosion of animal genetic diversity was the very first motivation, in the 1960s, to initiate conservation activities in Europe (Ollivier, 1998). Such efforts were initially scattered and isolated, and were undertaken in response to critical situations. Over many years, concern for the loss of genetic diversity led to development of specific programmes to maintain and support native breeds that were at risk of becoming extinct. Initially, most of the conservation activities were focused at the national/ provincial level and were led by enthusiastic individuals, public society institutions and non-governmental organisations. In France, the first conservation programme to save Solognote sheep was initiated in 1969, and other programmes for various species were undertaken shortly afterwards by breeders' organisations. The financial and statutory state support for the implementation of these programmes was granted in 1975 (Audiot et al., 1992). Similar activities supporting the maintenance of highly endangered breeds were undertaken in other Western European countries: the White Park, Kerry, Old Danish Black and White or East Finish cattle, are examples.

The need to co-ordinate national level conservation activities and programmes, and communicate the importance of local breeds, led to the establishment of several indigenous breed conservation non-governmental organisations. The UK Rare Breeds Survival Trust (RBST), which was initiated as a Working Party in 1968, and formally established in 1973, provided a first model of an umbrella private organisation that played an extremely important role in ensuring the conservation of indigenous breeds. Other examples include: SZH – the Dutch Foundation for Rare Breeds (1976), GEH – the Society for the Preservation of Endangered Livestock Breeds in Germany (1981), ÖNGENE – the Austrian National Society for Gene Reserves (1982), Pro Specie Rara in Switzerland (1982), and SERGA – the Spanish Society for Animal Genetic Resources (1989).

In Central and Eastern European countries (CEECs), conservation programmes to ensure the preservation of native breeds of livestock were also undertaken in the 1970s. Most of these conservation activities were formally established and financially supported by governments, with implementation being realised through research organisations and universities. For example, in Romania from 1967-1969, the state

E. Martyniuk

financed the collection of endangered chicken populations. 'Genotoque' was established to ensure the long-term survival of chicken genetic resources that were on the verge of extinction. In 1990, this Poultry Gene Bank comprised 94 varieties of 36 chicken breeds (Draganescu, 1992). In Hungary, the conservation of native breeds dates back to 1973, when the Ministry of Agriculture and Food entrusted the National Inspectorate of Animal Breeding and Nutrition with the implementation of compulsory conservation programme. These included both mammalian and avian species. In Poland, successful restoration programmes resulted in the conservation and restoration of populations of several native breeds, including: Polish Koniks, Polish Heath Sheep and Swiniarka Sheep.

Unfortunately, economic transformation in Central and Eastern Europe has not always been beneficial for AnGR conservation programmes, in spite of the fact that native breeds were considered as national heritage resources. Financial support for some state conservation programmes was reduced or stopped, and the introduction of the market based economy put pressure on farmers to increase efficiency and productivity, often resulting in the replacement of native breed with high performing commercial stock. This process was routinely accompanied by decreased state involvement and reducing financial support from the public sector for agriculture and AnGR.

Concerted action at the sub-regional and regional levels

The Nordic countries were among the first to initiate conservation programmes at the national level, with Iceland adopting a special Law for conserving Icelandic goats in 1965, and Sweden establishing *ex-situ* semen collection in 1967 (Maijala et al., 1992a). The Nordic countries also provide an excellent example of sub-regional governmental co-operation for AnGR. Their Council of Ministers, in 1979, established a Working Party on AnGRs to organise and coordinate activities across the region. This led, in 1990, to the establishment of the Nordic Gene Bank for Farm Animals (NGH) (Adalsteinsson, 1994).

Concerted action for AnGR was initiated across the European region in 1980, following recommendations of the FAO/UNEP Technical Consultation on Animal Genetic Resources Conservation and Management (FAO, 1981). The Commission on Animal Genetics of the European Association for Animal Production (EAAP) established its Working Group on AnGR which organised three successive surveys of European breeds of cattle, sheep, goats and pigs in 1982, 1985 and 1988, with the participation of 22, 17 and 12 countries respectively (Simon and Buchenauer, 1993). In 1986, the Department of Animal Breeding at Hannover Veterinary University (TIHO) was entrusted by the

17

EAAP with the task of creating the European Animal Genetic Data Bank (AGDB). Between 1988 and 1991, TIHO under the agreement with the United Nations Food and Agriculture Organisation (FAO), managed the Global Data Bank for AnGR. Establishment of the FAO global level Domestic Animal Diversity-Information System (DAD-IS), with support and data transfer from the AGDB, resulted in the Hannover database returning to its European status. Since 1994, the two databases have been developed separately using slightly different questionnaires for data collection. The databases are accessible at their respective web sites (www.tiho-hannover.de/einricht/zucht/eaap and www.fao.org/dad-is).

International non-governmental organisations (NGOs) have been instrumental in supporting collaborative actions in the management of AnGR. Activities are focused on the undertaking of joint conservation programmes, co-operation in research and various public awareness initiatives. The most active NGOs in the European region include, SAVE (Safeguard of Agricultural Varieties in Europe), established in the Netherlands in 1992; DAGENE (Danubian Countries Alliance for Conservation of Genes in Animal Species) founded in Hungary in 1989/1991; and Rare Breeds International (RBI) registered in England in 1991.

The motivation for conserving AnGR and conservation approaches

Early concerns regarding the loss of genetic diversity, which has been associated with the application of modern breeding methods, was accompanied by enhanced appreciation of the heritage values of native breeds, and their cultural and historical importance. A growing number of concerned individuals and organisations were becoming aware of the need to conserve unique characteristics of indigenous breeds, to enable farmers to address future challenges in the livestock sector. Support for conservation was enhanced by examples of conservation programmes originally implemented as an insurance measure, becoming economically sustainable. This has been achieved through, for example, utilisation of native breeds in commercial crossbreeding systems, promotion of speciality products and increased utilisation of native breeds in landscape management and agro-tourism activities (Gandini and Oldenbroek, 1999).

The early conservation programmes initiated in the 1970s, applied both *in-situ* and *ex-situ* methods to achieve the conservation of AnGR (Maijala *et al.*, 1992a). The EAAP Working Group on AnGR provided an analysis of possible advantages and disadvantages of maintaining live animals, frozen semen and frozen embryos in different farm animal species, and their guidance has been widely used in planning conservation activities (Maijala *et al.*, 1984).

E. Martyniuk

Where we are now? The state of AnGR conservation in Europe

Implications of the Convention on Biological Diversity

The Convention on Biological Diversity (CBD) provides countries with an international legal framework to address the wise management of their agricultural biological diversity. Conservation of plant and animal genetic resources has become one of the long-term objectives addressed within the scope of National Biodiversity Strategies. In many countries, the National Focal Point for AnGR (NFPs) contributed to the preparation of the national biodiversity strategy, as well as progress reports that were provided to the Conferences of the Parties to the CBD. The European Union has developed legislation responding to obligations under the CBD, and some countries, like the Netherlands, have adopted a national policy document exclusively addressing agricultural biological diversity (Sources of Existence, 2000). Implementation of the CBD has resulted in the undertaking of concerted initiatives; the establishment of the Platform for Biodiversity (The Netherlands); adoption of special financial measures, for example a Fund for Biological Diversity in Hungary; and initiation of joint activities including research, database development, and public awareness covering plant, animal and microbial genetic resources (Germany, Czech Republic and Lithuania).

The FAO framework for the Global Strategy for the Management of Farm Animal Genetic Resources

Development of the FAO technical programme, the Global Strategy for the Management of Farm Animal Genetic Resources in 1993, has proved to be an important milestone in support of AnGR conservation efforts. In order to participate in the programme, countries were invited to establish a formal structure, a National Focal Point for AnGR that is responsible for the co-ordination of conservation activities at the national level (Hammond, 1998). Countries of the European region were among the first to establish National Focal Points and nominate National Coordinators.

Establishment of National Focal Points for Animal Genetic Resources

In most countries, the NFPs are established within existing organisations, mainly research institutes, universities or within Ministries of Agriculture. In a few instances, new organisations have been established to serve as National Focal Points and carry out work on animal genetic resources. In France, BRG - Bureau des Ressources Génétiques founded in 1983, has undertaken responsibilities of the NFP with two full time scientists in charge of the AnGR sector. In Austria, the Ministry of Agriculture and

19

Forestry founded in 1997, an Institute focusing on biodiversity research (Institut für Biologische Landwirtschaft und Biodiversität) that is the NFP for AnGR. In Italy, ConSDABI (Consorzio per la Sperimentazione, Divulgazione e Applicazione di Bitecniche Innovative) was established to co-ordinate conservation activities as well as to maintain *ex-situ* live collections of endangered livestock breeds, and conduct research focused on the development of conservation biotechnologies. Table 1 indicates NFP affiliations across Europe.

Most European countries, following FAO guidelines, have established an Advisory Committee on AnGR as well as subsidiary Working Groups/ Task Forces to address specific issues, and prepare proposals for action. Several countries have also developed extensive networks, species- and subject-specific networks, ensuring the involvement and contribution of stakeholders in supporting conservation efforts.

Conservation efforts

In Western European countries, conservation activities are advanced and the necessary infrastructure is well established, including breeding societies and stakeholder networks. The Convention on Biological Diversity and the FAO Global Strategy, have further enhanced existing activities, and led to the initiation of new ones. Most Western European countries have secured significant financial resources to implement conservation programmes and enjoy significant public support for such activities. Public awareness of the heritage and other non-monetary values of indigenous breeds are gradually increasing in Western Europe, often as a result of the efforts of non-governmental organisations.

In Central and Eastern Europe, the situation is less uniform than in Western Europe. There are some countries with a long, successful, tradition of conserving their native breeds, like Hungary and Poland. There are also countries where such activities have been undertaken only recently, and carried out with a lot of enthusiasm but limited resources, both in terms of capacities and available financial resources. The situation regarding AnGR in the Balkan region is particularly difficult due to severe damages to the infrastructure, and the loss of livestock resulting from civil strife and war.

The introduction of the market-based economy in Central and Eastern Europe, beginning in 1989, created many challenges, especially during the early transition period. These required major changes in the structure of the agricultural sector. Unfortunately, during this period, AnGR were badly affected. The negative impacts were confirmed by an inventory of AnGR during the 1994/1995 FAO Project Identification Mission, which included 14 countries. The Mission identified the urgent need to save

Table 1.
AnGR National Focal Point affiliations across Europe.

Ministry of Agriculture (or equivalent)	Research Institutes	Universities	Specially established institutions	Others
Belgium	Albania	Bosnia Herzegovina	Austria – Institut für Biologische Landwirtschaft und Biodiversität	Iceland – The Farmers Association
Germany	Bulgaria	Croatia	France – Bureau des Resources Génétiques	Lithuania – Rural Business Development and Information Center
Hungary	Cyprus	Estonia	Italy – Consorzio per la Sperimentazione, Divulgazione e Applicazione di Bitecniche Innovative	Sweden – Swedish Board of Agriculture
Ireland	Czech Republic	Greece		
Latvia	Denmark	Macedonia FRY		
Luxembourg	Finland	Norway		
Malta	Israel	Romania		
Serbia and Montenegro	Slovakia	Slovenia		
Spain	The Netherlands			
Switzerland	Poland			
Turkey	Portugal			
UK				

endangered native breeds, and to initiate programmes among neighbouring countries. In total, 41 projects were developed, 30 of them addressing breed conservation; 6 projects proposed promotion of specific products; and 4 projects proposed a focus on education and training (Glodek and Ochs, 1995). Additional projects to conserve AnGR in Central and Eastern Europe have emerged from other processes. An FAO sub-regional workshop organised in Kaunus for the Nordic and the Baltic countries, and Poland, resulted in two proposals for concerted action in Northern Europe. The first project was to focus on the dissemination of information and knowledge on AnGR to increase public awareness and support for conservation programmes. The second project proposed an analysis of genetic diversity between cattle breeds in the region (Smith, 1998). Different priorities were identified at a similar workshop held in Thessaloniki for the Balkan countries. There were three project proposals prepared aiming at: development of networks on AnGR in the region; harmonisation of identification, recording and genetic improvement programmes for AnGR; and support for establishment of NGOs, namely breeders' societies and farmers' organisations to undertake activities in management of AnGR. All these project proposals further confirmed discrepancies in capacities between Western and Eastern Europe to manage AnGR, and indicate the long distances some countries in the east will need to travel to be able to effectively enhance the management of their AnGR. Although substantial resources were invested in the organisation of meetings and development of project proposals, the Author is unaware of any of the proposed projects being financially supported, illustrating a lack of interest by the donor community in financing AnGR projects.

Activities of the National Focal Points for AnGR

Since 1995, co-operation among the 37 countries of the European region has been facilitated through annual Workshops for National Coordinators organised jointly by the FAO and the EAAP, and hosted by annual EAAP meetings. National Coordinator Workshops provide an important forum for information exchange in the region, and enable discussion and development of common positions regarding specific AnGR issues. Each National Coordinator provides a short report on NFP activities during past year, and future plans. Annual national reports included in the final report from each Workshop have provided background material for the analysis in Table 2. Participation in the workshops varies, with some countries having attended each Workshop, while a couple of countries have never been represented. Financial constraints are the main reason for restricting participation in the annual workshops.

Table 2. Number of national reports in recent years.	Year and number of national reports	*National Coordinators Workshops in:*							
		1995	*1996*	*1997*	*1998*	*1999*	*2000*	*2001*	*2002*
		28	23	22	16	21	23	24	24

a. Policy and legislation development

Development of legislation addressing AnGR provides a framework for implementation of conservation programmes. In recent years, many initiatives were undertaken to develop new, or amend existing, legislation on animal breeding to facilitate conservation activities. In Belgium for instance, formal recognition of native sheep breeders' societies, and adoption of legislation in 1998 for the poultry and rabbit breeding sectors, facilitated the conservation of genetic resources of these species.

Legislative development has been most pronounced in Central and Eastern Europe, as countries needed to adjust their legislation to better reflect the principles of a market economy and ensure conformity with legislation in the European Union. In most countries, legislation is not solely devoted to AnGR, rather specific articles providing for the conservation of AnGR are included in breeding laws. For example, the Hungarian Breeding Law states in Article 11 that: 'Protected indigenous, native breeds have significant genetic value, preservation of them in their original form is a national interest and a duty of the State' and 'Maintenance of protected indigenous breeds is supported by the State' (Workshop Report, 2001). In Slovenia, a proposed new Law on Animal Husbandry contains several very specific articles on AnGR addressing conservation, monitoring and *ex-situ* banks (Workshop Report, 2002). References to AnGR conservation has also been included in breeding legislation in the Czech Republic, Estonia, Poland, Slovakia and Turkey.

b. National Strategies and Action Plans for the Conservation of AnGR

The development of national programmes for AnGR conservation is one of the key tasks for the NFP. In most countries, elaboration of a comprehensive national strategy has been carried out in a step-by-step manner starting with an inventory of native AnGR, establishment of breeding societies where necessary, and development of conservation programmes for endangered breeds. Conservation programmes had to be prepared for all breeds eligible for EU financial support within the Council Regulation 2078/92, and this requirement provided an additional strong incentive to complete this work.

The Nordic countries were the first to develop their strategies, with the Danish Management of Farm Animal Genetic Resources Strategy being implemented since 1997. In France, the National Charter for Management of AnGR was approved in 1999. In Germany, development of a National Programme on AnGR, within the framework of Genetic Resources for Food and Agriculture, is seen as a major outcome of the preparation of their Country Report as a contribution to the *First Report on the State of the World's Animal Genetic Resources* (Workshop Report, 2002). In Central and Eastern Europe, most countries developed their strategies before 2000, with the Czech Republic and Croatia being among the first ones (1997), followed by Slovakia, several Baltic countries, Poland and Yugoslavia.

Within the framework of national AnGR conservation strategies, many countries have undertaken restoration programmes of populations of native breeds, which had never been included in any conservation activities before. Examples include: Cachena cattle, Campaniça sheep and Bisaro pigs in Portugal, Krško Polje pigs in Slovenia, Galway sheep in Ireland, Leine sheep and Senner horses in Germany, Native Horse and quail genetic resources in Estonia, Vištines geese in Lithuania, and many others. In enhancing conservation activities, a step-by-step approach is most often employed. For example, in France, the scope of species included in conservation activities has been progressively broadened to cover poultry, horses and dogs. Conservation programmes tend to give considerable attention to maintaining within-breed genetic variation, especially in breeds with small population size. For instance, breeders of rare breeds in Austria and Germany are provided with information on inbreeding levels in their herds, available males, and expected inbreeding of progeny from mating within their female stock (Workshop Report, 2000).

c. Financial support for conservation

The European Union provided significant support for the conservation of animal genetic resources through Council Regulation No. 2078/92. This Regulation allowed *inter alia* provision of a special premium for farmers who 'rear animals of local breeds in danger of extinction'. The first evaluation report of the state of the application of this regulation indicated considerable differences between Member States in regard to both the level and scope of the support. The report also underlined that this *in-situ* conservation measure had a very positive effect on the conservation of animal genetic diversity (DGVI, 1999).

In 2000, a proposal to change the endangerment criteria within the new regulation 1750/99, raised concern among the European Union's National Coordinators. The proposed criteria only considered breeds with less than 1000 breeding females as being endangered. Many local breeds that were enjoying a renewed interest, and as a result of application CR 2078/92 had increasing population sizes, would no longer be eligible for financial support. If the resolution had been adopted, in France for example, the number of breeds eligible for financial support would have dropped from 17 to 12, in Italy, from 82 to 57. As a result of the efforts of the EU NCs, and technical support from the EAAP Working Group on AnGR (Danell et al., 2001), a new set of criteria to recognise endangered breeds has been developed and adopted by Commission Regulation 445/2002. This Regulation provides much higher threshold values for particular species, but these have to be applied at regional level. Two EU countries, United Kingdom and Denmark are not implementing this EU incentive measure, having their own system for supporting farmers carrying out *in-situ* conservation. In the UK, *in-situ* support is provided through private funding by the RBST. In Denmark, a national system was developed operating with a higher level of premium than in the EU (max. 400 EURO/LU) (Workshop Report 2002). Switzerland, Norway and Cyprus have more recently initiated farmer support for *in-situ* conservation.

In some of the CEECs, the state continues to provide substantial financial resources for the conservation of native breeds (e.g. the Czech Republic, Hungary, Poland, Slovakia, and Slovenia). In a few countries, financial support available before 1989 was temporarily stopped due to public financial constraints (e.g. Bulgaria, Romania, and Yugoslavia). In other countries, the financial involvement of the state has been initiated as a result of the activities of the NFP, and is slowly increasing to cover a wider number of breeds and species (e.g. Estonia, Latvia, and Lithuania). It is noteworthy that the level of state support is much higher now than it was in 1997, when only 11 European countries out of 26 countries responding to a survey indicated that they were applying subsidies for *in-situ* conservation (Martyniuk and Planchenault, 1998).

d. Conservation approaches

Sustaining traditional production systems

Management practices and animal husbandry systems are specific and highly diversified in Europe. They are based on local traditions, which have been developed to best utilise environmental conditions and respond to the specific needs of local communities. Utilisation of AnGR, especially in harsh environmental conditions and extensive marginal grazing areas, enable the management of ecosystems and maintain traditional farming practices as well as other activities including, tourism, hunting and fishing. Intensification of animal production has led in many instances to abandoning traditional production systems and the associated AnGR. This process was most pronounced in poultry, and resulted in the extinction of varieties suitable for floor/free range management systems or farmyard production. Most cage-adapted populations are not fit for floor systems. The Skalborg hen, a particular Danish line of White Leghorn, seems to be the only Danish line that still maintains production capacity to meet new welfare-associated market requirements (Sørensen, 1997). Renewed interest supported the conservation of the Settler hen in Iceland, a highly endangered breed that has increased recently to 3000 hens as a result of recognition of its potential role in farmyard production, and the high quality of its products (Hallgrimsson, personal communication). In pig breeding, there are trends towards less intensive systems to ensure animal welfare and provide higher product quality. Good foraging and maternal abilities, highly valued in extensive systems, have been preserved only in a limited number of local breeds like the Saddleback pig of the UK, and the Pulawy pig in Poland.

Marketing of speciality products

One of the best strategies to ensure the long-term maintenance of local breeds in a cost-effective manner is to successfully produce and market goods and services that are valued by consumers. This requires establishing direct linkages between specific breeds and products that are derived from them. Only a few breed-product links have been well documented in Europe (Gandini and Oldenbroek, 1999). One of the most successful examples is the Reggiana cattle breed, which increased from 500 individuals in the early 1980s, to 1868, by 2002 (Workshop Report, 2002). This rapid population increase can be directly attributed to consumer appreciation of a new brand of Parmigiano Reggiano cheese that is made exclusively from milk obtained from Reggiana cows, and commands a price about 16 % higher than the common Parmigiano.

Of the 65 cheeses produced in France, Italy, Portugal and Spain, which have Appellation d'Origine Contrôlée (AOC) status, 15 must be made of milk obtained from local breeds. Abondance, Beaufort, Comté, Reblochon and Fontina cheese, can be directly linked to specific cattle breeds (Bertozzi and Panari, 1993). Product links have been developed also in sheep and goats, especially in some Mediterranean countries where traditional processing and high quality of local or regional small ruminant dairy products are highly valued by consumers (e.g. Ossau Iraty, Roquefort, Pecorino Romano, Manchego, Serra da Estrela, Feta, etc.; Boyazoglu, 1999). Other successful attempts to link breeds with specific products have been made in the meat sector, with Mirandesa cattle in Portugal, Piedmontese, Chinina, Marchigiana and Romagnola cattle in Italy and Hinlerwälder cattle in Germany (Gandini and Giacomelli, 1997). In the United Kingdom, the Rare Breeds Survival Trust has established the Traditional Breeds Marketing Scheme to create speciality markets for discerning clientele through developing links among owners of traditional native breeds, accredited butchers, and specialist retail outlets (Alderson, personal communication).

Landscape management

There is growing recognition of the strong links between specific agro-ecosystems and AnGR, and the contribution of local breeds into managing ecosystems. For example, fragile mountain pastures and river valley meadows often require grazing by domestic animals to maintain their specific characteristics and wildlife habitats. Several factors, including EU policy for extensification, decreased food production, and economic transition in the CEECs, has resulted in increasing amounts of uncultivated land across Europe. These areas must be properly managed to maintain their ecological values, their biodiversity and biological resources, and to ensure opportunities for a variety of out-door recreational activities. Utilisation of grazing animals in landscape management is widely accepted in Europe as a tool to maintain ecological integrity and biodiversity and to ensure the sustainable utilisation of biological resources. In recent years in Poland, several Small Grant Projects financed by Global Environment Facility were carried out where local cattle and sheep breeds were used for vegetation control maintaining important nesting habitats for endangered bird species. Grazing is also used as a fire management tool, removing woody vegetation and preventing the accumulation of biomass that could result in the rapid spread of intense wild fires. This has proven successful in the dryland Mediterranean zone (Nastis, 1993). Controlled grazing has shown to be appropriate in the management of some protected areas in Europe, including national parks and nature reserves. Local and rare breeds could more effectively contribute in providing ecological services in protected areas and other landscapes, as has been shown in France, Hungary and Sweden (Maijala, 1995).

Ex-situ conservation

Ex-situ conservation measures have been undertaken in all European countries, but the scope of activities, number of breeds and species included in the cryo-conservation banks and the organisational infrastructure differ among countries. In most cases, *ex-situ* storage of semen and embryos are implemented as an accompanying measure to *in-situ* breed conservation programmes. Very often, semen or embryo collection was conducted within a specific research project, and such collections provided the beginning for the development of gene banks. This is common in the CEECs where research institutions have been storing substantial amounts of genetic material both from mammalian and avian species. Only recently well designed comprehensive *ex-situ* programmes been developed. The most advanced cryo-conservation banks are currently established in France, Germany and in the Netherlands. The National Cryobank collection in France covers 5 species with 20 breeds, and contains more than 26,000 doses of semen (Workshop Report, 2002).

d. Education, communication and public awareness

AnGR education, information, enhanced awareness and public support are necessary to ensure sustainable conservation of native endangered breeds. In Western Europe dialogue on this matter was initiated long ago, and has resulted in better knowledge and appreciation of conservation needs. Native breeds have been placed in agricultural schools, agricultural museums, nature reserves, farm parks, and even prison farms, to enable direct contact between rare animals and the public, promoting important conservation messages. In a few CEECs public awareness initiatives were initiated long ago and resulted in appreciation of native breeds as part of the national heritage; the Hungarian Grey cattle provide an excellent example. Appreciation of native cattle among farmers and the public in Iceland, following by widespread, long and ardent debate, and two votes by the Parliament, resulted in the abandonment of a plan to import Norwegian Red Dairy Cattle.

In recent years, a lot of public relations activities were undertaken to promote utilisation of native breeds. They include various publications from booklets to albums (most of the European countries); films (Czech Republic, Estonia, Germany), and other materials for TV; organisation of native horse races, animal shows and various exhibitions (like 3 day event - Genetic Resources Days in France); and receiving numerous visitors in the NFP (Italy). Farm Parks and Ark Farms are playing a special role in communication with the public. They are especially appealing to young people and provide an excellent opportunity to develop understanding and interest in indigenous livestock breeds. Recent initiatives in this area include: establishment of Geno Park for Istrian cattle in Croatia, and a sanctuary for the Cyprus Donkey; presentation of farm animals in an open-air museum in Ballenberg, Switzerland; foundation of four 'Conservation and Visitor Centres' in Denmark; establishment of Genetic Park in Czech Republic; and approval of over 100 Ark Farms in Germany (Workshop Reports 2000-2002). However, in most of CEECs, there is still limited public understanding of the roles and values of local breeds and consequently little support for conservation initiatives.

Development of the European Regional Focal Point

The need for the establishment of the European Focal Point for AnGR (ERFP) was expressed in 1996 during the second NC's Workshop. Participants underlined the multiple benefits that would be derived from co-ordination of the various AnGR conservation activities across the region. Since 1998, informal regional co-ordination has been supported by the Government of France to facilitate establishment of a formal

ERFP. Over the next couple of years discussions on objectives, terms of reference, modes of operation and financing continued and a Steering Committee was appointed to oversee establishment of the permanent ERFP. The Steering Committee consists of NCs from Greece, France, the Netherlands, Poland and the United Kingdom. It was agreed that in order to ensure sustainability of the ERFP, funding is to be provided through contributions made by donor countries that will be placed in a Trust Fund, which is to be administrated by the Steering Committee. In order to create the ERFP, at least 10 donor European countries must contribute to the Trust Fund with a maximum country contribution set at 10,000 EURO annually. In 2001, ten countries (the Czech Republic, France, Germany, Greece, Ireland, Italy, the Netherlands, Spain, Switzerland and the United Kingdom) committed supporting the activities of the ERFP thus enabling its formal establishment. The ERFP Secretariat is located in the BRG - National Focal Point for France and Dr. Dominique Planchenault is the Regional Coordinator. According to agreed terms of reference, the ERFP is established within the framework of the FAO Global Strategy for the Management of Farm Animal Genetic Resources with the following objectives:

- To assist and enhance activities of NFP at the European level and to assist in co-ordinating those activities within and between other European organisations such as EU;
- To develop and maintain regular contact and exchange of relevant information on AnGR, horizontally between European NFPs and vertically with the Global Focal Point in Rome;
- To stimulate the funding and organisation of regional projects, workshops and national programmes on AnGR within the European region; and
- To stimulate and co-ordinate the maintenance and further development of national databases within the European Region and encourage European networking on AnGR.

The ERFP structure consists of three main elements, annual meetings of the National Coordinators, a Steering Committee, and a Secretariat. The annual budget covers basic costs of the Secretariat, costs of the Steering Committee meetings and expenses for specific activities approved at annual Workshops. Further voluntary contributions to support specific programmes or projects co-ordinated by the ERFP are welcomed.

Where do we want to be? A vision for animal genetic resources

Development of the Global Strategy has been strengthened through the establishment of an intergovernmental mechanism to address both policy and technical issues related to AnGR. The Intergovernmental Technical

Working Group on AnGR recommended in 1998, the preparation of the *First Report on the State of World's Animal Genetic Resources* that will provide for the first time a global assessment of the state of AnGR and the state of capacities for their management. The Commission on Genetic Resources for Food and Agriculture endorsed this recommendation in 1999. As a result of this decision, countries are now undertaking preparation of their Country Reports, which will provide the basis for the global report. Country Reports will describe the state of each nation's AnGR, and indicate national priorities for the better management of these resources. Country Reports will catalyse development of a strategic vision for AnGR at both the national and international levels and hopefully encourage a transparent and objective review of existing strategies for conservation and use of animal genetic resources.

The experience so far indicates that most of the European countries have made progress in the conservation of their AnGR over the last few years, but that further developments are necessary and anticipated. Progress at the regional level is demonstrated by cooperation among the Nordic countries, where the strategy for the Nordic Gene Bank was developed, and is successfully implementing conservation, research, education and other activities (Danell *et al.*, 1998).

While the progress is significant, we cannot be complacent. It is not yet clear how we can enhance overall understanding and appreciation of roles and values of AnGR or how we can further mobilise required financial and other resources for their better management across all of Europe, and beyond. We also need to consider new pressures and changes in the sector such as evolving livestock husbandry, globalisation, internationalisation of breeding programmes, and the impact of the current livestock revolution on AnGR. If we consider some of the advancement achieved by the Plant Genetic Resources community, which includes completion of a first report on the State of the World's Plant Genetic Resources and recent approval to an update, adoption of the Global Plan of Action, and successful negotiation of the International Treaty on Plant Genetic Resources, it is very clear that international efforts for the management of AnGR are way behind those for plant genetic resources.

Unfortunately, while PGR are universally accepted as international assets, AnGR have never been widely regarded as such. There is a need to promote AnGR as a public good, and highlight the multiple benefits associated with their utilisation and exchange - now and in future. It seems that developed countries committed to conserve their national AnGR are somehow reluctant to accept responsibility for supporting international collaborative actions targeted at the countries in need. This can be only changed by better communication and increased awareness and understanding of the roles and values of AnGR, in contributing to sustainable agriculture development and food security.

It is clear that we must better articulate the importance of the management of AnGR, and advocate our issues and needs at international forums. The experience from the last session of the FAO Commission on Genetic Resources for Food and Agriculture, held in October 2002, illustrates that we are not yet ready to fully capitalise on opportunities to draw due attention to AnGR issues. In spite of enormous region-wise conservation efforts and extensive activities undertaken at national level, we failed to adequately share information on this progress with the Commission. We failed to elect to the Bureau of the Commission anyone with significant experience in animal genetic resources, and thus the Commission continues to be dominated by plant genetic resources experts and issues. In addition, the plant genetic resources community will have another international forum to address plant genetic resources issues when the International Treaty came into force.

Future activities of the European Regional Focal Point

During the 8th Workshop in Cairo in 2002, the NCs agreed that future activities of the ERFP would cover two main areas: regional level activities and implementation of projects submitted by groups of countries. Regional activities will focus on preparation of annual NCs workshops, development of the ERFP website and international co-operation and lobbying. The interaction with the EC is of a special importance in light of the development of new regulations related to AnGR. The ERFP will also participate in the EC financed project, EFABIS (European Farm Animal Biodiversity Information System) aimed at development of national databases for AnGR, and establishment of data transmission from national to European (AGDB) and global (DAD-IS) database using an Open Source Model approach.

The following priority areas were identified for collaboration over the next couple of years: I. Breed development and conservation - *in-situ*: to develop cooperation on breeding programmes that will ensure long-term, economically viable and sustainable *in-situ* conservation through concerted efforts of countries maintaining the same breed or breeds that are similar in origin. II. Breed development and conservation – *ex-situ*: to develop guidelines for establishment and management of cryo-conservation banks for AnGR that will cover specific aspects of development and management of *ex-situ* banks, including procedures for collection of various genetic materials, access regulations, conditions for use and replenishment protocols, and ownership and organisational structure, etc. III. Monitoring animal genetic resources – practices and approaches: to provide overview of ongoing AnGR monitoring programmes including, identifying best monitoring practices, dimension/ organisation of monitoring, and best ways and means to provide quality input for national databases. IV. Monitoring animal genetic resources - overview of available data and information: to provide analysis on available European databases concerning various aspects of AnGR and analysis of possible use of these data in evaluating animal genetic diversity at European level.

The European Regional Focal Point is the first global sustainable regional co-ordinating structure, which is based on the voluntary contributions of participating countries. The European Regional Focal Point has already proved to provide an extremely valuable forum to consider and lobby on issues related to AnGR conservation. Europe has a wealth of AnGR and a long tradition of ensuring their successful management and also significant technological and human capacities that will enable the development of new and exciting applications of AnGR. Given Europe's history in the management of AnGR and its extensive cadre of experts and enthusiastic professionals, Europe must continue to play a significant role in supporting global AnGR conservation efforts.

Challenges and opportunities

To ensure the optimal use of AnGR and to contribute to support global efforts to achieve food security and rural development, a number of challenges must be met and opportunities realised. There are many important and pending issues to address both at national, regional and international level, and Europe is well positioned to lead developments in some of these areas. Several of the key issues and challenges include:

- Continuing to develop capacity across Europe to ensure the conservation and sustainable use of AnGR, especially to support capacity building in economies in transition;
- Enhancing public awareness and awareness among policy-makers of the multiple roles and values of animal genetic resources to ensure sound investments in the conservation and sustainable use of AnGR;
- Being one of the most advanced regions in the conservation of AnGR, Europe must enhance and improve communication, sharing of experience across the region and beyond the region and responding to current and future global challenges;
- Continuing the policy and technical work of working groups and networks already established in Europe, and further engaging non-governmental interests and the private sector in AnGR conservation programmes;
- Continuing to increase understanding of breed diversity and breed numbers in Europe, and indeed the world, and improving native

breed characterisation to evaluate their uniqueness and their best utilisation;

- Enhancing understanding of the implications of disease eradication programmes on maintaining genetic diversity within breeds subjected to such programmes;
- As a result of rapid developments in *ex-situ* cryo-conservation methods, considering access to genetic resources and benefit sharing arrangements, especially as valuable traits are better understood and described at the molecular level;
- Ensuring that all countries in the region prepare Country Reports as a key input to *the First Report on the State of the World's Animal Genetic Resources*, and that resources are made available to assist countries in need within the region and outside the region to prepare their Country Reports; and
- Ensuring that Europe plays a stronger role in the Global Strategy for the Management of Farm Animal Genetic Resources through the Commission on Genetic Resources for Food and Agriculture, and especially over the next several years when AnGR will be central in the work of the Commission.

A vision for the future of AnGR has to be developed and the ERFP has got an instrumental role in facilitating and leading this process, ensuring inputs from all countries of the region. Conferences, such as this one organised by the British Society of Animal Science and partners, provides an excellent opportunity to share views, discuss and enhance specific elements of our vision on future management of AnGR.

References

Adalsteinsson, S. 1994. Conservation activities in the Nordic countries: Farm animals. Genetic Resources in *"Farm Animals and Plants - Report from Research Symposium."* 27-29 May, ThemaNord 1994: 603: 63-71.

Audiot A., Verrier, E., Flamant J.C. 1992. National and regional strategies for the conservation of animal inheritance in France. In: *Genetic Conservation of Domestic Livestock*. Edited by L. Alderson and I. Bodo, CAB International, Wallingford, UK. pp. 110 -127.

Bertozzi L. and Panari G. 1993. Cheeses with Appellation of d'Origine Contrôlée (AOC) factors that affect quality. *International Dairy Journal*, 3: 297-312.

Boyazoglu, J. 1999. *Livestock Production Systems and Local Animal Genetic Resources with Special Reference to the Mediterranean Region*. Invited Paper, VII Congress of the Mediterranean Federation for Ruminant Health and Production, Santarem, Portugal 22-24 April, 1999.

Danell, B., Distl, O., Gandini, G., Georgoudis, A., Groeneveld, E.,

Martyniuk, E., Ollivier, L., van Arendonk, J. and Woolliams, J., 2001. The development of criteria for evaluating the degree of endangerment of livestock breeds in Europe. *Book of Abstracts of the 52 Annual Meeting of EAAP,* Wageningen Academic Publishers, Wageningen, The Netherlands.

Danell, B., Vigh-Larsen, F., Mäki-Tanila, A., Eythorsdottir, E. and Vangen O. 1998. A strategic plan for Nordic co-operation in management of animal genetic resources. *Proceedings of the 6th World Congress on the Genetics Applied to Livestock Production,* Armidale, NSW, Australia. January 11-16, 1998, 28: 111-114.

DG VI, 1999. *State of Application of Regulation (EEC) No.2078/92: Evaluation of Agri-Environment Programmes.* DGVI Commission Working Document VI / 655 / 98. http://europa.eu.int/comm/ dg06/envir/programs/evalrep/text_en.pdf).

Draganescu C. 1992. Conservation of rare breeds of livestock in Romania. In: *Genetic Conservation of Domestic Livestock.* Edited by L.Alderson and I .Bodo, CAB International, Wallingford, UK. pp. 110–127.

FAO, 1981. *Animal Genetic Resources Conservation and Management.* FAO Animal Production Health Paper 24: 388.

Gandini, G. and Giacomelli, P. 1997. What economic value for local livestock breeds? *Book of Abstracts of the 48th Annual Meeting of EAAP,* Wageningen Academic Publishers, Wageningen, The Netherlands. p. 20.

Gandini, G. and Oldenbroek, J.K. 1999. Choosing the conservation strategy. In: *Genebanks and the conservation of farm animal genetic resources,* Edited by J.K. Oldenbroek ID-DLO, 1999, pp. 11-31.

Glodek, P. and Ochs H. K. 1995. Report of the Project Identification Mission. Conservation of Domestic Animal Diversity (CDAD) in Central and Eastern European Countries.

Hammond, K. 1998. Animal genetic resources and sustainable development. *Proceedings of the 6th World Congress on the Genetics Applied to Livestock Production,* Armidale, NSW, Australia January 11-16, 1998, 28: 43-50.

Maijala, K., Cherekaev, E.V., Devillard, J.M., Reklewski, Z., Rognoni, G., Simon, D.L. and Steane D.E. 1984. Conservation of animal genetic resources in Europe. Final report of an EAAP Working Group. *Livestock Production Science* 11: 3-22.

Maijala, K., Adalsteinsson, S., Danell, B., Gjelstad, B. Vangen, O. and Neimann-Sörensen A. 1992a . Conservation of animal genetic resources in Scandinavia. In: *Genetic Conservation of Domestic Livestock.* Edited by L. Alderson and I. Bodo, CAB International, Wallingford, UK. Chapter 3.

Maijala K. Kantanen J. Korhonen, T. 1992b. Conservation of animal genetic resources in Finland. In: *Genetic Conservation of Domestic Livestock.* Edited by L. Alderson and I. Bodo, CAB

International, Wallingford, UK. 128-142.

Maijala, K. 1995. Potential practical uses of genetic reserves. *Proceedings International Symposium on Conservation Measures for Rare Farm Animal Breeds,* Balice, Poland, May 17-19, 1994, 11-21.

Nastis, A.S. 1993. Effect of grazing small ruminants on landscapes and wild fires prevention in the Mediterranean zone. *Proceedings of the 44th EAAP Meeting,* Aarhus, Vol. II, 208-209.

Martyniuk, E. and Planchenault, D. 1998. Animal genetic resources and sustainable development in Europe. *Proceedings of the 6th World Congress on the Genetics Applied to Livestock Production,* Armidale, NSW, Australia, January 11-16, 1998, 28: 35-42.

Ollivier, L. 1998. Guest editorial in *Livestock Production Science* 54: 67-70.

Vares, T. 2000. A new approach for the global strategy of management and conservation of farm animal genetic resources. Discussion Paper – unpublished.

Workshop Report from 1995 till 2002.

Simon, D.L. and Buchenauer, D. 1993. *Genetic Diversity in European /Livestock Breeds.* EAAP Publications No 66, Wageningen Pers, Wageningen, The Netherlands, p. 581.

Smith, F., 1998. Proceedings of an International Workshop on the Development of Programmes for Co-operation on Animal Genetic Resources Management in Poland, the Baltic and the Nordic countries. Kaunas, Lithuania, 11-14 June, 1998.

Sources of Existence: Conservation and the sustainable use of genetic diversity. *2000 Policy document of the Government of The Netherlands.*

Sørensen, P. 1997. The population of laying hens loses important genes: a case study. *Animal Genetic Resource Information,* 22: 71-78.

3

Conservation of farm animal genetic resources - a UK national view

R.J. Mansbridge
Rare Breeds Survival Trust, NAC, Stoneleigh Park, Warwickshire, CV8 2LG, UK

Abstract

In 2003 the Rare Breeds Survival Trust (RBST) celebrated its 30th anniversary and is now widely recognised as the national non-governmental organisation responsible for rare breeds of farm livestock in the UK. No breed of farm animal has been lost since 1973 and there are now 72 breeds which meet the Trust's criteria for recognition as a rare breed. These criteria, which are regularly reviewed, take into consideration how long the breed has existed, the number of adult females and geographic distribution. With one or two notable exceptions, the breeds listed by the Trust have built up numbers and have become well distributed. However, all rare breeds still face an uncertain future but their greatest enemy is no longer immediate extinction but extinction by stealth. The sustained downward pressure on livestock farming in the UK, National and European government legislation, loss of genetic diversity and public indifference make a dangerous combination.

So what can an organisation like the RBST do, funded entirely by membership subscriptions, donations and legacies? The answer is a great deal! Firstly, we need to make people care, so that Governments and legislators listen and consider the implications of new (and existing) legislation on rare breeds. There is strong evidence that where people are seen to care, then government does listen to 'umbrella' organisations like the RBST lobbying on behalf of a small and sometimes fragmented sector of the industry. Widespread concern during the early days of the recent Foot and Mouth Disease (FMD) epidemic in the UK enabled the Trust to secure unprecedented exemption for rare breeds of sheep and pigs from contiguous culls. However, unlike captive zoo collections or wild populations, the individual cattle, horse/ponies, pigs, poultry and sheep/goats belong to individual breeders! Without their support, co-operation and participation, progress can be difficult and slow.

This national role, to be effective and comprehensive, must be grounded in sound science, reliable technical data and impartial information. The Trust now has the expertise and the tools to quantify the genetic diversity in rare breeds while offering practical solutions to some of the problems facing rare breeds. The RBST is doing a great deal but there is still much to do. The next 30 years promise to be as important in securing the future of the UK's rare breeds of farm livestock as the first 30 years were in rescuing them from extinction.

In 2003 the RBST celebrated its 30th anniversary. The origins of the RBST can be traced back to a letter in 1966 from the (then) Hon. Sec. for The Zoological Society of London, Lord Zuckerman (later regarded as the father of the rare breeds movement). He wrote to the Agricultural Research Council, National Agricultural and Advisory Service, Ministry of Agriculture, Fisheries and Food, Royal Agricultural Society of England and the University of Reading proposing a joint meeting. The aim of this meeting was to bring together all those who shared an interest in and concern for 'vanishing breeds of farm animals'. Seven years and many meetings later the Trust came into existence on 19 May 1973.

Thirty years on and the RBST is now widely recognised as the national non-governmental organisation responsible for rare breeds of farm livestock in the UK. During the intervening years, everyone connected with the organisation has worked hard to 'increase populations by recruiting new keepers of rare breeds and providing advice'. No breed of farm animal has been lost since 1973 and there are now 72 breeds (15 cattle, 12 horse, 7 pig, 9 poultry and 31 sheep and goats) which meet the Trust's criteria for recognition as a rare breed (see Table 1).

These criteria, which are regularly reviewed, take into consideration how long the breed has existed, the number of adult females and geographic distribution. Where a breed has been 'improved' by the introduction of genetic material from one or more other breeds, the Trust accepts only those animals where pedigree data confirms that they are free of introgression. Thus 'original populations' of Hereford and Aberdeen Angus cattle are recognised by the RBST, despite the fact that both these breeds are amongst the most numerous world-wide.

Today, the Trust has a membership approaching 10,000, a Council of elected and appointed Trustees and a managerial, technical and administrative staff of 12 based at the Trust's headquarters at the National Agricultural Centre in Warwickshire.

Enough of the past! No organisation, however successful, can afford to rest on its laurels. Instead it must look to the future, where our future will depend in large part on our ability not only to anticipate the problems and challenges but to identify achievable and sustainable solutions.

Table 1. RBST criteria for recognition of a rare breed.

Criteria For Recognition

Numbers of adult females?

	Category
	1. Critical
	2. Endangered
	3. Vulnerable
	4. At risk
	5. Feral
	6. Imported
	7. Traditional

Cattle	$<150 \rightarrow 1500$
Goats and Pigs	$<100 \rightarrow 1000$
Horses and Sheep	$<300 \rightarrow 3000$

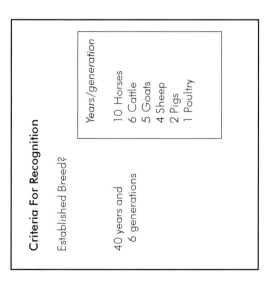

Criteria For Recognition

Established Breed?

Years/generation
10 Horses
6 Cattle
5 Goats
4 Sheep
2 Pigs
1 Poultry

40 years and
6 generations

With one or two notable exceptions, the breeds listed by the Trust have built up numbers and have become well distributed. While the dreadful events of the recent Foot and Mouth Disease (FMD) epidemic affected many, it decimated none of the breeds listed by the Trust. Some came close! The disease came within 4 miles of the only herd of Vaynol cattle left alive in the UK (at Temple Newsam, Leeds); the Chillingham herd of wild white cattle was surrounded by disease for many months, after the first outbreak 30 miles south at Heddon-on-the-Wall.

However, all rare breeds still face an uncertain future but their greatest enemy is no longer immediate extinction but extinction by stealth.

The sustained downward pressure on livestock farming in the UK, National and European government legislation, loss of genetic diversity and public indifference make a dangerous combination. Like all livestock keepers, owners of rare breeds look to recoup costs by selling breeding stock, surplus stock or produce in the form of meat, dairy products, wool or eggs. Extensive and exhaustive paperwork, movement restrictions, and poor returns are demoralising all but the most determined of rare breed keepers. Government legislation has greatly reduced the number of small and medium sized abattoirs, which were able to cater for the smaller producer.

The introduction of the National Scrapie Plan with its associated time-scales poses a huge potential threat to the genetic diversity of many rare breeds of sheep and could be catastrophic for some. Furthermore, some rare breeds, in an attempt to re-join mainstream livestock production, have introduced accelerated programmes of breed improvement, resulting in irretrievable loss of genetic diversity. Added to all this, the public perception of rare breeds is often limited to nostalgia rather than a conviction that maintaining genetic diversity may be a key to unlock the future.

So what can an organisation like the RBST do, funded as we are entirely by membership subscriptions, donations and legacies? The answer is a great deal!

Firstly we need to make people care. Our native breeds of farm animals are just as much a part of our heritage as Stonehenge, Hampton Court, Elgar, Kings College Cambridge, Ironbridge, etc. Numerically scarce breeds are even more precious and if we can make enough people care then we can avert the greatest threat to their future. This threat is not, as in the past, that the last flock or herd is about to be sold but, the insidious downward pressure of ever increasing European and National legislation on rare breed keepers. We need to make people care, so that Governments and legislators listen and consider the implications of new (and existing) legislation on rare breeds.

There is strong evidence that where people are seen to care, then government does listen to 'umbrella' organisations like the RBST lobbying on behalf of a small and sometimes fragmented sector of the industry. The Trust led a successful campaign of influential partners, including The Country Landowners Association and the Royal Society for the Protection of Birds (RSPB) urging Government to take the necessary measures to support small and medium sized abattoirs, which its own committees had already advised. While a number of these abattoirs have closed many more have continued operating as Government eased inspection criteria and therefore cost associated with their regulation.

Widespread concern during the early days of the FMD epidemic enabled the Trust to secure exemption for rare breeds of sheep and pigs from the contiguous culls. By working closely with the then Ministry of Agriculture, Fisheries and Food (MAFF), we secured remarkable and unprecedented recognition at Government level of the national importance of rare breeds. This 'sea change' in Government attitude, has been pursued with even greater vigour by the new Department of the Environment, Food and Rural Affairs (Defra), which is committed to maintain bio-diversity.

Defra listened to the concerns of the Trust and Breed Societies at the threat posed by the introduction of the National Scrapie Plan (NSP). We argued that there was simply not enough data on the prevalence of scrapie-susceptible genotypes in rare breeds of sheep and Defra responded by introducing the Rare Breeds Genotyping Scheme (RBGS). The Trust is only too well aware that arguments are best won with clear and unambiguous evidence and this scheme provides a unique opportunity to gain such evidence and work with government to secure the future of breeds shown to have a high proportion of scrapie-susceptible genotypes. However, our involvement in this often highly emotive issue has highlighted a truth fundamental to our aims and objectives. Unlike captive zoo collections or wild populations, the individual cattle, horse/ponies, pigs, poultry and sheep/goats belong to individual breeders! Without their support, co-operation and participation in initiatives such as this, progress can be difficult and slow.

While history will record that 2001 saw the worst ever recorded outbreak of FMD in the UK, it also saw an event which may have longer-lasting effects on rare breeds. The UK Government accepted an invitation from the Food and Agriculture Organisation (FAO) to participate in a report of the State of the World's Farm Animal Genetic Resources. For the first time, a UK government formally acknowledged the importance of farm animal genetic resources. The RBST was invited to join a National Consultative Committee comprising breeding and genetics experts and representatives from all sectors of the livestock industry. Under the

Chairmanship of Professor W G Hill, the Committee completed the UK Country Report in a remarkable 10 months. This has then been followed by a landmark meeting called by Defra to initiate the formulation of Government policy on genetic resources.

This national role, to be effective and comprehensive, must be grounded in sound science, reliable technical data and impartial (and wherever possible comparative) information.

The Trust now has the expertise and the tools to quantify the genetic diversity in our breeds. Developments in computer-based genetic analysis are facilitating a far greater insight into the effect of time (and breeding strategies) on the extent to which genetic diversity has been maintained. Early indications suggest that populations may have increased at the expense of genetic diversity. This effect can clearly be seen in one critically endangered breed of primitive sheep, on the Trust's lists from the mid 1980s. Although the breed still has less than 300 adult females, there has been a three-fold increase in population size during this time. However, this has been accompanied by a loss of over 40% of the original genetic diversity. The widespread use of DNA marker technology and significant advances in the field of cryogenics should ensure that further erosion of genetic diversity is minimised

During the early days and weeks of the FMD epidemic, it became only too clear that no-one, including the RBST, knew of the location and size of all herds and flocks of rare breeds throughout the country. The Trust, with the unstinting co-operation of the Breed Societies began to compile this information, enabling us to provide MAFF and then Defra with comprehensive lists of names and addresses of herd and flock owners by breed and size in each of the 'hot spot' areas of the country. These lists are forming the basis of a growing library of rare breed information currently under development by the Trust, with funding secured in 2002.

The Trust, a founder member of the Grazing Animal Project (GAP) Executive, has long recognised the role that rare breeds can play in extensive grazing systems. This is especially true where individual breeds are adapted to specific areas of the country, e.g. Norfolk Horn sheep, Exmoor ponies, Gloucester Old Spots pigs, Hereford cattle. The Traditional Breeds Incentive scheme recently launched by English Nature provides area-based financial assistance to farmers to support grazing by traditional breeds on a range of special sites. As the emphasis of the CAP shifts to 'second pillar' Agri-Environment support, schemes like this illustrate that conservation, environment and farming concerns can be successfully integrated into a holistic approach. They will provide an excellent opportunity to compare and contrast the ability of various breeds to deliver clearly defined and measurable results in a range of habitats and environments.

And last, but by no means least, the Trust offers practical solutions to some of the problems facing rare breeds. Niche markets for high quality, locally produced meat, milk products and woollen garments are rapidly emerging, thanks to the efforts of an increasing number of farsighted, committed and entrepreneurial individuals. Over a period of eight years, the Trust has invested approaching £0.25 million in a traditional breeds meat-marketing scheme (TBMMS). At a time when rare breeds keepers were open to derision and pitifully low prices for finished stock, the RBST introduced a scheme which established a direct link between producer and butcher. Through articles in the national press and lifestyle magazines, together with the endorsement of celebrity chefs, traditional breeds meat has a discerning and growing clientele who enjoy the taste and texture of animals reared slowly under extensive systems. The scheme has grown and is becoming increasingly independent of the Trust.

Further evidence of this practical support can be seen in the advice and funding which the Trust provides for networks of sales, Breeding Units, Conservation Centres and Conservation Breeding Groups.

In conclusion, the Trust is undertaking a great deal and yet there is still so much to do. We are funded through membership subscription, donations and legacies. In order that our income can keep pace with the increasing demand on the Trust's physical and financial resources, we launched the Regeneration Appeal in 2001 at the height of the FMD epidemic. We have raised £750,000 to date but are still a long way short of our target of £2.5million. But get there we must because the next 30 years promise to be as important in securing the future of the UK's rare breeds of farm livestock as the first 30 years were in rescuing them from extinction.

4

Evolution of Heritage GeneBank into The Sheep Trust: conservation of native traditional sheep breeds that are commercially farmed, environmentally adapted and contribute to the economy of rural communities

D. Bowles[1], P. Gilmartin[2], W. Holt[3], H. Leese[4], J. Mylne[5], H. Picton[6], J. Robinson[7] and G. Simm[8]

[1]CNAP, Department of Biology Area 8, University of York, P O Box 373, York, YO10 5YW, UK

[2]Centre for Plant Sciences, Leeds Institute for Plant Biotechnology and Agriculture, Faculty of Biological Science, University of Leeds, Leeds, LS2 9JT, UK

[3]ZSL Institute of Zoology, Regent's Park, London NW1 4RY, UK

[4]Department of Biology Area 3, University of York, P O Box 373, York YO10 5YW, UK

[5]Britbreed Ltd, 1 Airfield Farm, Cousland, Dalkeith, Midlothian, EH22 2PE, UK

[6]Academic Unit of Paediatrics, Obstetrics & Gynaecology, Leeds General Infirmary, Belmont Grove, Leeds, LS2 9NS, UK

[7]Scottish Agricultural College, Craibstone Estate, Bucksburn, Aberdeen, AB21 9YA, UK

[8]Scottish Agricultural College, West Mains Road, Edinburgh, EH9 3JG, UK

Abstract

The Foot and Mouth Disease (FMD) epidemic of 2001 clearly illustrated the fragility of the UK's farm animal genetic resources. In particular, millions of sheep were killed by the disease and by the 'stamping out' policy chosen for disease control. Loss of genetic resources was not evenly spread throughout the UK, nor throughout the many different sheep breeds that are native to the UK and for which the UK has a formal responsibility for protection to the United Nations. In fact, the FMD epidemic demonstrated for the first time that sheep breeds comprising large numbers of individuals which are commercially farmed, can nevertheless be at considerable risk of extinction. The breeds most affected were those restricted to geographical regions of the UK into which the FMD spread. These regionally important breeds are adapted to their particular regional environments, represent an important living heritage for the UK and are a key component in sustaining the rural economies of sheep farming communities.

The events of 2001 provided clear proof that there are two components of the UK's farm animal genetic resources demanding protection. One component is already recognised as a priority and is composed of the numerically rare breeds of all domesticated species: these are already under the protection of the Rare Breeds Survival Trust (RBST). The second component has not previously been recognised as a priority for protection. The FMD crisis proved that sheep breeds could exist as large numbers of individuals, but nevertheless face extinction due to their regional location. Urgent attention must be focussed on our Heritage Breeds of sheep. The UK has one of the greatest number of native sheep breeds of any country in the world. The Heritage Breeds provide potentially valuable genetic resources for environmental, low-input farming systems.

Heritage GeneBank was founded during the FMD epidemic specifically to protect sheep breeds at threat of extinction from the disease. A group of academic research scientists established a genetic salvage programme: collecting semen and embryos for protection in a gene bank. Germplasm from seven breeds is in long-term storage. Following the crisis, the scientists involved in the gene bank made a commitment to continue their conservation work in recognition that the Heritage Breeds of sheep in the UK continue to require protection.

This paper describes: (1) the work of Heritage GeneBank (HGB); (2) the threefold mission of The Sheep Trust, the new national charity that evolved from HGB (www.thesheeptrust.org); and (3) the ongoing urgent need for conservation of the UK's Heritage Breeds of sheep threatened by genetic erosion.

Heritage GeneBank is founded

The Herdwick breed of sheep in Cumbria comes under threat

Following the first diagnosis of FMD in the UK, the spread of the disease was worsened over the next days by continuing animal movements through auction sales. In particular, many thousands of sheep were continuing to be bought, transported, sold and re-bought at auction centres throughout the UK via the complex network of dealers buying in sheep from breeders and selling them on to other dealers in the UK and elsewhere. A major auction centre for sheep movements was Longtown, situated in Cumbria. This provided a focal point for disease build-up and spread, with the consequence that the epidemic was first strongest in Cumbria and the adjacent Lancashire and Scottish Borders regions of the UK.

At the height of the first stage of the epidemic in Cumbria, the Herdwick breed of sheep was losing numbers at a rate of several thousands daily. The breed is a distinctive feature of the Lake District, important to tourism as well as to environmental management of the fells and provision of income to the local farming community. Herdwick breeders contacted Professor Dianna Bowles to ask if, as a scientist, there was anything she could do to help, since they strongly believed the breed in entirety was being taken out. Professor Bowles kept a small (micro) flock of Herdwicks in Yorkshire, and was known to the breeders in Cumbria through tup (ram) sales.

A number of factors contributed to the heavy losses suffered by the Herdwick breed. (1) The breed is >98% situated in the one geographical region of the UK, where it is commercially farmed – this contrasts with rare breeds that are generally maintained by enthusiasts and are scattered in small numbers throughout the country. (2) The breed is farmed in Cumbria typically as 'fell flocks', often numbering 2,000–3,000 breeding ewes that are put to the fell (the high, unfenced mountains and moorlands of the Lake District) for most of the year, and 'heft' to specific regions of the fell that they regard as their home territory. (3) The landscape of the Lake District is such that open fells are contiguous and can house flocks from many different farms. (4) Gimmer lambs (female lambs prior to their first shear) are sent away from the fell farms to low country for their first 9 months.

The 'stamping out' policy was such that if one flock was diagnosed with FMD, all adjacent flocks and those within a 3 km radius were also killed. This led to killing across large swathes of the countryside, affecting not only animals on the producers' immediate land, but those on other holdings that he/she had visited and all those of his/her neighbours and neighbours to the other holdings. Some breeders lost all their animals; others 'saved' populations that they didn't visit; others lost all their young stock when FMD hit the low country, but saved their older stock through its isolation on fells beyond the FMD spread.

Neither the restriction of the breed to a single location and its adaptation to that environment, nor the large numbers (~80,000) that existed, protected the Herdwicks from the disease. In fact, these factors contributed greatly to its losses, and in late March 2001 led everyone to fear the breed could become close to extinction.

Heritage GeneBank - a science-led response to the crisis

Given the emergency that existed, the only means of salvaging the breed was to collect its germplasm. The academic network is such that it rapidly became feasible to gather together all the specialists necessary to undertake a field operation. Funding was sought from the Garfield Weston Foundation, and by return the charity provided £50,000 to enable the fieldwork to be planned and executed.

A team of scientists, together with a local breeder and a local vet in Cumbria, co-ordinated the interactions between the specialist team, the Herdwick breeders, and Defra, to ensure sanitary rules were in place, regulations were maintained, so that visits to farms outside the 3 km cull zones were permitted. A field team was operating within one week of the phone call requesting help.

It was agreed to undertake a dual strategy of semen collection and embryo collection, using farms most distant to known outbreaks to start the process of hormone treatment of ewes leading to ovulation and ultimately embryo harvest. March was considered to be the 'worst time of year' to undertake this work with sheep, but inclusion of world-class specialist vets and research scientists ensured success.

Sanitary regulations were such that all germplasm collected from a farm had to be maintained on that farm in liquid nitrogen containers. Eventually, Defra enabled it to be transferred from individual sites to a designated storage place at the local veterinary practice, which also held the liquid nitrogen reservoir. Offers of additional financial support came in from Defra, from the Countryside Agency and from a second charitable organisation, The Gatsby Charitable Foundation. This enabled all of the necessary storage equipment to be purchased and for the conservation work to extend to other sheep breeds as the FMD virus progressively spread into other regions.

The Rough Fell breed of sheep, also in Cumbria but closer to the Lancashire border, lost 75% of its breed numbers through the FMD. HGB helped this breed and also the Lonk breed, situated adjacent to the Rough Fells, but closer to Yorkshire. By mid-2001, FMD started new infection loci in Yorkshire, with the Dalesbred also coming under threat. The policy of HGB was essentially an open one, responsive to breed societies and individuals with bloodlines of particular significance. In consequence, HGB also collected germplasm from three 'rare' sheep breeds: the Lincoln Longwools, Whitefaced Woodlands and the Portland. This work was undertaken because no other organisation at the time was able to respond to the requests for help.

All of the germplasm collections are maintained in duplicated cryopreservation tanks at a single site in the UK. Essentially, they form the foundation of the first national gene bank for native sheep breeds.

Heritage GeneBank evolves into The Sheep Trust

Lessons learnt from the FMD epidemic: the importance of heritage breeds

The events of 2001 provided clear proof that there are two components of the UK's farm animal genetic resources demanding protection. One component is already recognised as a priority and is composed of the numerically rare breeds of all domesticated species: these are already under the protection of the RBST. The second component has not previously been recognised as a priority for protection. The FMD crisis proved that breeds could exist as large numbers of individuals, but nevertheless face extinction due to their regional location. In this context, urgent attention must be focussed on our living heritage of native sheep breeds. The UK probably has the greatest number of native sheep breeds of any country in the world. Most of these breeds exist in large numbers, constitute the greatest genetic biodiversity in the national flock and provide valuable genetic resources for environmental, low-input farming systems.

The UK Government has long since recognised the special case of the 'rare breeds'. A number of policy decisions are included in statutory requirements that are specifically targeted at the protection of rare breeds. For example, special clauses in the most recent (2002) Animal Health Bill. However, it is now very apparent that the attention of Government must be wider and must also incorporate measures to recognise and protect the regionally adapted 'Heritage Breeds'. These breeds already contribute to land management in many areas of the UK with high tourism interest – such as the Lake District, the Yorkshire Dales, the Scottish Borders, and the Welsh Mountains. With an increasing emphasis on the need for an environmentally sustainable agriculture, it is probable that breeds able to thrive in low-input, near-organic systems will become the sheep breeds of choice for many livestock farmers.

The mission of The Sheep Trust

The Sheep Trust was formally established in September 2001 as a charitable company and received Registered Charity status in November 2002. Through the continued support from the Garfield Weston Foundation and the Countryside Agency, a Co-ordinator for the Trust was appointed and The Trust's National Office was established to design new programmes of activity.

The Sheep Trust has a three-fold mission to: (1) conserve the genetic biodiversity of the 'Heritage Breeds' of sheep of the UK through *ex situ* and *in situ* measures; (2) help those breeds achieve greater economic and environmental sustainability; and (3) help ensure policy is underpinned by excellent science and the science is communicated to the public and farming communities in a readily accessible way.

The heritage breeds of sheep

Current priority breeds of The Sheep Trust

The Sheep Trust is working with a number of sheep specialists and organisations to define a priority breed list. Whilst the Trust's focus is native UK sheep breeds, it recognises that numerically scarce breeds are protected by RBST. Similarly, there are sheep breeds that are not geographically isolated and environmentally adapted to specific regions but rather are widely farmed throughout the UK. In recognition of the specific problems highlighted by FMD that face the regionally important pure-bred flocks of the UK, a priority list of 'Heritage Breeds' has been drawn up for immediate action. This list is shown in Table 1.

Three parameters have been examined in compiling the list of Heritage Breeds – the extent of their: (1) geographical isolation; (2) environmental adaptation; and (3) presumed genetic distinctiveness. Each breed has been allocated a 'score' under the three parameters from 1–8, with 8 representing the maximum. The Herdwick breed has the highest points, scoring maximum for each parameter.

Whilst the list provides a useful qualitative guide for prioritisation, it is essential that a scientific foundation is developed for the more precise quantification of each parameter.

Geographical isolation is relatively straightforward to define qualitatively, since the high-scoring breeds are indeed greatly restricted in their distribution across the UK. Breed enthusiasts may exist who maintain small flocks outside of the main area – such as the 'microflocks' of Herdwicks scattered in many regions. Nevertheless, even in this highly popular breed, >95% of individuals continue to be located and bred in Cumbria, where the hardiness and thriftiness of the breed continue to be encouraged by the indigenous fell-farming system.

D. Bowles et al.

Sheep Breed	Geographical Isolation	Genetic Distinctiveness	Envrionmental Adaptation	Total Score
Herdwick	8	8	8	24
Dorset Horn	4	8	6	18
Shetland	6	6	6	18
Rough Fell	6	5	6	17
Derbyshire Gritstone	3	6	6	15
Lonk	4	5	6	15
Romney	4	4	5	13
South Welsh Mountain	5	4	4	13
Welsh Hill Speckled	5	4	4	13
Badger Faced Welsh Mountain	4	4	4	12
Blueface Leicester	3	5	4	12
Dalesbred	4	3	4	11
Exmoor Horn	4	3	4	11
Swaledale	3	3	4	10
Devon & Cornwall Longwool	6	2	2	10
Devon Closewool	6	2	2	10
Black Welsh Mountain	2	5	3	10
Welsh Mountain	1	4	4	9
Cheviot	4	3	2	9
Brecknock Hill Cheviot	3	2	4	9
Blackface	1	3	4	8
North Country Cheviot	1	3	4	8
Clun Forest	2	3	2	7
Jacob	0	5	1	6
Lleyn	0	4	2	6

Table 1. Current Sheep Trust priority list.

Geographical isolation: based on the extent to which a breed is isolated to one geographical area and therefore placed at risk by a disease such as FMD and the resulting control policies

Genetic distinctiveness: based on presumed genetic distance, phenotype and performance criteria

Environmental adaptation: based on the extent to which the breed is restricted to a particular geographical region and may therefore be adapted to that environment and used in environmental habitat management

Environmental adaptation is a more complex parameter to define quantitatively. For example, it is unclear without further research, how much one hill breed could substitute for another, and therefore how adapted and dependent each breed is on a very specific environment. Could the fells of Cumbria survive in their full environmental biodiversity

51

if grazed by Blackface or Cheviots? Similarly, is it possible to really quantify the unique adaptation of the Lonk sheep, Dalesbred and Swaledale to specific regions of Lancashire and Yorkshire? Information from long-term grazing projects and ecological research into impact analysis of the different sheep breeds in the different habitats will be necessary before this parameter can be fully quantified.

However, in a related context, it is clear that rural communities are tied very closely to individual sheep breeds. A hill-farming family often will have specialised in breeding a specific sheep breed over many human generations. Just as the 'hill producers' have no interest in lowland sheep breeds, it is equally true that a Swaledale breeder will have limited knowledge of a Lonk, Rough Fell or Herdwick. Rural communities have grown around specific sheep breeds and this is particularly true for the remote hill-farming communities linked to many of the 'Heritage Breeds'. The sheep may be adapted to their environments, but just as significantly, the human communities responsible for caring for those environments are adapted to specific sheep breeds.

The parameter of genetic distinctiveness is perhaps the most readily accessible from a quantitative perspective. The advent of DNA-based molecular technologies, the development of new molecular markers and the increasing knowledge of the sheep genome, means research into the basis of genetic distinctiveness is now feasible. However, in the absence of this information for most breeds, some attempt has been made to assess genetic distinctiveness based on phenotype and breed history.

Genetic diversity at the level of the individual, the population, the breed and the species can now be characterised using molecular data (Frankham *et al.*, 2002). Typically, at present, much of the research relies on the use of microsatellites, but increasingly automated technologies involving the use of single nucleotide polymorphisms (SNPs) will revolutionise the sensitivity of the analyses. Recent research (such as described in Bruford's paper in this publication) has focussed on genetic variation between breeds, and perhaps more significantly, within breeds.

The question of relating molecular data on genetic diversity and distinctiveness to prioritisation for conservation is discussed in detail by Bruford. Some key issues raised in that paper are highlighted here due to their relevance to the work of The Sheep Trust and the heritage importance to the UK of the breeds prioritised by The Trust.

Bruford points out that in the UK there is a wide variety of livestock breeds, some of which may be comparatively ancient and extensively farmed over long periods of time, but many others that are much more recent and have originated less than 100 generations ago. For the latter breeds, it is highly unlikely that significant amounts of genetic diversity will have accumulated through mutation. In contrast, demographic fluctuations, such as imposed by disease outbreaks, in breeding, low effective population size in males and population isolation, are all highly likely to influence breed diversity. Bruford's major concern is that these demographic fluctuations, 'are inadequately incorporated in nearly all population and phylogenetic models for understanding, comparing and prioritising genetic diversity'.

In the context of this paper, two facts to emerge from his recent research are relevant: (1) there is a high level of genetic divergence between breeds even if they are geographically proximate; and (2) even a common breed may be under great risk of loss of diversity due to breeding practices. A consequence of the second point is that conservation policies cannot simply focus on the rarity status of a breed, but should take into account modern breeding practices, which he states: 'represent a threat to many breeds'.

In the above context, The Sheep Trust considers that a specific threat of major significance to the Heritage Breeds is the current selection pressure applied through the National Scrapie Plan (NSP) and associated European legislation.

Accelerated genetic erosion through the NSP

Today, the NSP represents a substantial threat to the UK's farm animal genetic resources. The national flock is being forced through a rapid artificial selection process based on the presence/absence of a single allele of a single gene. The gene encodes the prion protein, a rogue form of which is believed to be the causal agent of the family of diseases known as the Transmissible Spongiform Encephalopathies (TSEs). Amongst the TSEs, Scrapie has been a sheep disease known for several hundreds of years. The prion-related disease in cattle is Bovine Spongiform Encephalopathy (BSE) and amongst those in humans is variant Creuzfeld Jacob disease (vCJD). It is fully justified on food safety grounds to minimise the risks to human health from TSEs. However, the contribution of Scrapie-infected meat in the human diet to human TSEs is unproven and is considered by many to be unlikely, given the lack of causal relatedness over such a long timespan.

The principal driver in current Government policy is undoubtedly the potential for sheep to have contracted BSE from infected feed in the 1980s. The clinical symptoms of BSE and Scrapie in sheep cannot be distinguished readily. There is currently no EU validated laboratory test that can discriminate between Scrapie-induced defective prions or BSE-induced defective prions (although different TSE strains do show different incubation periods in mice). Similarly, there is no test to measure the level of defective prions in a live animal, nor any clear knowledge of the relationship of defective prion level to disease transmission.

New research is an urgent requirement to solve these unknowns and uncertainties. Similarly, it should be a priority to fund the development of new diagnostic tools. Instead of this science-based approach to the problem, policy has chosen to minimise the theoretical risk to human health by eradication of Scrapie from the national flock.

To achieve this eradication, a selection of animals based on their prion allele genotype is underway. This strategy has a number of drawbacks:

(1) Prion allele genotypes correlated with susceptibility to Scrapie may be different from those correlated with susceptibility to BSE – indeed there is emerging evidence that this is the case.

(2) There is no information on gene linkage. Eradication of prion alleles will lead to eradication of linked gene alleles. These may be genes encoding traits that are highly important, whether in terms of production, breed quality, environmental adaptation, disease resistance, and so on.

(3) Many native sheep breeds of the UK have a low to very low incidence of the ARR allele. This can be a low as 10% of the male population for many of the Heritage Breeds protected by The Sheep Trust. Through selecting only homozygous or heterozygous ARR genotypes, vast genetic resources are: (a) not being used in current breeding programmes; (b) threatened with extinction, given the NSP's requirement for slaughter or castration; and (c) threatened by substantially increased inbreeding and loss of genetic variation.

(4) The national voluntary programme, now coupled with the incoming EU legislation, imposes extraordinary selection pressure over a very short timeframe.

The reason(s) why sheep contract Scrapie remains unknown. The causal role of accessory proteins in conversion of prions to defective molecular species is unknown. The relatedness of prion alleles to clinical symptoms in the sheep and disease transmission to humans is unknown. Yet, even as this article is published, hundreds of thousands of healthy, high breed quality rams of superior production, adaptation and disease resistance traits, will be lost – simply because they do not possess the 'right' prion allele for the chosen policy.

The FMD crisis was extremely visible through the involvement of the media and the effects of the devastation on tourism. In contrast, the current crisis to sheep genetic resources is comparatively hidden. The farming community is anxious to promote food safety and since the NSP's principal raison d'être is to eradicate risk, successful marketing of meat will increasingly demand acceptance of the terms set out by the policy. The tragedy, however, is the lack of sufficient scientific evidence for that policy, the loss of irretrievable genetic resources and the long-term negative impact on the genetic and hence phenotypic qualities of the national flock.

Concerns such as these have recently led the UK Government to consider the formation of an NSP Semen Archive. Selected examples of rams due for slaughter/castration within the NSP will have their semen conserved in an archived gene bank. It is to be hoped that this initiative will not in future be described as too little, too late.

The Sheep Trust, in collaboration with other organisations including the National Sheep Association (NSA), Rare Breeds Survival Trust (RBST) and Rare Breeds International (RBI) are working to help ensure the Archive is put in place as soon as possible. As part of the ongoing discussions at national level on conservation of farm animal genetic resources, The Sheep Trust participated in the National Consultative Committee and in discussions to take forward the National Action Plan as part of the UK's commitment to the UN/FAO Convention on Genetic Biodiversity. Whilst all native animal species and breeds in the UK are deserving of conservation, The Sheep Trust remains particularly concerned at the immediate threat facing the genetic resources of the national flock, and particularly the Heritage Breeds.

Acknowledgements

We are grateful to Mr David Croston, Chief Executive of the English Beef and Lamb Executive, and Mr Joe Read, former Head of the Meat and Livestock Commission Sheep Improvement Services, for their valuable input to discussions on breed characteristics.

References

Frankham, R., Ballou, J.D. and Briscoe, D.A. 2002 *Introduction to Conservation Genetics*, Cambridge University Press, Cambridge, UK.

5

The UK Government policy on farm animal genetic resource conservation

M. Roper
Defra, 9 Millbank, c/o Nobel House, 17 Smith Square, London, SW1P 3JR, UK

Abstract

Historically the UK has not had or needed a defined Government policy on the conservation and utilisation of farm animal genetic resources. However, this situation has changed recently, partly as a result of international efforts, stimulated by the Convention on Biological Diversity, and led by the Food and Agricultural Organisation of the United Nations, to co-ordinate national strategies for conservation and utilisation of farm animal genetic resources. As part of this international effort, a National Consultative Committee was set up in the UK in 2001. This committee produced the UK Country Report on farm animal genetic resources, which was published in 2002 and submitted to FAO. This paper outlines the structure and recommendations of this report, and discusses government policy on farm animal genetic resources.

Introduction

Historically the UK has not had or needed a defined Government policy on the conservation and utilisation of farm animal genetic resources (AnGR). Unlike many other countries, genetic improvement of farmed species and breeds has been left to breed societies, commercial breeders and breeding companies with technical support from scientific institutes, University Departments and levy boards such as the Meat and Livestock Commission (MLC). British breeding companies and individual breed societies, supported by world renowned animal geneticists in the UK, have been extraordinarily successful in improving both native and exotic high performance breeds, and in some species, developing sophisticated hybrid breeding programmes from them. Such breeds and crosses have become the bedrock of the commercial breeding herd and flock replacement market and there has been little need for Government intervention, except in funding research and development in animal genetics in collaboration with industry and stimulating uptake of new breed improvement techniques principally through the levy bodies.

However, as performance levels of these mainstream commercial breeds, lines and crosses outstripped their counterparts – many of which were historic native breeds, the former became more popular at the expense of the latter. Concern at the potential loss in farm animal genetic diversity began to be expressed and the challenge of monitoring and conserving breeds at risk of extinction was picked up by non-Governmental organisations (NGOs) funded by charitable donations. Once again the private sector, led by breed societies and NGOs was very successful in rescuing breeds from extinction and setting up conservation programmes. In the absence of market failure there was no call on Government to intervene.

In the last 5 years however, there have been two principle drivers, which have led to Government reviewing its policy on AnGR:

1. International concerns over the loss of genetic diversity in farm animals at the global level, as improved mono-breed or crossbred production systems have evolved and spread, have led to the development of a global strategy for the conservation and utilisation of animal genetic resources in food and agriculture. This global strategy, fuelled by the Convention on Biological Diversity (CBD) and implemented through the Food and Agriculture Organisation of the United Nations (FAO) Commission on Genetic Resources in Food and Agriculture, has now extended to individual nation states. They are being actively encouraged to develop their own national strategies and action plans on AnGR through the submission of an official Country Report to the FAO as part of the *"First Report on the State of the World's Animal Genetic Resources"* (SoWAnGR). Furthermore, regional co-operation on AnGR issues is being fostered through the FAO global structure, which has included the establishment of Regional Focal Points for AnGR.

2. At the same time, the UK Government is indirectly encouraging the utilisation of a broader spectrum of breeds through policies on rural development, environmental management, tourism, cultural heritage, as well as for more biologically and environmentally sustainable food production. Also, as the demand for more consumer choice and variety in food products for different eating occasions grows, market opportunities for regional or breed-based food products are becoming evident which the Government is keen to encourage.

UK country report on farm AnGR

In August 2001 the Parliamentary under Secretary (Lords), Lord Whitty, accepted an official invitation from the Director General of the FAO to participate in the SoWAnGR Report. A 24-strong National Consultative

Committee (NCC) was set up. This included representatives from all the principal national farmed livestock species organisations (dairy cattle, beef cattle, sheep, pigs, poultry, equines), conservation NGOs, scientific institutes and University Departments with an interest in animal breeding and conservation, under the chairmanship of Professor W G Hill of the University of Edinburgh. The composition of the NCC was designed to ensure that a balance was struck in the Report between maintaining animal genetic resources for mainstream production in the UK and conserving those considered to be breeds at risk – especially rare breeds in danger of extinction. There was also a need to strike a realistic balance between the need to genetically improve individual breeds to meet the needs of an evolving market and the need to conserve genetic diversity in the UK's native breed populations.

Structure of the Report

A thorough review of the state of the UK's AnGR was conducted with contributions from all representative organisations. The consolidated Report:

1. Assesses the current state of farm animal biodiversity in the UK.

2. Analyses the changing demands on national livestock production and their implications for future national policies, strategies and programmes related to AnGR.

3. Reviews the state of national capacity to support the conservation and utilisation of AnGR in the UK and assesses where the gaps and needs are for capacity building.

4. Identifies national priorities for action on the conservation and use of AnGR.

5. Formulates recommendations for enhanced national and international co-operation in the field of AnGR.

It contains a number of useful appendices including background information on livestock and poultry production levels; a complete list of farm animal breeds in the UK, which will form the basis of an updated national AnGR database; the requirements for UK rare breed eligibility based on the Rare Breeds Survival Trust (RBST) list; and a review of minor species in the UK and AnGR in the UK's 15 overseas territories.

The completed report, adopted by Defra and devolved administration Ministers, has been forwarded to FAO to be incorporated into both a regional European Report and eventually a Global Report (by 2006) to form a basis for international action on the conservation and sustainable utilisation of the world's AnGR.

Categorisation of breeds in the Report

One of the key outcomes of the report writing process was to define more clearly the categorisation of breeds in the UK and their prioritisation for conservation. At present the UK Government only recognises numerically 'rare' breeds (as listed by the RBST) as being eligible for particular exemptions in, for example, the control of major disease outbreaks. However, there are other breeds, which may have a particular local adaptation (e.g. the Herdwick sheep breed which is a hardy hill breed from Cumbria, that demonstrates 'hefting' ability) or are genetically distinct populations in some way (e.g. Dexter cattle). These breeds may also be described as breeds at risk and worthy of specific conservation measures, although it was acknowledged that more work needed to be done in defining more scientifically the measures of local adaptation and genetic distinctiveness. After much discussion and debate a compromise categorisation was reached to ensure that the conservation requirement of all breeds at risk was addressed. This is shown in Figure 1.

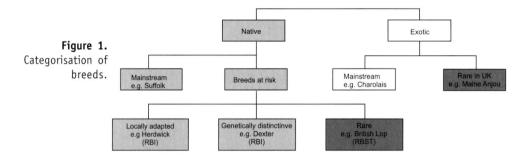

Figure 1.
Categorisation of breeds.

Breeds are initially split into 'exotic' or 'native'. They are then categorised as being either 'mainstream' or 'breeds at risk'. Breeds at risk are then subdivided into those which are 'rare', and those which are not numerically scarce but are judged to have a specific 'local adaptation' or are recognised as being 'genetically distinct'. These last two categories need to be much more clearly defined than is the case at present, particularly in assessing more scientifically the measures of local adaptation and genetic distinctiveness. However, they are very much part of our animal breeding heritage and have both cultural and potential economic value. Where either of these categories of breeds at risk are to be found concentrated in a particular geographical region or are particularly genetically susceptible to notifiable diseases then, even though they may not be numerically scarce, they may be very vulnerable to drastic, and threatening, population reduction in the event of disease epidemics such as FMD.

The revised official UK National Database held by the UK National Co-ordinator for AnGR at Defra, reflects these categorisations.

Recommendations of the Report

The Report makes a number of recommendations for enhanced co-operation at international, European regional and national level. It is current Government policy to encourage collaboration at international, and especially at regional European level. Defra plays an active role in global discussions at the Commission for Genetic Resources and in the development of the European Regional Focal Point for AnGR based in Paris. Co-operation in conservation of AnGR at European level is likely to deliver cost savings in the longer term.

The report also highlights the need to revitalise international trade in the UK's animal genetic resources, lost through BSE and FMD trading restrictions. This will assist in maintaining a threatened animal breeding infrastructure by bringing in much needed income to fund the implementation of evolving breeding objectives.

At national level a number of specific recommendations have been made, including:

1. Creating a more active information structure to enable all those active in animal breeding to benefit from the world-renowned expertise that exists in the UK in the management and sustainable development of our animal genetic resources.

2. Developing greater co-operation between Government, NGOs, private and commercial breeding organisations and scientific institutes. At national level there is a wide range of NGOs with various sectoral or species interests in the various categorisations of breed. Some of those interests overlap, whereas for some breeds there is no overarching body looking after their interests. Table 1 gives an example of the sort of matrix of NGOs that could be constructed to demonstrate the range of NGOs with leading interest in particular breed categories. The Report recommends that Government, through the UK National Co-ordinator for AnGR, should be active in facilitating co-ordination between these various NGOs and in a wider arena to include scientific institutes, private breeding companies, levy bodies and breed societies (Figure 2).

Table 1.
Example of matrix of lead NGOs (See Appendix 1 for Glossary of abbreviations).

Breed category	Cattle	Sheep	Pigs	Poultry	Equines
Mainstream	NBA RABDF	NSA	NPA	BPC	BHS NPS
At risk-locally adapted/ distinctive	NBA RABDF RBI	ST RBI	BPA	?	BHS NPS
At risk-rare	RBST	RBST	RBST	RBST	RBST

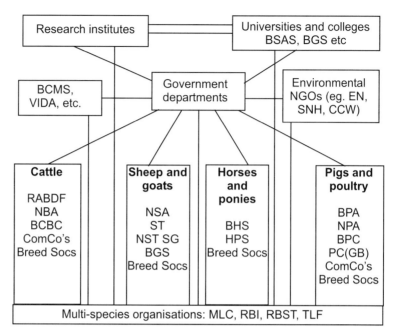

Figure 2.
Organogram of UK
organisations
involved in AnGR
(See Appendix 1 for
Glossary of
abbreviations).

Figure 2.
Organogram of UK
organisations
involved in AnGR
(See Appendix 1 for
Glossary of
abbreviations).

3. Providing dedicated resources for a range of *in situ* and *ex situ* conservation activities, such as the setting up of dedicated gene banks, the characterisation of breeds to identify their value for current and future utilisation and the development of national AnGR databases.

4. Creating a National Action Plan for Animal Genetic Resources based on the recommendations in the Report and the reconstitution of the existing National Consultative Committee into a Steering and Advisory Committee on Animal Genetic Resources to draw up the Plan and drive it forwards.

Ministers at Defra, and in the devolved administrations in Scotland, Wales and Northern Ireland, have accepted the UK Country Report. The recommendations contained in it now need to be assessed and responded to.

Defra policy review on genetic resources in food and agriculture

One of the problems in progressing Government policy on farm AnGR is the fact that there is no existing overarching genetic resource policy. Responsibility for AnGR lies currently within the Sustainable Agriculture and Livestock Products Directorate (SALPD) in Defra under the UK National Co-ordinator for AnGR. However, there is no specific policy unit covering it with an allocated budget. AnGR projects can only be engaged in on their individual merits and financed out of general livestock or research and development budgets. For a National Action Plan on AnGR to be developed and a Steering and Advisory Committee to be created, there is a need for a co-ordinated Defra (and devolved administration) policy on genetic resources in food and agriculture to justify the time and resources spent on it.

This is one of the reasons why Defra is currently reviewing policy on genetic resources across the Department – including all aspects of policy in the various Directorates that impact on their conservation and utilisation. The review, instigated by the Research Policy and International Division who have responsibility for plant GR, covers the plant, animal and microbial kingdoms and will be completed by the end of 2002. Stakeholders are being consulted as to the expectations they have of Government involvement in genetic resources. A Defra project officer has been visiting such stakeholders to canvass their views. Also a workshop was held at the University of Birmingham on 25 September 2002 attended by the Minister of State for the Environment, Michael Meacher, to develop collaboration between the three kingdoms and exchange ideas on a possible policy framework.

Government does also have certain obligations flowing out of the Convention on Biological Diversity to conserve biological diversity in food and agriculture. Up until recently the emphasis has been on wildlife biodiversity – indeed national biodiversity action plans rarely even make mention of farm animal biodiversity. However, at a recent meeting of the Convention of Parties (COP) of the CBD this issue was raised and Governments will now be more actively encouraged to give due consideration to the conservation of farm species and breeds in their future biodiversity plans.

If the current review results in a new policy for genetic resources within Defra approved by Ministers, then it will facilitate the setting up of a Steering and Advisory Committee to oversee the drafting and implementation of a National Action Plan for AnGR. The devolved administrations will develop their own policies on AnGR, which will reflect their own requirements but the UK National Co-ordinator in Defra will act as a focal point for UK co-ordination.

EU Commission Policy on AnGR

EU Commission support for AnGR has been limited in the past, but is increasing. There are two principal policy instruments that encourage the conservation and utilisation of AnGR:

1. *Rural Development Regulation (RDR)*. There is limited provision through the RDR (implemented through the various National Development Plans) to make headage payments to keepers of rare breeds as defined under criteria set out by the European Association of Animal Production (EAAP) and agreed by the ERFP. The breeding female population ceilings below which subsidy is paid are set on an EU-wide basis and the EAAP database in Hannover is used as a reference point. Subsidy is only payable where there is some additional rural development or environmental benefit derived from conserving the breeds concerned. This specific support for rare breeds was not implemented in the UK due to budgetary constraints at the time. Experience from other parts of the EU, notably Germany, indicates that such headage payment schemes have been expensive to administrate in proportion to the payments made to rare breed producers. In the UK, therefore, it is highly unlikely that such direct headage payments will be implemented. It is more likely that environmental or rural development schemes will be developed which indirectly benefit rare breed production. This may be particularly relevant in the context of the mid-term review of the Common Agricultural Policy and the decoupling of subsidy payments away from production support into environmental and other schemes which deliver 'public goods'.

2. *The Conservation, Characterisation, Collection and Utilisation of Genetic Resources in Agriculture Regulation.* This 5-year programme of funding which came to an end in 1999, was established under EC regulation 1467/94. It was criticised for a disproportionate allocation of funding in favour of plant projects. The Commission has since tabled proposals for a new version of this regulation, likely to be adopted during 2003. It is intended that the new regulation will give greater emphasis to projects related to animal genetic resources and *in situ* conservation, and will require several Member States to collaborate for projects to be accepted. This could be a significant vehicle for deriving additional funding for pan European conservation projects.

Conclusion

UK Government policy on AnGR is evolving more actively than at any time over the last 50 years. The emphasis of general livestock policy now is on environmentally, biologically and economically sustainable production systems. This will require a widening of breeding objectives to take into account environmental, animal welfare and wider diversification (e.g. tourism, rural development) requirements. The rich diversity of breeds in the UK is a reservoir of valuable genetic resources for breeders and the industry at large to call on. Government clearly has a role in ensuring that the conservation of those resources is well co-ordinated nationally and linked in to wider regional and global conservation programmes through the setting up of a Steering and Advisory Committee on AnGR to develop and implement a National Action Plan. The future effectiveness of that role will depend to a large extent on the establishment of an overarching national policy on genetic resources for food and agriculture.

References

UK Country Report on Farm Animal Genetic Resources 2002. (www. defra.gov.uk/science/publications)

Appendix 1

Glossary of Organisations:

BCMS:	British Cattle Movement Service
BGS:	British Goat Society
BPA:	British Pig Association
BPC:	British Poultry Council
Breed Socs:	Breed Societies
BSAS:	British Society of Animal Science
CCW:	Countryside Council for Wales
ComCo's:	Commercial Companies
EN:	English Nature
MLC:	Meat and Livestock Commission
NBA:	National Beef Association
NCAD:	National Cattle Association, Dairy
NGOs:	Non Governmental Organisations
NPA:	National Pig Association
NSA:	National Sheep Association
NSTSG:	Northern Short-Tailed Sheep Group
PC (GB):	Poultry Club of Great Britain
RABDF:	Royal Association of British Dairy Farmers
RBI:	Rare Breeds International
RBST:	Rare Breeds Survival Trust
SNH:	Scottish National Heritage
ST:	Sheep Trust
TLF:	Traditional Livestock Foundation
VIDA:	Veterinary Investigation Diagnostic Analysis

6

Genetic variation within and among animal populations

W.G. Hill and X.-S. Zhang
Institute of Cell, Animal and Population Biology, University of Edinburgh, West Mains Road, Edinburgh, EH9 3JT, UK

Abstract

Factors that influence variability between and within populations at levels ranging from the molecular to quantitative traits are reviewed. For quantitative traits, models of how levels of variation are determined and how they change have to be based on simplifying assumptions. At its simplest, variation is maintained by a balance between gain by mutation and loss by sampling due to finite population size. Rates of response in commercial breeding programmes and long-term selection experiments are reviewed. It is seen that rates of progress continue to be high in farmed livestock, but not in race horses, and that continuing responses have been maintained for 100 generations in laboratory experiments. Hence variability can be maintained over long periods despite intense selection in populations of limited size. The potential role of conserved populations is reviewed, and it is suggested that their role is unlikely to be as a useful source of variation in commercial populations but mainly to preserve our culture and to fill particular niches.

Introduction

Genetic variability can be observed at all levels from individual bases in the DNA, through individual genes, to polygenic quantitative traits. Within the DNA, many sites are non-coding and may indicate population genetic properties when there is no selection. We can use such information to describe the basic divergence between populations and individuals within populations, and draw inferences about the processes involved. These are restricted to population structure, genetic sampling and mutation. For quantitative traits, it is both harder to get adequate measures of genetic variability, because genetic and environmental differences may be confounded and because large samples are needed to disentangle genetic from environmental variation, and to explain the observations. The processes leading to within and between population variability have also to include both natural and artificial selection, and the operation of these depends on the genetic architecture of the trait, including gene effects and frequencies, which we do not see directly and on which we are only slowly accruing information. Much of our firm information on variability and on predictions of how it changes therefore comes from molecular genetic data, and its extension to genes affecting traits that are not under selection. The extension to quantitative traits is based either on simplifying assumptions, restricted models of the traits, or both (Falconer and Mackay, 1996; Lynch and Walsh, 1998; Barton and Keightley, 2002).

How variation changes with finite population size

Molecular genetic level (neutral genes)

Perhaps surprisingly, understanding of how variability in populations changes, at least in the short-term, is more adequate than those of what determines the current levels of variation that are maintained. Hence we first consider the influences on changes. The following basic theory applies to genes which themselves have no direct effect on fitness, induced either by natural selection, or by their effect on traits under artificial selection. Thus changes in the frequency of such genes are due to random sampling, with any response to selection due to changes in the frequencies of other genes.

Reduced population size induces a reduction in variability and heterozygosity (H, expected frequencies of heterozygotes with random mating, itself a measure of variability) within populations, as a consequence of the random Mendelian sampling of genes. It also produces inbreeding, because inevitably individuals must have ancestors in common (their genealogies coalesce) (e.g. Falconer and Mackay, 1996). In an idealised population, in which each of the N individuals is equally fertile and viable, so that family sizes differ only by chance, the loss of heterozygosity or variation in each generation is proportional to $1 - (1/2N)$. Similarly, the complementary increase in level of inbreeding (if low) is $1/2N$. In order to consider more general population structures the *effective population size* (N_e) is defined as the size of an idealised population with the same increment in drift or inbreeding per generation. Assuming N_e remains constant, the heterozygosity (H_t) at generation t is $H_0(1 - 1/2N_e)^t$, where H_0 is the initial heterozygosity. As the inbreeding coefficient (F) is proportional to $1 - H$ in a random mating population, $(1 - F_t) = (1 - F_0)(1 - 1/2N_e)^t$ and also $H_t = (1 - F_t)H_0$. If t/N_e is small, the increase in inbreeding is given by $F_t - F_0 = t/2N_e$, approximately.

The genetic sampling that causes the reduction in variability and increase in inbreeding occurs due to Mendelian segregation of genes within individuals, which breeders cannot simply control, and due to differences between individuals in their numbers of offspring, which breeders can reduce. The effective population size therefore depends on the number of breeding individuals, on the ratio of numbers of females to males that are mated and on the variation in family size (Falconer and Mackay, 1996). For example, with 25 breeding males and 25 breeding females, and no differences among animals in fitness so that family sizes are randomly distributed, the effective size is 50. If, however, there are 10 sires and 40 breeding females, the effective size is 32. With random family size, the number of offspring of each individual is Poisson distributed, with mean and variance of 2. With equal numbers of males and females, the effective size is reduced by 1/3 if the variance of family size is 4, for example. The effective number can be larger than the actual number if steps are taken to equalize family size; indeed with 25 males and 25 females, with each giving one offspring of each sex, the effective size is almost 100.

If population size varies over generations, $H_t = H_0(1 - 1/2N_{e1})(1 - 1/2N_{e2}) (...)(1 - 1/2N_{et})$. Therefore those periods in which population size is very low have proportionately bigger effect: a bottleneck of 10 individuals followed by three generations of 100 induces the same loss of variability as 4 generations at a constant size of 30 individuals.

The rates of change in inbreeding or variability are expressed per generation, so the rate of change per year is inversely proportional to the generation interval. If breeding animals are kept longer, the annual rates are therefore reduced, but only if the increase in generation interval is not accompanied by a greater increase in variation in family size due to the longest living individuals having disproportionately more progeny.

Natural selection or artificial selection based on individual performance results in families being represented unequally because high performing individuals tend to have high performing relatives, particularly when the heritability is high. Effective population size is therefore reduced by this increased variability in family size. With index or BLUP selection using family performance, weight is given to the family mean and so there is increased co-selection of family members. As most weight is given to relatives' performance for low heritability traits such as reproductive rate, the effects on effective size are then greatest when the heritability is low. As the parents of highest merit are also likely to have the best offspring, the effects of selection are exacerbated by a correlation of family sizes across generations. Wray, Caballero, Thompson, Woolliams and colleagues have developed theory over a number of years to predict the influence of selection on effective population size and thus the rates of inbreeding and loss of variability (see e.g. Bijma et al., 2001; Villanueva et al., in this publication). The assumptions needed in order to utilise this fully in breeding programme design are discussed later.

The previous discussion assumes individuals mate at random in the population. We now need to distinguish between the two effects of finite population size: reduction in variation, which is due to chance change of frequency of genes (genetic drift), leading to more extreme frequencies and ultimately fixation; and inbreeding, which is associated with the increased probability that individuals are homozygous. With non-random mating, the rate of inbreeding and drift may be altered in different ways. For example, matings between related individuals cause a rise in inbreeding and concomitant fall in frequency of heterozygotes, but gene frequencies and variability in the population are little affected. Similarly, if a population is subdivided into small non-intermating groups, inbreeding is increased and variability reduced in each. In the long term, however, such subdivision is theoretically an effective way to maintain variability in the population comprising the whole set of lines, for by considering all possible crosses between lines, the overall loss of heterozygosity is minimised. In practice, however, such sublines would be less fit due to inbreeding, and both random and selective loss of lines would be hard to avoid.

Variation among replicate populations started from a common base increases initially for neutral genes or DNA sequences at the same rate as that within populations declines, i.e. there is redistribution of the variance: as noted previously, the heterozygosity within populations (which can be regarded as the variance in gene frequencies among individuals in the population) falls in proportion to $H_t = H_0(1 - 1/2N_e)^t$, approximately $H_0(1 - t/2N_e)$, while that between populations rises in proportion to $1 - (1 - 1/2N_e)^t$, or approximately $t/2N_e$ if t/N_e is small. Over longer periods of time, the rate of population dispersion does not depend solely on the level of variability in the populations at the time they divided, but increasingly on the occurrence and fixation of mutations arising subsequently. In the very long term the rate of divergence becomes equal to the rate of occurrence of selectively neutral mutations.

The analysis of population dispersion can be taken further, for example using Wright's F statistics to describe variation among populations and among their subpopulations, e.g. breeds and herds within breeds (Falconer and Mackay, 1996). At its simplest, with random mating within subpopulations, the divergence, termed the genetic distance, is described by the inbreeding required to generate this amount of population dispersion. Thus two populations from the same base and maintained to inbreeding levels F_1 and F_2 would have distance $F_1 + F_2$. Note that genetic distance is increased both by time and by small population size. Migration between populations reduces their dispersion. Indeed even low rates of migration, of the order of one individual per generation, are sufficient to prevent appreciable population subdivision.

Quantitative traits

In order to predict the effects of restricted population size, inbreeding, selection and other forces on quantitative traits, it is necessary to make several assumptions. For the case of a finite population in the absence of selection, the basic necessary assumption is that there should be additivity of gene action (i.e. no dominance or epistasis, such that performance is a simple sum over the effects of individual genes, regardless of their genotypic combination). In order to apply the results for finite populations in the presence of selection a further simplifying assumption is needed, that of the infinitesimal model: in this all genes act additively, are unlinked, and are of infinitely small individual effect, such that their changes in frequency under selection are negligible. (As there are assumed to be so many such genes, their combined effects are not negligible, however, otherwise selection would be ineffective.)

Predictions of rate of loss of variability within populations and of dispersion between them are then simple extensions of the results for individual neutral genes. The additive genetic variance within populations, V_{At}, declines in proportion to the heterozygosity at individual loci, i.e. $V_{At} = (1 - F_t)V_{A0}$, whereas that between populations rises at twice the rate, $V_{Bt} = 2F_tV_{A0}$, where V_{A0} is the initial additive genetic variance. (This factor of two arises because the finite size induces both a variation among lines in gene frequency and a covariance in frequency of the two genes carried by each individual.) Hence the total variability, within plus between lines, is expected to rise, in proportion to $V_{Bt} + V_{At} = (1 + F_t)V_{A0}$, showing population dispersion creates genetic variance in quantitative traits but not in heterozygosity. Under the infinitesimal model, the predictions of effects of selection on effective population size discussed previously can immediately be used to predict the rates of change in genetic variance and inbreeding, and how these are influenced by changes in design of the breeding programme. Hence programmes can be designed so as to maximise the selection response for a minimum acceptable value of N_e or to optimise the trade off between response and N_e by defining a cost of reduced variability and increased inbreeding. There has been extensive work by Woolliams and colleagues (e.g. Bijma et al., 2001; Villanueva et al., in this publication) and by other groups to understand the process and facilitate the dynamic computation of optimal design.

With additive gene action, small population size does not lead to any expected change in mean performance due to inbreeding depression, but by chance individual lines may perform better or worse than their non-inbred founders. If there is non-additive gene action, there is an expected mean change in performance (usually seen as inbreeding depression) and variability departs from that shown. At low levels of inbreeding, however, the formula $V_{Bt} \approx 2F_t V_{A0} \approx V_{A0}/N_e$ holds for quite general gene action. The change in variance within populations is less predictable; indeed it may rise if rare recessive genes are 'uncovered' by the inbreeding. It is always important to distinguish between the effects of small population size on variability (with which we are primarily concerned in this paper) and mean performance, both because their impacts are different and because they are influenced differently by non-random mating.

Levels of standing variation

Neutral genetic variability

At sites in the DNA at which there is no selection (if there really are such: even different codons for the same amino acid are used somewhat non-randomly) and genes at which selection can be ignored, the variability maintained in the population depends on a balance between input from mutation and loss by sampling (genetic drift) through finite population size. In a closed population of constant size, if there are an infinite number of alleles (as is effectively the case for microsatellites) the heterozygosity (proportion of heterozygotes) at steady state is given by $4N_e u /(4N_e u +1)$, where u is the mutation rate. If the mutation rate is low, for example at the individual DNA base level, this equation simplifies to $4N_e u$. The level of variation observed at single DNA sites in man is around 0.001. Whilst this could be explained by a neutral mutation rate of 10^{-9} and an effective size of 10^6, it is important to note that N_e has clearly not been constant in human or indeed any livestock populations, so the actual level is a function both of historical population size and its rate of change.

Whilst there is extensive information for livestock on variation at microsatellite loci, these are chosen to be highly heterozygous for use in QTL mapping studies, for example, rather than to be representative. There is now a little information on variation at random sites in the DNA (single nucleotide polymorphisms – SNPs), although again there may be some upward bias due to selection of regions for sequencing around genes of interest or for mapping or parentage analysis. The first data on poultry indicate very high levels of DNA polymorphism in mixes of commercial layer and broiler populations, almost 1% of sites (Smith *et al.*, 2002). There was substantial overlap of polymorphic loci between the two types of bird. For sites around nine cytokine genes in a pooled set of breeds of beef cattle, Heaton *et al.* (2001) identified one SNP every 143 sites on average, with any two individual's chromosomes differing every 443 sites. In a further study of a group of SNPs selected to be useful for paternity identity, heterozygosity within the Angus breed in the USA was lower, but not substantially so, than in a composite of 17 beef breeds (Heaton et al., 2002). Much more data are needed, particularly on variability within populations, but what is very clear is that substantial variability is present at the DNA level within and among our livestock populations.

For genes that are individually under selection, it is not possible to make general statements as to how their frequencies are determined for it depends on the nature of gene action at that particular locus. The simplest model is that of heterozygote superiority; examples of it are rare, however, but include sickle cell anaemia and warfarin resistance genes, where there is, respectively, a balance between selection against sickle cell anaemia and susceptibility to malaria, and a balance between vitamin K deficiency and susceptibility to warfarin. Examples also occur for genes affecting quantitative traits, which have such large effect that they can be regarded as major genes. For example, the frequency of the double muscling gene in cattle breeds depends on the weight given to meat content and conformation relative to that on reproductive fitness.

Quantitative traits

Our understanding of what determines the actual level of variation in quantitative traits is really very poor at both the phenotypic and the genetic level. There is no clear theory, for example, which explains why the coefficient of variation of body size is typically close to 10% for most species at most ages, nor why that of reproductive rate is higher. Nor do we understand why growth rate has a heritability (h^2) of around 25% in most species, even though it formally depends on genes segregating in a particular population at a particular time, or why, given the consistency, it is typically 25% rather than say 5%. Although we are not surprised that the heritability of growth rate is higher than that of reproductive rate, as simple arguments suggest that natural selection will remove additive genetic variation faster from the latter than the former, it is not clear why the genetic coefficient of variation (genetic standard deviation/mean, termed 'evolvability') of reproductive rate is typically as high as or higher than that for growth rate.

There has been extensive analysis of the forces maintaining quantitative genetic variation in natural populations, but there is not yet any fully accepted theory that can predict the actual magnitudes of variability observed. Limited population size and most forms of selection lead to loss of genetic variability, and the main factor that increases it in the long-term is mutation. There have been many estimates of the rate of mutation to quantitative traits in laboratory animals and plants. Somewhat surprisingly, in view of the diversity of laboratory species (Drosophila melanogaster, Caenorhabditis elegans, mouse) and of traits (bristle number, body size) studied, the estimates are fairly consistent. They show a new variation (V_M) typically of the order of 10^{-3} times the environmental variance (estimates are usually made in highly inbred populations, so as to separate signal from noise) (Keightley, 2004). This increment or 'mutational heritability' of 0.1% or more each generation is small relative to the typical heritability of 25% for existing variation, but is cumulative over generations: $V_{At} = V_{A,t-1}(1 - 1/2N_e) + V_M = V_{A0}(1 - F_t) + 2N_e F_t V_M$ (Hill, 1982). In the absence of any selective forces and assuming additive gene action, a balance would be obtained in a finite population of increase in variance by mutation and loss by drift, to maintain a heritability value of the order of $0.002N_e/(0.002N_e + 1)$ if the mutational heritability is 0.1%. Thus a population size of 250 would maintain h^2 of the order of 1/3, with the actual value highly dependent on population size. Although this value of 0.1% for mutational heritability may be an underestimate for livestock with long generation intervals if mutations accumulate in the germ cells throughout life, many mutants are likely to be deleterious with respect to fitness, perhaps even lethal in homozygous form. Hence the neutral calculation almost certainly overestimates the variability that can be maintained and utilised.

Whilst an obvious force for maintaining genetic variability is heterozygote superiority, examples are rare. Although there is plenty of evidence for inbreeding depression, which can be explained by heterozygote superiority, there is far more evidence for its cause by fully or partially deleterious recessive genes. Another factor that can maintain variability is differential selection in separate environmental niches, with migration sufficiently rare that gene frequency differences between the subpopulations can be maintained. This is seen in livestock where artificial selection has generated extreme artificially selected lines for growth such as broilers and layers, and breeds differing for visual traits such as hair colour. The differences in environment may be spatial or temporal, in that changes in, for example, climate may produce periods when one or other type is favoured. To maintain variability, however, it is necessary that these cyclical changes span several generations, otherwise selection effects are just averages within cycles.

There are other features that have to be explained. Population means remain reasonably constant over time and space, not drifting haphazardly as expected if they were solely reflecting a mutation-drift equilibrium. Stabilising selection, whereby intermediates are favoured, has often been identified in field and laboratory experiments (Falconer and Mackay, 1996). There is not, however, agreement as to how strong such selection is in natural populations, and its strength may have been exaggerated in many studies. Stabilising selection is usually modelled as a fitness function of approximately quadratic form, with an intermediate maximum at the optimal value of the trait. Such selection removes variability from the population, and hence it is necessary to invoke some balance between the increment in variation from mutation and its loss by selection and drift. There has been extensive analysis of the levels of variability to be expected from this balance, with different models producing significant differences in outcome, although earlier models indicating that large amounts of variability could be maintained are now dismissed as requiring unreasonably high per locus mutation rates (see e.g. Zhang and Hill, 2002).

A critical assumption of the classical stabilising selection model is that the relationship between fitness and phenotype for any particular trait is due to natural selection acting directly and exclusively against extremes for it. Thus an analysis of mutation-selection balance would involve only the direct effects of fitness on that trait. This is obviously unrealistic: for example the observation that survival of human babies at birth is (or at least *was*, prior to improvements in medical technology) highest for those of intermediate birth weights is not solely due to the effects of birth weight *per se*, as small babies may be premature or otherwise abnormal. Hence a multi-trait formulation becomes necessary, and in principle it is not possible to construct theory for any single trait.

In a model that has been suggested to overcome this simple single trait/fitness model, mutant genes may affect, to varying extents, both the trait being observed and overall fitness through pleiotropic effects on all other traits associated with fitness, with the selection on the mutant depending not on its effect on that trait but on overall fitness. The apparent superiority of intermediates arises because, as all mutants are assumed to reduce fitness and some may affect the trait, extreme individuals for the trait are more likely to be carrying mutants and therefore be less fit. Whilst this pleiotropic model can maintain variation, it neither explains stability of means of traits over long time periods nor predicts enough apparent stabilising selection.

We have recently been constructing models in which mutants have pleiotropic effects on both the trait, which is subject to some stabilising selection, and overall fitness due to directional or stabilising selection on all other traits (Zhang and Hill, 2002). Mutations are assumed to have a distribution of effects on both the trait and fitness, i.e. some mutants may have very small effects on both, some a large effect on one and not the other, and some a large effect on both. These distributions may be highly leptokurtic, i.e. a small proportion of the mutants have a very large effect. By suitable choice of parameters it is possible to construct models that can give both apparent stabilising selection and maintain variation. Basically these indicate that the stabilising selection observed is mostly due to that directly on the trait, but genes of very small effect on fitness largely maintain the variation. Even so, the model requires a more strongly leptokurtic distribution of mutant effects on fitness than on the trait, whereas it is easier to explain a more leptokurtic distribution on the trait (few genes affecting it substantially) than on fitness (most genes likely to affect it). If pleiotropic effects on many traits with stabilising selection are included in the model so as to describe the pleiotropic effect on fitness, the distribution of fitness becomes less leptokurtic and the model less tenable. If mutations are more likely to be recessive for fitness, it might go some way to resolve this restriction, however (Zhang *et al.*, 2004).

Variability in populations under artificial selection

In natural populations the main problem is to predict and explain the level of genetic variation that is present; but in populations under artificial selection the problem is to predict how much the mean and variability of the population will change. Frequently predictions are made assuming the infinitesimal model, such that variability at individual loci does not change as a consequence of selection, as described above. Mutation is not expected to compensate fully for the loss of variation by drift if the selected population is substantially smaller than the natural or founder population from which it was drawn.

The infinitesimal model is useful for predicting genetic change and variability over short time periods, and is the basis for calculations of optimal design (e.g. Bijma *et al.*, 2001), but there is no reason to believe it should work well over the long term. More elaborate models are needed, which include information on the distribution of effects of genes influencing the trait, their pleiotropic effects on fitness, and their frequencies. Whilst some such information is being acquired through QTL mapping experiments, it is still very limited, certainly too much so for accurate prediction theory. Of course as these analyses actually reveal QTL of substantial effect, they 'disprove' the infinitesimal model; but although it was never 'correct' in the first place, it has remained useful and, for example, has been shown to give a very good description of change up to 20 generations in some small selected populations of mice (Martinez *et al.*, 2000). More sophisticated analyses in which models of genes of larger effect are incorporated show that the loss of variability is likely to be faster than predicted by the infinitesimal model, when it takes $1.4N_e$ generations to lose one half of the initial variability, but that the loss of potential response from small populations is less (foundations for this were established by Robertson, 1960). This apparent contradiction is because favourable genes of larger effect are likely to be fixed more quickly by selection, and the consequent reduction of variability reflects, in a sense, a success. Genes that have a deleterious effect on fitness and hence remain at low frequency in natural populations, but have a large effect on the trait and are at strong advantage under artificial selection, also contribute to an increase in

W.G. Hill and X.-S. Zhang

variability as their frequency rises, but with a concomitant fall in fitness of the population.

The variation contributed by mutations each generation (V_M or mutational heritability) is not dependent on population size, but the subsequent rate of loss is proportional to $1N_e$ (see equation above). Hence more standing variation from mutations is expected in large populations, and in unselected populations the final level is independent of the distribution of mutant effects. Mutations of large effect, however, if they occur, can contribute substantially much earlier to variability than those of small effect; but again there may be concomitant pleiotropic effects on fitness. Predictions of the effects of selection on initial variation and on that arising by mutation and its interactions with population size are therefore seen to be very dependent on the model that is assumed to describe the genetic basis of the trait.

An important source of information on how variability is maintained in populations under artificial selection comes from the many long-term selection experiments that have been conducted in laboratory species and the few in livestock and crop plants. Arguably, there has not been enough theoretical input into interpreting experiments so as to inform us about the underlying genetic architecture.

Responses in long term selection experiments in vertebrates have recently been reviewed (Bünger et al., 2001; Hill and Bünger, 2004) and the results updated by Dudley and Lambert (2004) of the classic maize oil and protein experiment at Illinois, which has been run for over 100 yearly generations in a closed population (for results of the first 76 generations, see Falconer and Mackay, 1996). Each generation seed from about 30 plants is selected for the next generation. Response for increased oil and protein content (in separate lines) has continued to the present, from its initial values of 5% and 11%, respectively. In the downwards direction, a plateau has indeed been reached, but at essentially 0% oil and 6% protein, with lower levels of protein probably not providing a viable seed. Also, in new lines drawn from both the high and low lines after 50 generations, the direction of selection was reversed; and these have responded back to the initial population means. Although heritability has fallen during the long duration of the experiment in the Illinois lines, variability has not therefore been lost to the extent that they cannot respond to selection, nor such that selection cannot be totally reversed. Selection lines for increased growth rate in closed lines of species of poultry have been maintained for very many generations and have shown continuing, albeit slowing, responses: for 100 generations in quail (Marks, 1996), and 40 in chickens (Dunnington and Siegel, 1996) and turkeys (Nestor et al., 1996). Likewise there have been long continued responses in lines of chickens and turkeys selected for egg production. Many more experiments have been conducted in mice, usually for growth rate to or size at about 6 weeks, an age near sexual maturity. As expected, the duration of responses have depended on the population sizes adopted, but in the biggest experiment, that at Dummerstorf, Germany (Bünger *et al.*, 2001), responses have continued for 100 generations. Mice of the selection line are now almost 3-fold heavier than in the unselected control population. Although there is evidence that response slowed, it has not actually ceased due, presumably, to the occurrence of mutations that have been utilised in the selection.

In addition to the continued responses in the mouse lines, albeit usually at declining rates as genetic variation has been utilised, phenotypic variances have risen in the high selection lines, presumably explicable in terms of scale because coefficients of variation have typically been little affected. This applies not just to the Dummerstorf experiment, but also to a whole range of selected lines that have been maintained or brought to our lab in Edinburgh. These have also been inbred, and again there is little change in CV (for further details see Hill and Bünger, 2004). Indeed, there is a need to consider models which include terms for the effects of selection on variance *per se*, rather than just through the route of changing genetic variance as a consequence of changing gene frequency. It can be shown, for example, that intense directional selection favours genes that increase levels of phenotypic variability (Hill, 2002).

Direct evidence for the role of mutations comes from selection experiments started from a highly inbred base population. One of these has been conducted in the mouse (Keightley, 1998), in which a high-low divergence of 30% was obtained after 50 generations of selection from two lines maintained with only 12 pairs of parents, a result concordant with a mutational heritability of over 0.1%.

Analyses of long-term selected commercial livestock populations also give us information on how much response and variation can be maintained (Hill and Bünger, 2004). Of course, the populations are not closed, but the more extreme they become, the more likely that significant external contribution, if any at all, comes only from other commercially selected populations. Recent analyses of broiler populations show that, despite over 50 years of intensive selection, rapid responses continue, not just for growth to market weight but also for yield of meat; and because selection pressure has also been applied to minimise fitness associated effects, such as leg weakness, declines in such fitness traits has been reduced or reversed (McKay *et al.*, 2000). Furthermore, the heritability of growth rate (broiler weight) in these populations does not seem to have changed appreciably: recent analyses of selected populations show a heritability of 25% or more (Koerhuis and Thompson,

1997), a value that has always been typical for the trait. Similarly, responses continue in improvement programmes for poultry for eggs, pigs and dairy cattle, albeit with the evidence that heritability in the latter has increased, presumably mainly through better management and parameter estimation techniques. In contrast, there appears to have been little improvement in classic events over the last 50 years in the winning times of either Thoroughbred racehorses (Gaffney and Cunningham, 1988) or greyhounds (Hill and Bünger, 2004). Although the Thoroughbred was founded from a limited number of animals, such a plateau also implies that there are no useful mutations appearing. Perhaps all mutations are deleterious with respect to speed after such long-term selection, such that they are 'running faster just to stand still', merely eliminating the mutations that reduce performance.

Variation among populations

Estimates of genetic distance are made using neutral markers, most usually microsatellites, such that they specifically exclude loci under selection and are attempting to describe the background divergence. Evolutionary trees of breeds can also be constructed using these distances, although because of intercrossing, a simple branching structure is unlikely to be a complete description. We do not propose to give further detail here. There are estimates of distance for many breeds and regions for our livestock species. The amount of variability is shown by extensive recent analyses of European pig breeds (but including imported American and Chinese breeds) by San Cristobal *et al.* (2002) (see also Blott *et al.*, in this publication), in which genetic distances (equivalent to sums of F values down the paths from common ancestor) of up to 0.4 were observed, with over 0.3 among solely European breeds. There have also been major analyses of breeds of cattle (Blott *et al.*, 1998) and of sheep (see also Bruford, in this publication). These and other data on microsatellites, enzyme or blood group polymorphisms, or SNPs in the DNA (see comments above) are all indicative of large amounts of variation both within and among breeds.

Estimates of variation among populations in quantitative traits are simply obtained from mean performance, *providing* comparisons can be freed of environmental confounding. The results may not be very useful, however: the fact that the egg layer-broiler difference is many standard deviations for juvenile body size tells us little about the underlying genetic architecture. What is more important is the extent to which there is useful variability that can be brought from one population into another, and the extent to which there are local adaptations or, equivalently, genotype x environment interaction.

Variation in synthetic populations

If genetically distant lines are crossed, the genetic variation between them translates into increased variation within the population founded from their cross. Only under limited situations can we predict, however, how much this increase will be. The simplest, and indeed almost only explicit formula, applies to a group of small populations that have diverged from a common base population without selection. As noted previously, assuming that within each line random mating is practised, the heterozygosity within lines decreases in proportion to $(1 - F_I)H_0$. If lines are crossed at random, the heterozygosity within any two-way cross rises to $(1 - F_I/2)H_0$, and in a synthetic population constructed by putting many such independent lines together again, the base population is reconstituted, with heterozygosity H_0. Variation has merely been temporarily redistributed. Similarly, inbreeding accumulated in the parental lines is recovered in the F1 individuals of any cross or in the synthetic formed from a multi-line cross, and one-half is recovered in the F2 of a two-way cross. Of course these calculations assume either no selection or the infinitesimal model and consequently are only a reference point if selected breeds are considered.

Whilst it is simple to write down formulae for the variance in a cross or synthetic obtained from a pair or many parental lines in terms of gene effects and frequencies in the lines, evaluating these requires information on the actual gene frequency differences between them. For a single locus, for example, the additive genetic variance in the F2 of a cross is $V_{AX} = (V_{A1} + V_{A2})/2 + (p_1 - p_2)^2 a^2/2$, where V_{A1} and V_{A2} are the variances in the parental lines corresponding to their gene frequencies p_1 and p_2, and a is the effect at that locus on the trait. In general, if many loci contribute to the difference in means, and their frequency differences are all in the same direction, the increment in variance in the cross becomes proportionality less; whereas random (from drift) rather than directional changes (from selection) can lead to large increases in variance in the synthetic if the lines are genetically very distant. With many loci involved, calculations based on genetic distance computed from neutral markers may therefore be a better predictor of increment in variance than differences among the populations in the traits of interest.

Conserved populations

Conserved populations could potentially be useful in a number of circumstances. One is to bring variation into a population that is under selection but in which variation is thought to be becoming exhausted. We have seen, however, that this does not happen quickly. To be useful in a simple cross, the conserved population would need to have at least moderate performance, otherwise the resultant cross would be so far behind the commercial population as to be useless over finite time, or contribute so little to the commercial population, no more than a quarter, that the additional variation would be small. The second possibility would be to incorporate QTL (or genes if known) from that population using marker-assisted introgression. This is a lengthy process, however: the QTL have to be identified in these particular populations, and then several generations of backcrossing are needed. If a different environmental niche is to be exploited, then we have to consider some population that will be useful there. If the major breed or line already occupying it is competitive, the same catch-up requirement in incorporating useful variation from other populations applies. The third possibility is to exploit in a particular niche the commercial population from elsewhere by marker-assisted introgression. The difficulty with dealing with particular local environments is to decide whether the interaction of genotype and environment is sufficiently large that a separate population is necessary or whether the same can be used more generally. It is notable that more or less the same broiler strains and the same breed of dairy cattle are being used world-wide. In many cases the environment is modified to cope with them, by providing housing and appropriate diets; and whether or not this will be more efficient than aiming to further improving locally adapted breeding stock will depend on circumstances. It requires substantial resources and is unlikely to be cost effective to sustain improvement programmes for each of many different environments each with limited numbers of commercial stock. With a single large population and by appropriate recording and selection, perhaps incorporating molecular methods, greater response in local environments may well be made than in a population selected specifically for local adaptation unless the GxE interaction is very substantial.

The great majority of UK animal production is of poultry, pigs and dairy cattle, in which global improvement programmes operate and there is evidence of continuing variability and response. There are inadequate resources to operate highly effective independent breeding programmes in all of our many native breeds of beef cattle or sheep. In neither case, therefore, is it likely that conserved unselected populations in the UK will contribute much to commercial animal improvement. Our native breeds are, however, an important part of our landscape and culture, and there is a case for maintaining and improving them in so far as funds allow, so as not to reduce their competitive position too much. We do not yet know how much the consumer will pay for niche products that are really different (uniquely textured meat perhaps, but slow growing chickens and slow horses?); but if such are developed it will become worthwhile to practice selection in the populations rather than leave them unimproved.

Conclusions

Ironically, in a discussion on conservation as a source of variability, it is important to point out that with modern transgenic technology it is becoming possible both to transfer genes between populations and to make new variants of genes, or even new genes (for further discussion, see Hill, 2000). Lack of variability over the long term is not likely to be the problem; utilising it will be harder. Hence the cultural argument dominates, at least in temperate environments. Issues may be different in tropical production with particular diseases and extremes of environment, but in places where animal breeding is or soon becomes well organised, a continued fall in the number of breeds in commercial application is likely. The extant diversity may reflect lack of success, not success in commercial animal breeding programmes, and so we retain diversity in museums as our history.

Acknowledgements

We are grateful to the BBSRC for financial support and to Beatriz Villanueva and to colleagues for helpful comments.

References

Barton, N.H. and Keightley, P.D. 2002. Understanding quantitative genetic variation. *Nature Reviews Genetics* 3: 11- 21.

Bijma, P., Van Arendonk, J.A.M. and Woolliams, J.A. 2001. Predicting rates of inbreeding for livestock improvement schemes. *Journal of Animal Science* 79: 840-853.

Blott, S.C., Williams, J.L. and Haley, C.S. 1998. Genetic relationships among European cattle breeds. *Animal Genetics* 29(4): 273-282.

Bünger, L., Renne, U. and Buis, R.C. 2001. Body weight limits in mice - Long-term selection and single genes. P. 337-360. In: *Encyclopedia of Genetics*. Edited by E.C.R. Reeve. Fitzroy Dearborn Publishers, London, Chicago.

Dudley, J.W.and Lambert, R.J. 2004. 100 generations of selection for oil and protein in corn. *Plant Breeding Reviews* 24: (in press).

Dunnington, E.A. and Siegel, P.B. 1996. Long-term divergent selection for eight-week body weight in White Plymouth Rock chickens. *Poultry Science* 75:1168-1179.

Falconer, D.S., and Mackay, T.F.C. 1996. *Introduction to Quantitative Genetics. 4th ed*. Longman, Harlow, UK.

Gaffney, B., and Cunningham, E.P. 1988. Estimation of genetic trend in racing performance of thoroughbred horses. *Nature* 332: 722-724.

Heaton, M.P., Grosse, W.M., Kappes, S.M., Keele, J.W. Chitko-McKown, C.G., Cundiff, L.V., Braun, A., Little, D.P. and Laegrid, W.W. 2001. Estimation of DNA sequence diversity in bovine cytokine genes. *Mammalian Genome*. 12: 32-37.

Heaton, M.P., Harhay, G.P., Bennet, G.L., Stone, R.T., Grosse, W.M., Casas, E., Keele, J.W., Smith, T.P.L., Chitko-McKown, C.G., and Laegrid, W.W. 2002. Selection and use of SNP markers for animal identification and paternity analysis in U.S. beef cattle. *Mammalian Genome*. 13: 272-281.

Hill, W.G. 1982. Predictions of response to artificial selection from new mutations. *Genetical Research* 40: 255-278.

Hill, W.G. 2000. Maintenance of quantitative genetic variation in animal breeding programmes. *Livestock Production Science* 63: 99-109.

Hill, W.G. 2002. Direct effects of selection on phenotypic variability of quantitative traits. *Proceedings of 7th World Congress on Genetics Applied to Livestock Production*. CD-ROM. Communication No. 19-02.

Hill, W.G. and Bünger, L. 2004. Inferences on the genetics of quantitative traits from long-term selection in laboratory and farm animals. *Plant Breeding Reviews* 24: 169-210.

Keightley, P.D. 1998. Genetic basis of response to 50 generations of selection on body weight in inbred mice. *Genetics* 148: 1931-1939.

Keightley, P.D. 2004. Mutational variation and long-term selection response. *Plant Breeding Reviews* 24: 227-247.

Koerhuis, A.N.M., and Thompson, R. 1997. Models to estimate maternal effects for juvenile body weight in broiler chickens. *Genetics, Selection, Evolution* 29: 225-249.

Lynch, M. and Walsh, B. 1998. *Genetics and Analysis of Quantitative Traits*. Sinauer, Sunderland, MA, USA.

McKay, J.C., Barton, N.F., Koerhuis, A.N.M. and J. McAdam. 2000. The challenge of genetic change in the broiler chicken. p. 1-7. In: *The Challenge of Genetic Change in Animal Production*. Edited by W.G. Hill, S.C. Bishop, B.J. McGuirk, J.C. McKay, G. Simm, G. and A.J. Webb, Occasional Publication of the British Society of Animal Science, No 27.

Marks, H.L. 1996. Long-term selection for body weight in Japanese quail under different environments. *Poultry Science* 75:1198-1203.

Martinez, V., Bünger, L. and Hill, W.G. 2000. Analysis of response to 20 generations of selection for body composition in mice: fit to infinitesimal assumptions. *Genetics, Selection, Evolution*. 32: 3-21.

Nestor, K.E., Noble, D.O., Zhu, J. and Y. Moritsu. 1996. Direct and correlated responses to long-term selection for increased body weight and egg production in turkeys. *Poultry Science* 75: 1180-1191.

Robertson, A. 1960. A theory of limits in artificial selection. *Proceedings* of the Royal Society of London B153: 234-249.

SanCristobal, M. and 27 others. 2002. Genetic diversity in pigs. Preliminary results on 58 European breeds and lines. *Proceedings of the 7th World Congress on Genetics Applied to Livestock Production*. Communication No. 26-14.

Smith, E.J., Shi, L. and Smith, G. 2002. Expressed sequence tags for the chicken genome from a normalized 10-day-old white leghorn whole-embryo cDNA library. 3. DNA sequence analysis of genetic variation in commercial chicken populations. *Genome* 45: 261-267.

Zhang, X.-S. and Hill, W.G. 2002. Joint effects of pleiotropic selection and stabilizing selection on the maintenance of quantitative genetic variation at mutation-selection balance. *Genetics* 162: 459-471.

Zhang, X.-S., Wang, J. and Hill, W.G. 2004. Influence of dominance, leptokurtosis and pleiotropy of deleterious mutations on quantitative genetic variation at mutation-selection balance. *Genetics* 166: (in press).

7

Managing populations at risk

J.A. Woolliams
Roslin Institute (Edinburgh), Roslin, Midlothian, EH25 9PS, UK

Abstract

The procedures outlined by the Food and Agriculture Organisation of the United Nations (FAO) guidelines for managing small populations at risk are reviewed. These cover identification of breeds at risk, prioritising and deciding upon actions, managing in vivo populations at risk, and managing gene banks of cryoconserved material.

Introduction

In the U.K., the Foot and Mouth Disease (FMD) epidemic and the surrounding crises have brought to the forefront a wide range of issues concerned with managing risks associated with agriculture, and not least among these has been the way that we conserve and utilise our farm animal genetic resources (AnGR). The profile of this issue had already been raised internationally by the Convention on Biological Diversity (CBD, known as the Rio Convention) and management of AnGR was taken up as a global action programme by FAO in 1992. Part of this programme was to document guidelines for the management of small populations at risk to help underpin national management strategies. These guidelines are now available in book form (FAO, 1998) and on the Internet (http://dad.fao.org/en/Home.htm) and this paper is intended to provide an overview of their content. Many of the guidelines are straightforward and are not technically demanding, but their proper completion and co-ordination underpin the successful management of AnGR.

The guidelines are intended for global use in planning strategies for both developed and developing countries. It is undoubted that strategies and priorities for action in developing countries will be very different from those in developed countries, as will be the resources to implement the plans, but the underlying processes in arriving at a strategy will be very similar. In broad terms, the ends and the means will be very similar, although different countries may be at different stages in their action plans and may need to respond to different environmental, social, cultural and economic pressures. Therefore it is reasonable to argue that guidelines to promote (i) sound analysis of conservation needs, and (ii) synthesis of an achievable conservation strategy that is informed, considered and prioritises the variety of activities that are needed, may aspire to have a global perspective.

In some aspects of what follows, the UK and other EU countries are more advanced than most countries, particularly developing countries. Consequently the text may seem less relevant to readers from UK and the EU, and yet in very few aspects covered by this paper does the UK (for example) achieve best practice. This is for various reasons to do with policy, infrastructure, stakeholder involvement, technical awareness, and, not least, cost. There are no easy solutions, but all stakeholders need to consider how practices can be improved within the very limited resources available.

Identifying populations at risk

Identifying populations at risk is an essential pre-requisite to the whole process! Unless an inventory is made of populations within a country, defined by breed, then the extent of AnGR within the country will be unknown and populations at risk will remain unnoticed. Whilst the need is clear, in very many countries, this information is often unavailable. An initial problem is defining what is meant by the term 'breed': FAO recognises this term as a cultural term i.e. a 'breed' is defined by the usage of the human population. This is a pragmatic approach, but it is compatible with the CBD, in that each country has sovereignty over its AnGR, and helps in identifying stakeholders from the outset. Bruford, in this publication, examines alternative approaches to defining conservation units.

The primary sources of information for identifying the populations at risk are from censuses and surveys. These two terms can be confused. A census is a counting of head, nationally or regionally, with a properly designed sampling strategy to avoid the requirement to count and account for all animals, and is likely to be an infrequent event e.g. once a decade. The term survey is more applicable to targeted sampling on a smaller scale that is aimed at finding out information on specific breeds and is conducted at regular intervals. However, what is important is that the information obtained is dynamic: (i) the surveys are sufficiently regular to provide up to date information e.g. every third year, and (ii) ancillary questions are asked that help assess, along with previous survey information, whether the breed is increasing or decreasing in number, and what are the likely future trends.

In consequence, the quality and usefulness of a survey for a breed will depend on the questions that are asked. Such questions should cover general information, breed development, management conditions, breed performance and special qualities, along with other relevant details. It is important to recognise that responses obtained in a survey remain subjective and cannot be used as a substitute for a properly designed scientific study, although they may serve as a good way of developing hypotheses for such studies to test.

General information

This includes synonyms of the breed, location, primary and secondary uses, and physical appearance (approximate weight, colouring, visible characters such as horns and other discriminating features).

Breed development

This includes the numbers of breeding males and females, juveniles intended for breeding, any apparent trends in these numbers, and the use of artificial breeding. A very relevant question is the extent of crossbreeding that is practiced, since this determines the risk that the breed will be unable to produce sufficient straightbred replacements to maintain itself (depending on the lifetime reproductive rate of the female). In addition, the existence of a selection programme for the breed or organised conservation measures (*in situ*, *ex situ* live, or cryoconservation) is important for establishing the degree of activity and involvement of stakeholders within the breed. Where selection programmes exist, information on the objectives can be informative on the concerns of and intentions of the stakeholders. Where cryoconservation has been carried out, the numbers of doses of conserved semen and embryos and the numbers of donor males and females helps prioritise future conservation actions across the different breeds at risk.

Management conditions

Information on management conditions is critical since breeds often suffer from misleading comparisons with other breeds and crosses. Thus information on whether the breed is given less preferential treatment in any form compared to other breeds kept on the same holdings is valuable, as is whether the breed plays a particular role in the production system e.g. grazing specific types of pastures. The primary stressors (e.g. climatic, disease, forage availability or quality) of the production environment should also be listed, particularly when the environment may be characterised as being of low or medium input.

Breed performance and special qualities

This is concerned with characterisation of the breed and should attempt to assess overall lifetime economic performance and not solely the ability to produce commercial products. The lifetime economic performance depends on age at first breeding, breeding intervals, litter size, neonatal and juvenile survival, and longevity in the breeding population within the production environment. Special qualities, e.g. product qualities, should also be covered.

Additional relevant information

Additional information can sometimes be very helpful in assessing future trends. One such item is the age of the owner in relation to herd/flock size. If a breed is only kept by the older farmers then its population is likely to decline as those farmers retire from farming, whereas a breed kept by a young farmer may represent a positive choice in its favour.

Breed societies that record and maintain pedigrees can readily provide accurate and up-to-date numbers of breeding males and females within their breed, past trends in numbers and information on what the breed is 'good' at. This seems *a priori* to ensure that countries with many breed societies (such as the UK) would have a clear idea on the populations at risk and the degree of risk faced. However, in the UK there is no account taken of the numbers of individuals in the breed that are outside the breed society and some breeds such as the Scottish Blackface do not have a flock book (although it is clear that the Blackface is not at risk!). As a result there remains an important degree of uncertainty over the population size of some of the more rare UK breeds. Whilst it is completely unreasonable to expect that numbers of individuals in each breed be known precisely, it is reasonable to ask:

(i)   What percentage error can we assume our estimates have?
(ii) What is a reasonable percentage error to have as an achievable goal, particularly for breeds with small to moderate census size?
(iii) What can be done in practice to decrease such errors?

In addition, the UK has only very rudimentary information, often anecdotal, on breed performance and special qualities, in particular on those qualities that make a breed peculiarly fitted to dealing with stressors in the production environment in which the breed was developed. As discussed below, such information relevant to phenotypic characterisation is important in developing conservation strategies.

Assessing the risk

From a conservation standpoint, one of the most important outcomes of the census and the survey is the categorisation of risk status for the extinction of the breed. Risk assessment is the difficult task of identifying the risk factors that are present and how strongly they apply. These factors are clearly dependent on census size (the numbers of breeding males and females) but there are more pertinent factors that need to be accounted for.

Genetically, the risk may best be described by the rate of inbreeding (ΔF) in the population. ΔF determines how quickly gene frequencies may change in a population, e.g. of deleterious recessive alleles that may threaten population fitness. ΔF will primarily depend upon the numbers of parents used in the population, and is often expressed by reference to effective population size (N_e), which has the strict definition of $N_e = 1/(2\Delta F)$ i.e. ΔF goes down as N_e increases. Recently Sanchez *et al.* (2003) have shown that in populations that are managed to minimise the loss of genetic variation, $N_e \leq 6\,Min(N_{sire},N_{dam})$ for random mating where N_{sire} and N_{dam} are the number of sires and dams used in a generation. This upper bound for N_e is only achievable under stringent conditions (i.e. following the mating system of Sanchez *et al.* and requiring a fertile offspring of the correct sex to be obtained from each mating), and in practice N_e *will be decreased dramatically* by aspects of management e.g. how parents are selected, the influence of breed fashions, how matings are managed. It is often quoted, incorrectly, that $N_e = 4N_{sire}N_{dam}/(N_{sire}+N_{dam})$; this too will often greatly overestimate N_e and underestimate ΔF as it based upon specific assumptions that *rarely hold in real populations*.

There is an additional issue in whether or not a management tool based on ΔF should be concerned with the magnitude of ΔF measured /year or /generation. These differ in considering the fractional loss of genetic variance /year or /generation respectively, and the latter is approximately L times greater than the former, where L is the generation interval. Treating all species equally, a deontological perspective argues for the management of ΔF /generation (ΔF_g) since this accounts for the different biological timescales of the different species. An empirical observation is that managed populations where N_e /generation ($=1/(2\Delta F_g)$) is ≥ 50 can be successfully maintained over long periods, and this is argued by FAO (1998). Mace and Lande (1991) argue that a value of 500 is more suitable for feral populations.

Demographic factors can have much more impact over short timescales than genetic risks e.g. geographical distribution, numbers of herds, and sociological factors e.g. the age of the farmers keeping the breed. This was brought sharply into focus only recently with the FMD crisis where the geographical concentration of the Herdwick in the Lake District was responsible for placing this breed in a very high degree of risk. Such factors will often be more closely related to the census size of a breed rather than N_e, or the census size of the number of farmers keeping the breed.

The risk categories that are used by the FAO (1998), classifies breeds into one of seven categories: extinct, critical, critical maintained, endangered, maintained, endangered maintained, not at risk and unknown. The definitions listed below are taken from the World Watch List (3rd Edition; FAO, 2000). These differ slightly from FAO (1998) and may be revised further as experience grows. The categorisation is based on census size (the total number of breeding males and females) and the trend in census size, i.e. whether it is increasing, decreasing or stable. This is why it is important to organise a regular survey to enable informed categorisation and the appropriate prioritisation.

Extinct. A breed is categorised as extinct if it is no longer possible to easily recreate the breed population. This situation becomes absolute when there are neither breeding males (semen) and breeding females (oocytes) nor embryos remaining. In reality extinction may be realised well before the loss of the last animal, gamete or embryo.

Critical. A breed is categorised as critical if: (a) the total number of breeding females is ≤ 100 or the total number of breeding males is ≤ 5; or (b) the census size is ≤ 120 and decreasing and the percentage of females being bred straight is below 80%.

Critical maintained. As for Critical, but for which active conservation programmes are in place or populations are maintained by commercial companies or research institutes.

Endangered. A breed is categorised as endangered if: (a) the total number of breeding females is > 100 and ≤ 1000, or the total number of breeding males is ≤ 20 and > 5; or (b) the census size is > 80 and < 100 and increasing, and the percentage of females being bred straight is above 80%; or (c) the census size is > 1000 and ≤ 1200 and decreasing, and the percentage of females being bred straight is below 80%.

Endangered maintained. As for Endangered, but for which active conservation programmes are in place or populations are maintained by commercial companies or research institutes.

Not at risk. A breed is categorised as not at risk if: (a) the total number of breeding females and males are > 1000 and 20, respectively; or (b) if the census size is > 1200 and increasing.

Unknown. Self-explanatory, but also a call to action: find out!

If categorisation of a particular breed is borderline, further consideration is given to factors such as the number of animals actively used in artificial insemination, and/or the amount of semen and number of embryos stored, and/or the number of herds as collected in the other parts of the questionnaire. The use of AI is an indicator that the effective population size may be smaller than that indicated by the number of breeding males alone and so the category would increase in priority. Existing storage of suitably sampled germplasm would reduce the priority for conservation.

Categorisation carried out using the FAO system when applied within countries is not without its absurdities; for example, Piedmontese cattle are listed as Endangered by FAO (2000) within the UK whereas it is clearly not endangered in Italy (EAAP, 2003). Likewise Duroc and Hampshire pigs and Montebelliarde cattle are all considered Critical or Endangered within the UK. The need for rationalisation of classification systems to account for cross-border populations (towards the wider idea of 'conservation unit' discussed by Bruford in this publication) is acknowledged and widely recognised, but basing classifications upon countries has the operational benefit of organised communication, and is in accord with the CBD which recognises the sovereignty of countries over their AnGR.

The FAO system is not the only classification system in use for domestic livestock. EAAP has developed a system of endangerment based upon ΔF (see http://www.tiho-hannover.de/einricht/zucht/eaap/factors.htm#methods). In this system ΔF /year is the critical factor rather than /generation. EAAP has also proposed a system called 'NFN', to help identify endangered breeds that may be eligible for EU support. NFN is the product of: (a) the number of breeding females; (b) the proportion that are straightbred; (c) a factor for the trend in females, which is 0.7 if the population is decreasing, 1 otherwise; (d) a factor for the number of herds, which is 0.5 if < 10, 1 otherwise.

In the UK, the Rare Breeds Survival Trust (RBST) operates a further system. Breeds that RBST consider to have been established in the UK for a sufficiently long period are classified primarily by reference to the number of adult females registered with the Breed Society. A breed qualifies as being 'Rare' if the number of adult females is ≤ 3000 for horses and sheep, 1500 for cattle, and 1000 for goats and pigs. The status is amended to 'At Risk' 'Vulnerable', 'Endangered' or 'Critical' if the numbers falls below 1/2, 3/10, 1/6 and 1/10 of the qualifying number for 'Rare' respectively.

It is clear that defining a comprehensive system of endangerment for livestock is a complex task if it is to allow for the different demographic factors, trends and unexpected crises caused by disease or economics. The different systems can lead to a confusion of terms e.g. the Cleveland Bay horse is Endangered-Maintained (FAO, 2000), Critical (RBST as cited in Defra, 2002) or Not Defined (EAAP, 2003; considered to be a composite of foreign breeds!) on very similar population data. Importantly, there is broad agreement among systems on relative priorities, but it would be desirable if some standardisation were to occur particularly as more experience is obtained over time in identifying the important demographic risks factors and quantifying their likely impact.

Making action plans

The CBD places an obligation upon its signatories to manage biological resources so as to conserve diversity (*within as well as between species*) in such a way that their actions are sustainable and allow for equitable sharing of benefits. The relative affluence of the EU and the degree of food security that it enjoys makes the equitable sharing of benefits less of an issue within the EU compared to outside the EU. The major constraint on sustainability is the economic cost of conservation activities, requiring plans that will be self-financing to a large degree in the medium- to long-term.

The CBD clearly ascribes priorities to different types of actions: *in situ* conservation has a higher priority than *ex situ* conservation, irrespective of whether the latter is with live animals or with cryoconserved material. This maintains the relationship with the environment in which the breed was developed and encourages continued co-evolution. An additional argument that is particularly relevant to livestock populations is the maintenance of the relationship between the livestock and the human societies in which a breed was developed. This emphasis on *in situ* is relevant to UK livestock resources, even though it is tempting to assume that the environmental extremes found in the UK are not especially demanding of adaptive fitness. For example, the North Ronaldsay sheep is a breed that became adapted to eating seaweed in its native habitat, and many died of Cu poisoning when removed from the island even on pastures that were classified as Cu deficient, suggesting that the *ex situ* live population may have been subject to strong natural selection pressure. Effective strategies will often have an element of *in situ* conservation combined with cryoconservation (gene banks, a form of *ex situ* conservation).

The assignment of risk status to the different breeds gives a basis for prioritising conservation actions among them. In practice, action plans will depend not only upon the risk status, but also upon some consideration of the relative value of the particular population at risk, especially where funds are limited. FAO (1998) recognises that in some situations 'no action' is a legitimate outcome in the context of an overall conservation plan dealing with many breeds and limited funds, providing this decision is informed and properly considered.

The most difficult task is to develop sustainable *in situ* conservation schemes for breeds at risk, since sustainability without continual subsidy will depend on making the breed attractive to a significant number of farmers. There are a number of approaches to raising the profile of a breed at risk, but the first step must be to ask why the majority of farmers no longer favour it. This is very likely to stem from the view that the breed is not as profitable as other breeds, which may happen for three reasons: (i) poor performance relative to other locally available breeds; (ii) products are competitively produced elsewhere; (iii) demand for the products is declining. The approach to securing sustainability will depend in part on identifying which of these is true. The first reason is the most common cause of breeds declining in popularity and it emphasises the need, described in 'Identifying populations at risk', for phenotypic characterisation of breeds.

Determining the facts about economic performance

This requires evaluating the overall lifetime profitability of the breed within the appropriate environment and production system. This overall profitability includes an evaluation of the fitness and longevity of the breed, and the management inputs that are required, as well as the yield of primary products. The key question is: *'has the economic value of the breed been underestimated?'*

Incorporation into a crossbreeding scheme

The holistic assessment of the economic performance of the breed at risk may indicate that other breeds are superior (there is no genetic reason why a breed developed for one environment must perform better in that environment than another breed that had originally been developed for an alternative environment). Nevertheless the breed at risk may still have an important economic value providing crossbreds are superior. Crossbreeding is an emotive issue in conservation, and two forms of crossbreeding must be distinguished: (a) sustained crossbreeding in which the breeds involved need to be maintained as straightbred populations, as is the case with all the breeds involved in the stratified sheep production system operating within the UK; and (b) development of new breeds, where it is possible that one or more of the straightbred populations involved will not survive in their current form. The first type of crossbreeding may be very valuable for maintaining a breed at risk. The second may be desirable in the light of alternatives of extinction or the existence of the breed solely in a gene bank, since the genes of the breed are maintained even though the breed itself (and its gene combinations) may not. The key question is: *'does the crossbred have better economic value than the imported exotic?'*

Selection within the indigenous breed

Improvement may be obtained from within the breed itself through selection. Where the product demand is in decline, small changes in the product quality might halt or reverse this trend and this may be achieved easily through selection. Selection lends itself to incorporation into nucleus operations, which can be an effective means of managing small populations at risk. The benefits of selection are enhanced through publicity on the improvement scheme promoting the advantages of the breed and breeding goals of the scheme as this raises the profile of the breed as being dynamic and developing, which helps to secure and expand ownership.

This option is most practicable with breeds that are either not yet at critical levels or have recovered in numbers. A sustainable conservation programme with $N_e < 50$, when managed to reduce the rate of loss of genetic variation, should be concentrating on multiplication and spreading the genetic base before embarking upon selection, and this will limit the opportunity for selection. If the breed is very uneconomic then selection alone will only remove the problem in the long term, since rates of progress from well-run selection schemes may only achieve 1–2% improvement per year. Nevertheless, for a breed that is suffering gradual erosion in profitability and numbers, a selection scheme is a very effective option since genetic gain is permanent and cumulative (so 5–10% improvement after 5 years may be possible). The selection may have consequences for other traits that are considered valuable in the breed and may compromise the adaptation of the breed, which may undermine the motivation for its conservation! Therefore selection objectives and recording need to be carefully considered to avoid deterioration of important secondary traits associated with adaptive fitness. However, faced with the eventual extinction of the breed, or its existence solely in a gene bank, maintaining viability *in situ* is a gain in biodiversity. The key question is *'will development by selection restore profitability to the breed?'*

Niche markets for high quality products

The product from the breed at risk may be capable of being marketed at a higher price than the comparable product from other breeds. For endangered breeds, the products already have a rarity value, but this needs to be capitalised upon to increase both local and export markets. This added value may arise simply because the local breed is traditionally reared and the products traditionally processed, but it may require the breed to become more closely associated with its region, to emphasise the management of the breed and to market the products accordingly. There are other opportunities for adding value to the product through highlighting additional qualities in the product from the breed e.g. larger proportions of fat as polyunsaturated fatty acids rather than saturated fatty acids. The key question is *'can the product be differentiated to make it command a higher market price?'*

Developing novel products

This depends on the imagination. One route to explore is the identification of a traditional product that is no longer produced, and if this is identified, competition from other breeds may be overcome by the kind of marketing associated with niche products. One kind of novel product is the breed itself: generating tourist income from maintaining a breed by incorporating it into a farm park or similar venture, perhaps in combination with wildlife indigenous to the region. This is an approach that has been explored widely within the UK. The key question is *'can the breed be marketed to generate income from non-traditional products?'*

Incentive payments

The EU has implemented this option in the past (although the UK did not participate), but the scheme is no longer in operation. Such schemes can cause problems of sustainability: for example if incentives are based directly upon the endangerment status, there is an incentive for farmers to maintain the breed at small numbers. Furthermore when the scheme stops the breed is at risk of a rapid population collapse. The key question is *'can a progressive and sustainable incentive scheme be devised and funded?'*

Improving the management

Improving the management, depending on the circumstances, can make significant improvements in profitability in a short period. As with selection, changes in management may compromise the traits in the breed that are valued. It may therefore be advisable to limit management changes, providing such limitations can be enforced. However, faced with the eventual extinction of the breed, or its existence solely in a gene bank, maintaining viability through this means is a gain in biodiversity. Because of the modification of the environment implied by this option, it has a lower priority compared to other options that would allow the maintenance of a live population within the region. The key question is *'where other options are limited, can cost-effective improvements in management be made?'*

After considering the conservation options, difficult choices need to be made on the form of conservation actions, (including no action). This will inevitably depend on the funds available and the degree of stakeholder support. For reasons given above, *in situ* conservation has the highest conservation value and, where the options allow, the lowest long term costs. Cryoconservation requires facilities for gene banks and staff to manage and monitor the facility, whereas ex-situ live conservation requires little in the way of start-up costs but will require costly maintenance of the population over time particularly if the population is not managed as a livestock production enterprise (although there may be opportunities for keeping such populations on farms belonging to government institutions or similar organisations). Table 1 gives a perspective on the relative costs and benefits.

Table 1. Comparative advantages between the different forms of conservation. Reproduced from FAO (1998).

Form of conservation	Contribution to biodiversity	Maintenance of adaptive fitness	Sustainability	Cost of establishment	Cost of maintenance
Cryoconservation	+	+ +	+	- - -	- -
Ex-situ live	+ +	+	+	-	- - -
In-situ	+ + + +	+ + +	+ + +	-	-

Managing live populations

If it is decided to maintain a conserved breed as live animals then many of the issues relating to management of the population as a whole (i.e. excluding funding, marketing, etc.) will be the same for *in situ* and *ex situ* conservation. Some of the important issues are (i) establishing a viable population, (ii) coping with the history of the population; (iii) the genetic structure of the population; and (iv) the physical management of the population. This section will not be described in detail since it is thoroughly addressed by Villanueva *et al.* in this publication.

Establishing a viable population

FAO (1998) argue that the first objective is to establish a population that has $N_e \geq 50$. Thus the numbers of parents should be increased to achieve such a population size. It is a classical result of population genetics that for a total of N parents the value of N_e is maximised by having equal numbers of male and female parents. However, in many livestock species, it is common to have far fewer breeding males than breeding females, partly for reasons of cost and partly because of the difficulty of managing mature males. Sanchez et al. (2003) show that N_e is at most 6 Min(N_{sire},N_{dam}), and must be assumed to be much less (see above), and therefore conservation schemes for livestock should seek to increase the number of breeding males as far as is practicable.

In addition, the mating of relatives should be avoided wherever possible, and this can be facilitated by creating circles of farmers keeping the breed that are willing to exchange breeding males in a managed way. In particular, the mating of half-sibs and full-sibs not only creates a longer term risk of losing genetic variation but also creates offspring that are notably more inbred than contemporaries, and such offspring will be more likely to show inbreeding depression. This depression in performance occurs in traits related to fitness, such as survival and reproductive capacity, which are precisely those traits that are important to secure and multiply up the population at risk.

Where a nucleus of conserved animals is being formed it may be possible to do a limited amount of sampling among farmers donating stock, and this offers an excellent opportunity to sample the gene pool widely to secure as much genetic variation within the managed population as possible.

Coping with the history of the population

The history of the population may bring genetic problems that need to be tackled. Bottlenecks (a period of time during which very few parents left offspring that went on to breed) that have occurred in the past cannot be removed but recent bottlenecks may be ameliorated by selecting animals that are as unrelated as possible. This can be achieved using the methods described by Villanueva et al in this publication.

A further problem is related to lethal or unfavourable recessive alleles segregating in the population taking the form of a recognised syndrome or inherited disease. *These can be dealt with effectively providing: (a) pedigree is recorded; (b) incidence is recorded; and (c) such data is made available for analysis using modern software tools.* Where the exact mode of inheritance is unknown, application of modern software tools will allow the data to be tested for different modes of inheritance. Whether or not the alleles are identified, the risks from each candidate for breeding may be estimated, e.g. in simple lethal recessive models, the carrier status can be predicted, and the breeding risk for the population as a whole can be managed. With such syndromes present in the population an important initial breeding objective is to remove the syndrome, and this can be done by selecting parents from among those that have minimum breeding risk. This problem is particularly relevant to the UK since the National Scrapie Plan is intended to remove all susceptible haplotypes from the sheep population, including rare breeds.

Such selection must be managed, however, as it could result in a large loss of genetic variation (depending on the frequency of the unfavourable alleles) due to removal of all offspring from carrier ancestors. The good news is that the controlled removal of such syndromes can be managed effectively and efficiently using the selection algorithms described by Villanueva *et al* in this publication, with minimal cost to ancestral family lines. Such algorithms may be easily implemented at little cost and *not to use such tools to help remove deleterious alleles known to be segregating in a breed at risk runs counter to the principles of the CBD.*

The genetic structure of the population

This topic includes which animals to select and which to mate together and is dealt with by Villanueva *et al* in this publication.

The physical management of the population

An important aspect is individual identification in the population. This serves several purposes, for recording performance and other events, tracing location, and for recording pedigree. In practice this requires the best available system combining ease of attachment, ease of reading minimal loss of identifiers, and low cost, when applied in the relevant production system.

Pedigree recording is the most powerful tool for managing genetic variation in the population. To achieve this cost- effectively it is necessary to: (i) manage the matings either individually or through careful management of breeding males, and (ii) to identify mothers of offspring at birth or soon after. A system that leads to pedigree errors of < 10% will be effective. Error rates can be checked occasionally using DNA markers if costs allow.

Dispersal of the conserved population is important since a single location is very vulnerable to fire and disease, or climatic disasters. If adaptive fitness is important, then wide dispersal can be disadvantageous since it begins to depart from *in situ* conservation. If a strictly defined management is associated with the breed then the greater the number of holdings, the more difficult it is to control management. These considerations of adaptive fitness and management need to be set against the need and importance of securing the small population and increasing the numbers of farmers interested in sustaining the breed. Catastrophic risks, coupled with genetic risks present in small populations (such as deleterious alleles making an impact upon the population) argue for a cryoconservation back up where this is technically feasible and affordable. The considerations for cryoconservation are discussed below.

Cryoconservation

Cryoconservation means gene banks. Gene banks require planning, facilities for obtaining samples, expertise to obtain and prepare the germplasm, duplicated and secure facilities for storage of samples, continuous and sustained management and monitoring of the facility to maintain the integrity of the bank, and agreements with stakeholders on ownership, access, use and replenishment. Consequently cryoconservation costs money and commitment in both the setting up and the maintaining the gene bank, but, if carried out properly, it can provide a valuable means for reducing the risks of managing AnGR.

Nevertheless an implication of the large cost and commitment involved is that if gene banks are to be established, they should be done after rigorous planning, starting off with clear statements on the objectives of the gene bank (i.e. what purpose is the gene bank meant to achieve?). From the earliest planning stage the answer to the question should not be left vague, e.g. 'protecting against future risks', since achieving simple tasks involving the use of gene banks can very easily imply large numbers of samples and, consequently, large costs. If the size of the gene bank is found to be too small to make a useful impact when the gene bank is called upon to overcome problems, then the enterprise will have been a costly failure. Furthermore, important conservation decisions may have been taken between the establishment and the use of the gene bank that were predicated upon the existence of the gene bank and its expected success in protecting against particular risks. It is therefore important to establish clarity at the outset, under what circumstances the gene bank will be used, what size of impact is it expected to cater for, and hence how big does the gene bank need to do the job?

An additional consideration when considering the design of the gene bank is that the small sizes of populations at risk, and the limited objectives that can be expected from gene banks, implies that many projects will involve only small numbers of animals. When numbers of donors are small and/or the numbers of offspring desired from utilising the bank are small, random chance may have a large effect on whether or not the intended outcomes are achieved e.g. the offspring obtained may be fewer in number, or have an unfavourable sex ratio, or are all full-sibs. Therefore from the outset, resource planning should not be based on expected success rates, but should consider the question what resources are required to make 90% certain (for example) of achieving a particular objective. Gene banks are often designed with the aim of protecting against unpredictable risks, and it would be incongruous if failure arose from not properly considering entirely predictable risks, resulting in a waste of cost and effort.

These considerations form the background to the guidelines of FAO (1998). The design of the gene banks requires to consider: (i) forms of germplasm; (ii) achievable objectives; (iii) numbers of donors and samples; (iv) facilities; and (v) access, use and replenishment. The text below discuss the first three of these items, and full details on all five items are given in FAO (1998).

Forms of germplasm for cryoconservation

There are a number of forms of germplasm that can be considered: semen, embryos, oocytes, somatic cells and DNA. The ease with which the germplasm can be obtained, handled and used varies with species. The routine nature of many techniques in cattle, which have been developed to be minimally invasive or non-invasive, makes it easy to underestimate the limitations in applying such techniques in other species. At present, the two main routes for cryoconservation are semen and embryos.

Semen has **strengths** in that it has the most widely established techniques of collection and use across the species, techniques are generally least invasive, and it is the *only* established form of cryoconservation for poultry. The **weaknesses** are that it is haploid and therefore does not store an intact genome. Embryos have **strengths** in storing intact genomes, including cytoplasmic DNA, and potential disease threats within stored material can be more readily identified from screening than with semen. The **weaknesses** are the species limitations in collection and storage of embryos, longer collecting procedures, and the highly variable recovery rates among females following superovulation.

Of the other sources of germplasm, oocytes are becoming a feasible route in cattle. Ovum pick-up is increasingly reliable and a significant benefit of the technique in practice is that it has a less variable response across donors compared to embryo recovery. However, the long-term storage of oocytes as haploids (with the cytoplasmic DNA) is still difficult.

have its own objectives, which will most likely be a mix of the objectives described below, but what is important is that the objectives are clearly defined and realistic, and the numbers of samples stored are sufficient to achieve them.

Re-establishing an extinct breed

This objective uses the gene bank as an 'Ark'. Whilst this use is close to the ethos of the gene bank it is reasonable to ask: (i) how often will a breed have a sufficient need for re-establishment to justify the considerable resources that will be required to achieve a large population (especially for species with low reproductive rates); and (ii) whether a breed that only exists in a gene bank will be lost from memory. Therefore FAO (1998) only considered achieving this objective on a small scale, basing calculations on re-establishing 12 males and 12 females). To reduce the numbers further was considered to establish a population with a built-in bottleneck, unnecessarily introducing genetic risks into the complex management process of multiplying up whilst securing the genetic variation. An important point to note is that with this objective samples used from the bank can ultimately be replaced after re-establishment.

The re-establishment may take two forms: ideally, from embryos in which a population of straight breeding males and females are immediately produced; or in a more cumbersome fashion, from semen alone in which several generations of backcrossing (i.e. grading up) are necessary before the population becomes straight-breeding. The re-established breed will never be totally free of the base breed used to provide initial females for the backcrossing process. Therefore an informed but otherwise arbitrary number of backcrosses has to be decided upon to arrive at some form of minimum requirement. Hill (1993) shows how the number of generations of backcrossing will alter the mean and variance of the proportion of the genome coming from the cryoconserved breed and FAO (1998) recommended 4 generations of backcrossing, since this will give a 95% chance that 90% of the genome of the individuals produced derives from the cryoconserved breed. Additional generations prolong the process and the dependence on the gene bank, whereas fewer generations may be regarded as forming a new breed (discussed below). Re-establishment via embryos, where feasible, is considered to be very much more desirable than by semen.

New breed development

An important contribution of cryoconserved material is for new breed development to fit new production circumstances. Breed development has been a part of animal breeding from its origins (see Hall and Ruane (1993) for an analysis of livestock breed development in relation to human demography). In this case the objective is to use the unique genes and gene combinations of a stored breed to establish a new population with desirable properties. The need for such a population may arise from: new selection objectives possibly due to changed environmental conditions or disease threats, or to replace or improve a breed that is seen to be genetically inadequate maybe through excessive inbreeding depression, or a lack of performance or variation in desirable traits. One route to achieve this is the use of stored semen for a number of generations of backcrossing. The use of embryos in this instance would not be warranted (unless cytoplasmic inheritance was very clearly implicated in the desired trait, which is very rare). Defining this as an objective requires stating how a stored breed may contribute to a new breed, how large the contribution from the gene bank would be, and how often the process can be repeated. A new breed cannot replenish the germplasm from the original breed that were withdrawn from the gene bank.

FAO (1998) considered the storage requirement by the needs for semen to obtain a 100 breeding males of a crossbred containing 75% of the stored breed. In global terms, where developing countries are seeking to achieve food security that is sustainable, this may constitute a very important objective for a gene bank, although within the EU this objective may have a much lower priority.

Supporting in vivo conservation

The *in vivo* population may acquire lethal recessives over time. In a small population these may very quickly reach gene frequencies that are difficult to remove from the population. Therefore individuals that have been cryopreserved may aid the maintenance of the population through semen or embryos. In some cases, depending upon the extent of the inbreeding depression, semen may be sufficient to aid the removal of the recessive. Before introducing the cryopreserved semen, consideration should be given to the possibility that some of the cryopreserved individuals may also be carriers and so should be avoided (see Section 'Coping with the history of the population'). An additional use may be to help maintain genetic variation by re-introducing ancestral lines that had been lost or have become under-represented (although the question must be asked whether this lack of representation was chance or natural selection!), or simply to help with the dissemination of sires across a dispersed population. Sonesson *et al.* (2003) considered the genetic management of the animals that were conserved alive when there was regular availability of cryopreserved semen in very large amounts, and found that inbreeding was minimised when cryopreserved semen from the founding generation and the first generation of males was used for all generations.

FAO (1998) defined the storage requirement by considering the need for semen to mate 100 females per generation over two successive generations. Assuming a successful outcome for the population the semen withdrawn from the gene bank can be replenished.

Research into identifying single genes of large effect

The inclusion of research into the identifying single loci of large effect (QTL) as a legitimate claim on cryo-preserved stocks is justified by the observation that the primary purpose of *ex-situ* conservation is to conserve a set of loci that may be valuable in the future. The initial objective of this type of research, often conducted in the form of genome mapping studies, is to tag such loci with genetic markers whose location within the genome is already known. This knowledge will lead to breeding programmes that can more efficiently utilise the QTL, either directly through the use of the cryoconserved breed or from subsequent research in other breeds that builds upon the results of the mapping study. This activity will encourage the use of the conserved stocks.

FAO (1998) defined the requirements for storage for this objective based upon power calculations for detecting QTL of large effect. The samples withdrawn would not be replenished. The sole objective of comparative examination of DNA sequences would not have demanding requirements relative to those for mapping QTL described in the last paragraph.

Numbers of donors and samples

The numbers of samples required will depend on the objectives. The number of donors should be sufficiently large so that the gene bank 'generation' will not represent a bottleneck, and in general the more donors that are represented in the cryo-conserved material the better. Care should be taken to sample all ancestral lines, and the methods of Villanueva *et al* (in this publication) can help to maximise the genetic variation among individuals stored. The number of donor males used for semen collection, and donor females for embryo collection must consider not only the number that are selected but also the likely failure rate in obtaining samples for storage. This is particularly challenging for embryo recovery where the variability among individuals means that single individuals may easily dominate (see FAO (1998) for a solution to this problem).

Discussion

The guidelines of FAO (1998) take an approach to the issues of managing AnGR at risk that is centred upon breeds and individual breeds at risk: which breeds are at risk, is conserving a particular breed a priority, what actions may be needed for managing the risks for a breed, how should the breed be managed *in vivo*, and what role might cryoconservation have for the breed. It assumes that effective conservation actions will require the involvement of stakeholders that are interested in particular breeds and that conservation of livestock diversity should strive to be inclusive of all breeds. This may not be possible for economic reasons and choices may need to be made.

Eding and Meuwissen (2001) explored an alternative approach to cryoconservation that is more centrally organised. This attempts to define objectively (i) which breeds would or would not be stored, (ii) how many individuals to sample from each breed, and (iii) which individuals should be sampled, so as to maximise the genetic variation that can be recreated from a gene bank. This method depended upon deriving relationships among breeds based upon DNA markers. Other related approaches use genetic distances (again based on markers) to define alternative measures of diversity rather than the concept of stored genetic variation that was used by Eding and Meuwissen. Weitzmann (1992) provided rules for defining what may or may not be a legitimate utilitarian measure of diversity. d'Arnoldi *et al.* (1998) and Caballero and Toro (2002) provide interesting discussions on these rules and the consequences of the different strategies for conservation of diversity, both within and between breeds.

The perspective of FAO (1998) is that action is required, and that such actions should be based upon the best information available at the time, using the best practice that is feasible at the time, and involving stakeholders at all stages.

Acknowledgements

I would like to thank BSAS for inviting me to address the conference in Edinburgh in November 2002, and to Beatriz Villanueva for a number of helpful comments on earlier versions of this manuscript.

References

d'Arnoldi, C.T., Foulley, J.L. and Ollivier, L. 1998. An overview of the Weitzman approach to diversity. *Genetics Selection Evolution* 30: 149-161.

Caballero, A. and Toro, M.A. 2002. Analysis of genetic diversity for the management of conserved subdivide population. *Conservation Genetics* 3: 289-299.

Defra, 2002. *UK Country Report on Animal Genetic Resources.* Defra, United Kingdom

EAAP, 2003, *EAAP-Animal Genetic Data Bank.* http://www.tiho-hannover.de/einricht/zucht/eaap/

Eding, H. and Meuwissen, T.H.E. 2001. Marker-based estimates of between and within population kinships for the conservation of genetic diversity. *Journal of Animal Breeding and Genetics* 118: 141-159.

FAO, 1998. *Secondary guidelines for development of national farm animal genetic resources management plans: management of small populations at risk.* FAO, Rome, Italy

FAO, 2000. *World watch list for domestic animal diversity, 3rd edition.* FAO, Rome, Italy

Hall, S.J.G. and Ruane, J. 1993. Livestock breeds and their conservation -a global overview. *Conservation Biology* 7: 815-825.

Hill, W.G. (1993). Variation in genetic composition in backcrossing programs. *Journal of Heredity* 84: 212-213.

Mace, G.M. and Lande, R. 1991. Assessing extinction threats: towards a re-evaluation of IUCN threatened species categories. *Conservation Biology* 5:148-157.

Sanchez, L., Bijma, P. and Woolliams, J.A. 2003. Minimising inbreeding by managing genetic contributions across generations. *Genetics* 164: 1589-1595.

Sonesson, A.K., Goddard, M.E., and Meuwissen, T.H.E. 2002. The use of frozen semen to minimize inbreeding in small populations. *Genetical Research (Cambridge)* 80: 27-30.

Weitzmann, M.L. 1992. On diversity. *Quarterly Journal of Economics* 107: 363-405.

Woolliams, J.A. and Wilmut, I. 1999. New advances in cloning and their potential impact on genetic variation in livestock. *Animal Science* 68: 245-256.

8

Experiences from plant GR conservation

M.J. Ambrose
*John Innes Centre, Norwich Research Park, Colney, Norwich,
NR4 7UH, UK*

Abstract

*There is a long history in the UK of procuring and maintaining plant
genetic resources for curiosity, novelty, taxonomic reference or direct
utilisation. This paper describes the evolution, the current structures
and the processes involved in plant genetic resource activities in the
UK, and discusses similarities and differences in the issues in and
approaches to plant and animal genetic resources conservation and
utilisation.*

The UK has a long history in procuring and maintaining plant genetic
resources for curiosity, novelty, taxonomic reference or direct utilisation.
This has involved many prestigious academic institutions and much of
this institutional landscape is still evident today in collections at
universities, botanic gardens, museums and research institutes. While
these collections remained small scale, their conservation was never a
major issue. As the collections grew and overheads for their maintenance
increased, decisions were required as to what was to be kept and what
was to be ditched or left unsupported.

The suggested formation of a National Crop Plant Genetic Resources
Committee to address such issues first arose out of discussions within
an Agricultural Research Council (ARC) working party on data
management for plant breeding and plant genetics in 1978. The idea
took further shape following discussions held by UK delegates at an
International Board for Plant Genetic Resources (IBPGR) meeting on
information handling systems for genebank management (Konopka and
Hanson, 1985). The first meeting of what was called the UK Plant Genetic
Resources Group (UKPGRG; http://ukpgrg.org) was convened in March
1985. Membership of the Group at that time was drawn from research
institutes, the universities of Reading and Birmingham, the Royal Botanic
Gardens Kew and the Ministry of Agriculture Fisheries and Food or
MAFF (predecessor to Defra).

In the late 1980's a concerted effort was made to raise awareness of
PGR conservation and biodiversity issues in the media (Sattaur; 1989a

and b). Following lobbying and questions in Parliament, the Government of the day commissioned MAFF and SOAFD (Scottish Office, Agriculture and Fisheries Department) to examine the UK's involvement in the *ex-situ* conservation of plant genetic resources, taking into account the likely scientific and public good requirements in the field over the coming ten years (MAFF, 1991). The UKPGRG maintained a close dialogue with MAFF and was involved with others in the review process. One of the recommendations of the review was to encourage the UKPGRG to formalise its terms of reference so as to offer advice to Government, to promote a UK network of plant genetic resources, and to develop contacts with other stakeholder organisations. Within a year, the Group had met to discuss this proposal and formally adopted terms of reference (Table 1). Under these terms MAFF supported the Group by providing secretarial support at meetings. Every sector and collection within PGR is different and offers opportunities for others to learn. In addition to two regular meetings each year, the Group organises technical visits to individual institutions on a regular basis covering the formal and informal *ex-situ* collections as well as the commercial sector. These are run as open meetings with invitations being extended to other interested parties and have proved particularly useful for administrators and governmental officers in developing an appreciation of the practicalities of conservation management for different scenarios. Occasional *ad hoc* meetings of the Group enable more detailed discussion on specific issues. Examples of topics include the exploration of issues and protocols for the *ex-situ* conservation of GM material, 'Linking Resources', a meeting held in conjunction with the Henry Doubleday Research Organisation which brought together a range of governmental and non-governmental organisations working in the field of PGR to explore the common ground between the two sectors and how they could integrate and complement each other more effectively.

Table 1.
UKPGRG terms of
reference.

To act as a forum for discussion on PGR issues and exchange of information between curators of *ex-situ* collections, industry and other interested organisations within the UK;

Provide technical and policy advice (at UK and International levels) to Defra, DFID, SEERAD, DARD and other Government Departments;

Address and exchange information on scientific and technical issues related to PGR;

Encourage the co-ordination of *ex-situ* conservation through networking of collections and other initiatives;

Encourage the utilisation of collections by industry and other customers;

Develop contacts and co-operation with botanic gardens and informal groups in pursuance of *ex-situ* conservation

The Group's strength lies in its breadth of technical expertise. While the focus of the Group remains predominantly *ex-situ*, the complementarity with *in-situ* and on-farm are increasingly being addressed. This is not just in terms of the practical issues of conservation, characterisation and exchange of materials but also in terms of information management, outreach programmes and user contacts. Membership of the Group has been broadened to include representatives of all the main stakeholder communities with practical involvement in PGR including industry and the informal sector (Table 2). The range of formal *ex-situ* collections held within the UK covers a wide range of different forms from seed collections such as the forage grasses, cereals and vegetables to vegetative tissue collections such as potatoes through to living collections such as the National Fruit collection held by Brogdale Horticultural Trust. While each individual collection has its own operational procedures and programme of characterisation and evaluation work, issues such as standardised formats for passport data, pedigree information and housing of off-site safety duplicates are common for all collections. The existence of the Group has been responsible for many examples of bilateral and multilateral collaborations and offers of help. Significant assistance in terms of regeneration of material, verification and development of characterisation programmes can be cited over the years. The Group maintained an email server that acts as a notice board on which issues can be raised and advice sought.

Collections

National Vegetable Gene Bank: HRI Wellesbourne
Commonwealth Potato Collection: SCRI
Dundee Forage Grasses: IGER, Aberystwyth
Cereals and Grain Legumes: JIC, Norwich
National Fruit Collection: Brogdale
Millenium Seed Bank: RBG Kew, Wakehurst Place
Arabidopsis Stock Centre: University of Nottingham
Willow Collection: IACR, Rothampsted
Statutory Collections: NIAB, SASA, DARD
Heritage Seed Library: HDRA, Ryton
BGCI: Kew NCCPG
Wye College, University of London
University of Reading
Forestry Commission
Joint Nature Conservation Committee
Institute of Terrestrial Ecology
British Society of Plant Breeders
BBSRC

Training

University of Birmingham

Observers

Defra, DIFD, SEERAD, DARD

Table 2. Membership of the UK Plant Genetic Resources Group.

Increasingly, the Group is called on to offer technical advice and support. It offers suggestions of nominees to act as technical contacts on a range of PRG initiatives and has been called on to offer detailed technical input in support of government Departments with respect to PGRFA negotiations at the UK, European and International level (Table 3). These include the International Undertaking, now known as the International Treaty on Plant Genetic Resources for Food and Agriculture (FAO; http://fao.org/ag/cgrfa), EU programmes and the European Cooperative Programme for Crop Genetic Resources (ECP/GR: http://ecpgr.cgiar.org/Introduction/aboutECPGR.htm). The Group's profile is such that it has become the primary forum and focal point for ex-situ PGR information in the UK. The Group is currently participating in a EU Framework V programme which aims to deliver a European PGR search catalogue with passport data on ex *situ* collections across European institutions (EPGRIS; http://ecpgr.cgiar.org/epgris). Meetings and information exchange between UK institutions is being organised through the UKPGRG. The European Inventory will be called EURISCO, which stands for European Internet Search Catalogue. The catalogue will be frequently updated automatically from National PGR Inventories and will be accessible via the Internet. The PGRFA community across Europe has benefited greatly over the past thirty years through organising itself into coherent groupings such as ECP/GR and EPGRIS that have both gained in credibility and enabled them ot take advantage of European funding initiatives. Their very existence has provided important technical and communication networks through which Europe and individual countries can work to meet their own and wider international obligations.

Table 3.
UK Plant Genetic Resources Group. Technical inputs and interactions.

Convention for Biological Diversity
International Treaty
EU Community programmes in the area of PGRFA
ECP/GR Steering Committee
DEFRA Policy Review
National Focal Point for PGR
EPGRIS National inventory of PGR
UK response to the Global Strategy
National List and Seed Committee

Across the European PGR community, there are a series of coordinated programmes addressing key issues relating to the documentation of PGR both in terms of passport and characterisation data and in the long term looking to foster a sharing of responsibilities across European ex-*situ* collections. Both are seen as adding value to the germplasm resources. The availability of high quality data is considered essential to the wider awareness of these resources and their effective utilisation.

This is increasingly important in the era of molecular characterisation of collections and of emerging bioinformatics.

PGR collections aim to hold and document their resources, whether they are collected from the wild, as ruderal populations or developed and cultivated in the UK in terms of crops going as far back as possible. Statutory reference collections are maintained to underpin legisliation regarding national listing, plant breeders rights and seed certification. These specialist collections form an important component of PGRFA within the UK but are not complete without the resources held in other formal and informal *ex-situ* collections. The immense value in bringing together these different types of collections into one grouping and working through a common agenda should not be underestimated. I would add that there is added value as a result of the plurality represented within the Group. While there is clearly commonality between the crop plant and farm animal genetic resources communities in terms of objectives for the conservation and characterisation of their material, there are a number of fundamental differences that separate them.

I would like to state my primary reflection on the difference between the two communities on attending the Farm Animal Genetic Resources meeting in Edinburgh. This rests with the fundamental difference in the relationship between a farmer and their livestock as opposed to their crops within the Western Europe. A lot may just come down to basic biology or the higher monetary value invested in each animal, but for those involved with rearing, maintaining or improving different breeds, there is the clear sense of farmers being the custodians of bloodlines and thus part of an unbroken tradition with agricultural practices that trace back to the earliest days of domestication. Thus many of these resources and their associated expertise reside with a large number of individual practitioners and their strong sense of commitment to these activities become immediately clear. The same cannot be said for crops where the regulation of plant breeding and the marketing and provision of crop seed that has occurred since the early 1900s (Wellington and Silvey, 1997) has broken that link. It can be argued that the introduction of national listing, seed certification and plant breeders rights, while unquestionably transforming productivity and providing the necessary incentives and security of products, have redefined the role of the arable farmer to that of grower. There are examples of farmers acting as maintainers or custodians of old crop forms but these are predominantly restricted to marginal regions where the strong adaptation and multiple use in these forms cannot be met with modern types (Jarman, 1996), or niche markets such as straw for thatching or seed for health markets. The proportion of farmers actively involved in the development of crops through active selection of superior forms from landrace crops is never likely to have been that great. Having said that, there are a good number of documented examples of crop varieties that were developed in this way that survive in *ex-situ* collections today, particularly in the areas of fruit and vegetables but there are no accounts from recent times in terms of our major crops. Interestingly these activities are still practiced in the amateur grower market and long may they continue (Stickland, 2001).

References

Jarman, J.R. 1996. Bere barley - a link with the 8th century. *Plants Varieties and Seeds*, 9: 191-196.

Konopka, J. and Hanson, J. 1985. Information handling systems for genebank management. *Proceedings of a workshop held at the Nordic Gene Bank*, Alnarp, Sweden, 21-23 November 1984. IBPGR.

MAFF, 1991. *Review of UK policy on ex-situ conservation of plant genetic resources.*

Sattaur, O. 1989a. The shrinking gene pool. *New Scientist*, 29: 37-41.

Sattaur, O. 1989b. Genes on deposit: saving for the future. *New Scientist* 5: 41-44.

Stickland, S. (2001). Back garden seed saving- Keeping our vegetable heritage alive. HDRA- the organic organisation. eco-logic books, Bristol, UK.

Wellington, P.S. and Silvey, V. 1997. *Crop and Seed Improvement; A History of the National Institute of Agricultural Botany 1919 to 1996*. The Dorset Press, Dorchester, UK.

9

Managing genetic resources in selected and conserved populations

B. Villanueva[1], R. Pong-Wong[2], J.A. Woolliams[2] and S. Avendaño[1]

[1]*Scottish Agricultural College, West Mains Road, Edinburgh, EH9 3JG, UK*
[2]*Roslin Institute (Edinburgh), Roslin, Midlothian, EH25 9PS, UK*

Abstract

Managing the rate of inbreeding (ΔF) provides a general framework for managing genetic resources in farmed breeding populations. Methods for managing ΔF have been developed over the last five years and they allow the attainment of the greatest expected genetic progress while restricting at the same time the increase in inbreeding. This is achieved by optimising the contribution that each candidate for selection must have to produce the next generation. The methods take into account all available performance and pedigree information and use Best Linear Unbiased Prediction (BLUP) estimates as a predictor of merit. Importantly, these tools give at least equal, but more often more gain than traditional selection based on truncation of BLUP estimated breeding values when compared at the same ΔF. Deterministic predictions for the expected gain obtained with optimised selection with ΔF restricted are now available. The optimisation tool can be also applied in a conservation context to minimise ΔF with restrictions to avoid loss in performance in valuable traits. Information on known quantitative trait loci or on markers linked to them can be incorporated into the optimisation process to further increase selection response. Molecular genetic information can also be incorporated into these tools to increase the precision of genetic relationships between individuals and to manage ΔF at specific positions or genome regions.

Introduction

Breeding programmes have been important contributors to the enormous advances in biological and economic efficiency of farmed livestock

113

species in many countries over the last 50 years. However, much of the emphasis in the past has been focussed rather narrowly on increasing performance in production traits (e.g. milk yield in dairy cattle and growth rate in meat-producing species). Narrow breeding objectives and intense and accurate selection have given rise to concerns over substantial increases in the rates of inbreeding and consequent decreases in genetic variability and, most importantly, reductions in fitness-related traits. Indications of negative genetic consequences of selection for production on health and reproductive traits together with changes in consumers' demands (e.g. quality, welfare) has led to broadening of the breeding objectives for livestock species (e.g. Simm *et al.*, 2001). Methods that control inbreeding without concomitant losses in genetic gain are also starting to be implemented in livestock breeding programmes (Avendaño *et al.*, 2003a).

Managing the rate of inbreeding (ΔF), or equivalently the effective population size ($Ne = 1/2\Delta F$), provides a general framework for managing genetic resources as it determines levels of neutral genetic variability and the relative effects of genetic drift and selection on non-neutral loci. The control of ΔF could avoid or alleviate the reductions in viability and fertility which eventually can make further progress imposible despite the presence of genetic variance for traits under selection. Other processes influenced by ΔF such as increased probabilities of losing beneficial rare alleles (Caballero and Santiago, 1998), developmental instability (e.g. Fernández *et al.*, 2002) and risk of extinction (Saccheri *et al.*, 1998) would be also controlled. There is also a relationship between ΔF and the variance of response to selection, another component of the risk in breeding programmes (Nicholas, 1987), and in some circumstances, a restriction on ΔF can lead automatically to an acceptable coefficient of variation of genetic gain (Meuwissen and Woolliams, 1994a). Thus, ΔF provides a measure of genetic risk in improvement schemes (Woolliams *et al.*, 2002). This paper outlines the application of methods developed over the last five years for managing ΔF in farmed breeding populations.

Methods for managing the rate of inbreeding

Two main decisions need to be taken in the operation of breeding programmes: 1) which animals should be used for breeding and how widely they should be used (i.e. selection decisions); and 2) how the selected animals should be mated (i.e. mating decisions). A considerable amount of research in the past has dealt with the development of selection and mating strategies for restricting the increase of inbreeding. Selection strategies included increasing the numbers selected, restricting the number of individuals selected per family and reducing the emphasis given to family information in the selection criterion (e.g. Toro and

Perez-Enciso, 1990; Villanueva et al., 1994). Mating strategies included factorial designs, minimum coancestry matings and compensatory matings (Caballero et al. 1996). These methods considered rates of gain and inbreeding separately and although they were succesful in controlling the increase in inbreeding, they also led in general to losses (modest in most cases though) in selection response.

The first attempt to simultaneously manage genetic gain and inbreeding in selection decisions was by Wray and Goddard (1994) and Brisbane and Gibson (1995) who developed dynamic selection rules by placing a direct constraint on cumulative inbreeding while the usage of selected candidates was optimised for maximising genetic gain. These methods were improved by Meuwissen (1997) and Grundy et al. (1998) for achieving a constraint on ΔF. The application of these tools only require information that is commonly available in practical breeding programmes; i.e. pedigree information and estimates of breeding values (EBVs) for candidates for selection, which are now routinely obtained using BLUP in most species.

Selection that applies optimisation tools (referred to here as 'optimised selection') differs from the traditional method, usually referred to as 'truncation selection', in the information used when deciding which animals should be selected and how widely they should be used. With truncation selection, candidates are ranked according to the best estimates of breeding values available and only the candidates above (or below) a given value (the truncation point) are chosen as parents and they all contribute equally to the next generation. Thus selection decisions are taken independently of the genetic relationships between candidates and therefore, in extreme cases, individuals selected can come from very few families. In contrast, optimised selection takes into account all genetic relationships, optimises the number of inividuals selected and allows them to contribute differently, resulting in higher gains at the same inbreeding, or to lower inbreeding at the same gain. The idea of allowing unequal contributions of selected candidates as an alternative to truncation selection was first proposed by Toro and Nieto (1984) who showed that lower ΔF with no loss in response could be obtained by allowing more contributions to those candidates with higher EBVs without compromising selection intensity. A more detailed description of optimal methods is given in the next section.

Optimised selection

Selection decisions can be optimised in order to manage ΔF without implying any loss in genetic gain (Meuwissen, 1997; Grundy et al., 1998). The problem to be solved is concerned with the allocation of contributions of the candidates to selection so as to maximise genetic gain with restrictions on ΔF and can be formulated as:

Maximise $\mathbf{c}^T\mathbf{g}$ subject to $\mathbf{c}^T\mathbf{Ac} \leq C$ and $\mathbf{Q}^T\mathbf{c} = \frac{1}{2}\mathbf{1}$

where $\mathbf{c}$ is a vector of solutions (i.e. contributions or proportions of total offspring left by each candidate), $\mathbf{g}$ is the vector of BLUP-EBVs (or the best estimate available of breeding values) of the candidates, $\mathbf{A}$ is the additive genetic relationship matrix (e.g. Lynch and Walsh, 1998), C is the desired ΔF, $\mathbf{Q}$ is a known incidence matrix for sex and $\mathbf{1}$ is a vector of ones of order 2. The first inequality ensures that the constraint on ΔF is met (note that, with fully random union of gametes, $\mathbf{c}^T\mathbf{Ac}$ is twice the inbreeding coefficient of the next generation) whereas the second inequality ensures that half of the contributions come from males and half from females. The problem can be solved using Lagrangian multipliers by maximising the function:

$$H_t = \mathbf{c}_t^T\mathbf{g}_t - \lambda_0 (\mathbf{c}_t^T\mathbf{A}_t\mathbf{c}_t - C_t) - (\mathbf{c}_t^T\mathbf{Q}_t - \frac{1}{2}\mathbf{1}^T)\boldsymbol{\lambda} \qquad [1]$$

where subscript t refers to generation number, $C_t = 2[1-(1-\Delta F)^t]$ and λ_0 (scalar) and $\boldsymbol{\lambda}$ (a vector of order 2) are Lagrangian multipliers (Meuwissen, 1997; Grundy et al., 1998).

Table 1 illustrates the benefits expected from using the optimised method relative to standard truncation selection for an example with 50 males and 50 females as selection candidates and a heritability of 0.2. With truncation selection, a fixed number of individuals (N sires = N dams) with the highest estimated breeding values were selected to be parents of the next generation. The number of parents and the family sizes were fixed across generations and each dam produced 50/N males and 50/N females. With optimised selection, the numbers of individuals selected and their contributions were not fixed but optimised each generation for maximising genetic progress while restricting ΔF to a specific value. This value was equal to the corresponding value obtained for ΔF with truncation selection to compare gains from the two types of selection at the same ΔF. The effective number of parents per sex was computed as $1/(4\sum_{i=1}^{N_o} c_i^2)$, where N_o is the actual optimised number of parents per sex.

Except for the extreme restriction on ΔF (i.e. 0.25%), for which no gains were obtained with either method, optimised selection always gave higher gains than truncation at the same ΔF. This was achieved by selecting less individuals and allowing them to contribute differentially, as indicated by the values for V_{no}. More stringent restrictions on ΔF led to larger numbers selected and more equal contributions (Table 1 and Avendaño et al., 2003a) and to within family selection (Villanueva and Woolliams, 1997).

Table 1.
Cumulative genetic gain (G) at generation 10, actual (N) and effective (N*) number of parents selected of each sex and variance of number of offspring of each sex among selected parents (V_{no}) under truncation (T) and optimised (O) selection that maximise gain under different ΔF. These values for ΔF are only illustrative rather than recommended.

	$\Delta F = 15\%$		$\Delta F = 8\%$		$\Delta F = 1.5\%$		$\Delta F = 0.25\%$	
	T	O	T	O	T	O	T	O
G	2.45	2.73	2.43	2.72	1.62	2.11	0.00	0.00
N	5.0	4.9	10.0	5.9	25.0	20.8	50.0	50.0
N*	5.0	3.9	10.0	3.9	25.0	14.3	50.0	50.0
V_{no}	0.0	294.9	0.0	252.9	0.0	11.9	0.0	0.0

When the benefits of these techniques have been evaluated in real livestock populations, large improvements in rates of gain over conventional truncation BLUP at a given ΔF have been suggested. Avendaño et al. (2003a) predicted extra gains of the order of 20% and higher when the methods were applied to sheep and beef cattle populations. This contrasts with the conclusion of Weigel and Lin (2002) that genetic gain might be sacrificed by imposing constraints on inbreeding after applying dynamic optimisation methods in dairy cattle. The comparison of optimised *versus* traditional selection of Weigel and Lin (2002) was however unfair as gains from both strategies were compared at different ΔF. A fair comparison would be to compare the predicted average genetic merit from the optimisation to that obtained with current selection at the current observed inbreeding. Furthermore, these authors applied a direct constraint on the absolute expected level of inbreeding in the next generation, which does not guarantee a constant ΔF in future generations.

In depth investigation of how the optimised method works clearly show that it is by allocating the contribution of selected candidates according to their Mendelian sampling terms, rather than their breeding values (Avendaño et al., 2004; Figure 1). The relevance of this observation is that individuals are managed, and will contribute to future generations, in relation to their unique superiority (or inferiority) with respect to the parental average. Therefore, dynamic selection manages the 'new' genetic variation (i.e. the Mendelian sampling variance) that is created each generation. This has the underlying concept that sustained genetic gain is related to the creation of a covariance between long-term contributions (r) and the Mendelian sampling terms (a) of selection candidates (i.e. $\Delta G = \Sigma(r_i a_i)$), put forward by Woolliams and Thompson (1994).

Although it is clear that the optimisation algorithms constitute powerful operational tools, it has only been possible to predict their benefits from stochastic computer simulations since deterministic predictions have not been available. Grundy et al. (1998) derived an expression for the expected gain when ΔF is constrained but this prediction was only an

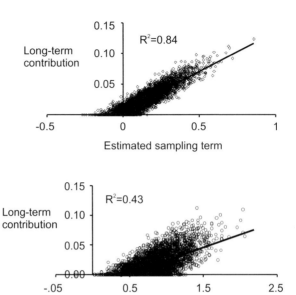

Figure 1. Relationship between long-term contributions and estimated Mendelian sampling term or estimated breeding values. R^2 is the proportion of the variance of long-term contributions explained by the regression (reproduced from Avendaño et al., 2004).

upper limit, impossible to obtain in practice as it assumed that Mendelian sampling terms were known without error. More recently, Avendaño et al. (2003b) have developed a fully deterministic prediction of gain under constrained inbreeding by using index selection theory – the pseudo-BLUP index of Wray and Hill (1989). This approach relies on predicting the accuracy of the Mendelian sampling terms and, when compared with simulation results, has proved to accurately predict the rate of gain for a range of heritabilities, inbreeding constraints and population sizes. Thus it is possible now to predict in advance how much gain it is possible to obtain in practice.

The function to be maximised for optimising contributions of candidates (equation 1) implies that no restrictions are imposed on the reproductive capacity of the candidates for selection. With the exception of species with high reproductive capacity in both sexes (e.g. fish species) this may be unrealistic in practice, in particular for female candidates for which high reproductive rates are less feasible. More flexible constraints may be needed to allow full implementation in practical breeding programmes. These may include a minimum number of sires, a fixed contribution for a particular set of males, or a maximum contribution for females. In theory, any number of additional constraints can be incorporated into the optimisation process but the problem to be solved may become complex. Also, the optimisation tools already available have dealt mainly with selection decisions, assuming that the selected parents are mated at random. Non-random mating systems can be applied subsequently to selected animals resulting from the optimisation, to further increase gains without increases in inbreeding (Sonesson and Meuwissen, 2000). The simultaneous optimisation of selection and mating strategies, in addition to including extra constraints, complicates the optimisation process further. Evolutionary and genetic algorithms may provide a more flexible framework for obtaining the solutions (e.g. Kinghorn et al., 2002).

The optimised methods allow us to restrict the increase in inbreeding to the desired levels, but the value at which ΔF should be fixed still remains a question. Meuwissen and Woolliams (1994b) gave rates of inbreeding to be chosen in order to achieve a balance between the depression in fitness due to inbreeding and its improvement due to natural selection. Rates of inbreeding allowing fitness not to change after ten generations of natural selection ranged from 0.002 to around 0.016, depending on the coefficient of variation and the heritability of fitness and on the number of loci affecting this trait. The critical factor to decide the desired rate was inbreeding depression in fitness and thus the consequences on genetic variability and inbreeding depression in artificially selected traits was ignored. Past estimates indicated that an effective population size of 50 (which corresponds to a $\Delta F = 1\%$ per generation) was sufficient to avoid a significant inbreeding depression and to retain reproductive fitness. However, more recent data have suggested that this figure was optimistic and that the required size is much larger than 50 (Frankham et al., 2002). Much higher effective sizes are required for balancing genetic variation lost by drift and that gained by mutation (thus retaining the evolutionary potential of a population) and for avoiding the accumulation of new deleterious mutations (Frankham et al., 2002). Active management can do something to reduce the numbers required providing that breeding goals include fitness and that goals are backed up by appropriate records. Although there is clear evidence that inbreeding reduces fitness in wild and farmed populations (e.g. Borrow, 1993) and contributes to the extinction of small and isolated populations (Saccheri et al., 1998), there are also highly inbred populations that exist with no decline in fertility or viability. This has been explained by the fact that, when combined with selection, inbreeding may eliminate deleterious alleles. Most deleterious mutations appear in a recessive state and then they are only exposed in the homozygotes. As inbreeding increases the frequency of homozygotes it would also increase the opportunity for natural selection to remove deleterious alleles, particularly if their effect is large (e.g. lethal). This phenomenon, known as 'purging', may explain why some highly inbred populations do not show inbreeding depression in fitness. An example of this is the Chillingham cattle kept in a park in northern England with no immigration for at least 300 years. The effective population size is estimated to be 8.0 (which implies a rate of inbreeding of 6.25% per generation) and the population is genetically uniform for a variety of markers (Visscher *et al.*, 2001). Still the population has not shown any decline in fertility or viability indicating that natural selection has been efficient in preventing random fixation of deleterious alleles (Hall and Hall, 1988). However, the impact of purging appears to be relatively small and most efforts should be devoted to avoid inbreeding.

There may be some reticence to change selection policies in favour of optimised selection in some commercial populations where current ΔF is of no concern. However, current programmes all impose, in one way or another, restrictions on ΔF (e.g. by deciding on the numbers of breeding males and females and avoiding matings between close relatives). The difference is that these restrictions are *ad hoc* rather than optimal and gains obtained are lower than those possible with optimised selection at the same ΔF.

Selection *versus* conservation programmes

There is not a clear distinction between selection and conservation programmes in farmed populations. The difference between the two types of programmes is in the relative emphasis given to rates of gain and inbreeding. The method described above was developed for selection programmes where the aim is to maximise the increase in performance for economically valuable traits but imposing restrictions on ΔF. This method is also valid for conservation programmes where the aim is to minimise ΔF but imposing restrictions to avoid decreased performance in traits that make the breed valuable. In the conservation scenario the problem can be formulated as:

Minimise $\mathbf{c}^T\mathbf{Ac}$ subject to $\mathbf{c}^T\mathbf{g} \geq K$ and $\mathbf{Q}^T\mathbf{c} = \frac{1}{2}\,\mathbf{1}$

where K is the desired rate of gain. The inequality with K can also be omitted, implying no concern with gain (i.e. K can be made large and negative, offering no restriction). In a similar way to the problem of maximising gain with a restriction on ΔF, the problem can be solved using Lagrangian multipliers. In the conservation context the following function is minimised:

$$H_t = \mathbf{c}_t^T\mathbf{A}_t\mathbf{c}_t - \lambda_0(\mathbf{c}_t^T\mathbf{g}_t - K_t) - (\mathbf{c}_t^T\mathbf{Q}_t - \frac{1}{2}\mathbf{1}^T)\lambda$$

and solutions (vector **c**) are obtained as described in the Appendix.

An example of the advantage of the optimisation tool in comparison to truncation selection is shown in Table 2. For this particular example (50 males and 50 females as selection candidates and a heritability of 0.2), the inbreeding coefficient with optimised contributions was nearly half of that obtained with truncation selection and in general this was achieved by effectively selecting a smaller number of parents. It should be noted that the solutions from both types of formulations (minimise c^TAc with a constraint on c^Tg or maximise c^Tg with a constraint on c^TAc) are technically equivalent.

Table 2.
Average inbreeding coefficient (F, in %) across generations (t) and actual (N) and effective (N*) number of parents selected of each sex under truncation (T) and optimised (0) selection that minimise ΔF with restrictions on cumulative genetic gain (G).

	G	F		N		N*	
t	T and O	T	O	T	O	T	O
1	0.16	0.0	0.0	25	33.2	25	22.3
2	0.33	1.6	0.7	25	30.0	25	22.1
3	0.51	2.9	1.5	25	29.8	25	21.0
4	0.67	4.4	2.4	25	28.5	25	21.8
5	0.84	5.9	3.3	25	29.2	25	34.7

With no concern over gain, the optimal solution is to select all the candidates and to mate them to contribute equally to the next generation. These models ignore mutation and it has been argued that when mutation is taken into consideration, equalising contributions may be counterproductive as it may lead to accumulation of mild deleterious mutations due to a reduction in the intensity of natural selection (Schoen et al., 1998). More realistic models have shown however that equalising contributions does not imply a lower fitness in the population but it is a recommended method in conservation practices (Fernández and Caballero, 2001).

Incorporation of molecular genetic information in breeding programmes

The developments in algorithms that optimise selection decisions while restricting ΔF were initially proposed for the infinitesimal model, which assumes that there are many (strictly an infinite number of) genes each of infinitely small individual effect and that these genes act additively and are unlinked. With the tremendous advances in molecular genetics, a large amount of information at the molecular level is now available. This information is being integrated with existing quantitative genetics techniques in order to increase genetic gains from selection and the precision of genetic relationships between individuals and to eliminate deleterious identified alleles. These applications are discussed below.

Use of molecular information for increasing response to selection

Information on identified single genes that affect quantitative traits (quantitative trait loci or QTLs) or on markers linked to them can be used to increase the accuracy of selection and response to selection. Statistical methods are available for identifying QTLs and estimating their effects (Lynch and Walsh, 1998; Weller, 2001) and for using marker information in BLUP genetic evaluations (Fernando and Grossman, 1989). However, given the fact that, in general, strategies that increase gain tend also to increase ΔF, the application of optimised selection extended for using molecular information is advisable.

Initial studies evaluating the potential benefits of gene and marker assisted selection (GAS and MAS, respectively) in mixed inheritance models, where an identified QTL is segregating together with polygenes (e.g. Gibson, 1994; Larzul *et al.*, 1997; Pong-Wong and Woolliams, 1998) focussed on potential gains ignoring the consequences on ΔF and assumed fixed contributions of selection candidates and equal emphasis on estimated breeding values for the QTL and polygenes in the selection criterion. The general finding was that extra short-term gains were obtained with both GAS and MAS but these extra gains were not maintained and, in the long-term, the response was lower than that obtained when ignoring molecular information. The lower accumulated long-term response when using genotype information was mainly due to a decrease in the selection pressure applied to the polygenic background due to the high short-term selection pressure imposed on the QTL.

More recently, there has been some controversy over whether there is always a loss in the long-term gain by using information on the QTL or on (linked) markers, relative to selection schemes that ignore gene or marker information and over whether the use of BLUP and optimised contributions avoid the problem. One factor that may have contributed to this controversy is the arbitrary definition of what constitutes 'long term'. If 'long-term' is defined as the time where the favourable QTL alleles are fixed in all the schemes compared, then for schemes under truncation mass selection (i.e. EBVs are estimated from phenotypes of candidates alone), the long-term loss is only avoided when the gene is completely recessive and has a large effect (Pong-Wong and Woolliams, 1998). With BLUP truncation selection, there is also a reduction in the selection intensity applied to the polygenes but the long-term loss of using the genotype information can be reduced relative to mass selection and even avoided. Several factors contribute to this, including an extra bias with BLUP ignoring QTL information that does not occur with mass selection, the greater accuracy of polygenic EBVs with BLUP, the reduced bias in the EBVs as a consequence of the linkage disequilibrium between the QTL and the polygenic effects induced by selection and the bias in the heritability used in BLUP ignoring QTL information (Villanueva *et al.*, 1999). Other parameters such as QTL effects and frequencies, population size and heritability also determine whether or not long-term response with GAS-BLUP is lower than that with BLUP ignoring QTL information, hence generalisations are not straightforward. The use of BLUP-EBVs (with equal emphasis given to the QTL EBV and the polygenic EBV) and optimised contributions with a restriction on ΔF per se does not avoid in general the long-term loss (Figure 2).

Figure 2.
(see text below figure).

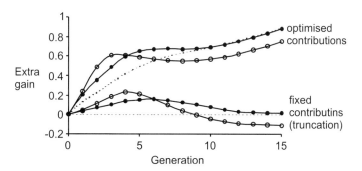

Total accumulated genetic gain over generations from truncation and optimised BLUP selection for three types of scheme: phenotypic selection which ignores QTL information (- - -), standard GAS (o) and GAS with optimal relative weight given to the QTL in the selection criterion (•). Results are for an additive QTL and are expressed as deviations from gains from truncation selection ignoring genotype information.

There is a way of consistently eliminating the detrimental long-term effect with GAS and this is by optimising the relative weights given to estimated breeding values for the QTL and polygenes in the selection criterion. Methods for this type of optimisation were developed by Dekkers and Van Arendonk (1998) and by Manfredi et al. (1998). These studies however assumed fixed contributions of candidates and infinite population sizes with no accumulation of inbreeding. It seems logical then to combine the optimisation of genetic contributions for maximising gain while restricting ΔF with the optimisation of the relative emphasis given to the QTL over generations for further increasing the benefits from GAS. By doing this, Villanueva et al. (2002a) observed increased gains both in the short and in the long term when molecular information was incorporated and the restriction on ΔF was still achieved. Most of the increase in gain was produced by the optimisation of contributions of selection candidates. It is notable that optimised selection ignoring the genotype can be superior to standard truncation (i.e. with fixed contributions) using the genotype both in the short- and in the long-term (Figure 2), indicating that the optimisation of contributions has a higher impact on genetic response than the use of QTL information. The optimisation of the relative emphasis given to the QTL over generations had, however, a higher impact in avoiding the long-term loss usually observed in GAS schemes (Figure 2).

Results when selection was on markers rather than on the QTL itself follow the same trends although the benefit of MAS is only a small proportion of the benefit of using GAS (Villanueva *et al.*, 2002b). MAS with a marker bracket as close as 0.05 cM away from the QTL achieved only half the benefit observed with GAS, indicating that it may still be worthwhile to find the gene even when it has been finely mapped. However, the use of prior information about the QTL's effects substantially increased genetic gain, and, when the accuracy of the priors was high enough, the responses from MAS were practically as high as those obtained with direct selection on the QTL. This highlights the potential improvement still possible from further developments of the methodologies for using marker information in genetic evaluations.

Optimised selection has been applied also in more realistic scenarios where the breeding objective includes several traits. Avendaño *et al.* (2002) considered GAS and optimised selection on an index that includes two negatively correlated traits and a QTL affecting only one of the traits. The combined use of genotype information and optimal selection achieved the highest gains in the breeding objective although this scheme was not the most efficient for improving each trait individually. As with the single trait scenario, for a gene of small effect, the optimisation of genetic contributions had a higher impact on the aggregate genotype than the use of the QTL information.

Use of molecular information for increasing precision of genetic relationships

Selection and mating decisions are heavily based on the knowledge of the additive relationship matrix (the matrix **A**) which describes the genetic variances and covariances of the complete population. Estimates of breeding values in effective selection programmes are obtained from BLUP methodology for which **A** is the fundamental parameter and knowledge of genetic relatedness is essential to decide matings of selected individuals. Also, as described above, **A** is a key parameter used in the optimised methods for managing genetic variability.

Normally, **A** is calculated exclusively from pedigree information (A_p) and therefore gives only expectations of the proportion of genes that two particular animals have in common; for example, it implicitly assumes that half of the genes of a pair of full-sibs are identical by descent from their parents. However, these expectations can differ from the exact proportions not only for full-sibs but for any other type of relationship, with the exception of parent-offspring, ignoring sex chromosomes (Christensen *et al.*, 1996). Molecular DNA markers can be used to estimate these proportions with a high degree of precision and the genetic covariances may be re-calculated to reflect the true proportion of the genome in common.

A potential application of marker-based matrices is to increase the accuracy of BLUP evaluation under the infinitesimal model and to apply subsequently the optimisation tools for managing ΔF. Ideally, the **A** to be used in BLUP for obtaining the highest accuracy of estimated breeding values would be that describing the additive genetic relationships for those loci affecting the trait under selection. Nejati-Javaremi *et al.* (1997) found substantial increases in accuracy and in selection response when using the relationship matrix computed exclusively with QTL information ($\mathbf{A}_q$). The advantage of $\mathbf{A}_q$ was partly due to the fact that variation in the additive relationships among a particular type of relatives (which is ignored with $\mathbf{A}_p$) is accounted for. Another characteristic of $\mathbf{A}_q$ is that it allows the use of information from pedigree-unrelated individuals. The computation of $\mathbf{A}_q$ assumes implicitly that genotypes for all the loci affecting the trait are known and BLUP is used to estimate the unknown allelic QTL effects. The assumption of having all QTLs affecting the selected trait identified is unrealistic and, even if this was possible, it could be argued that direct estimation of their effects would be possible and BLUP genetic evaluation would be redundant (i.e. QTL information could be used directly in selection).

A more realistic scenario is to have genotypes for markers linked to QTLs rather than QTL genotypes themselves and to compute an **A** based on both pedigree and marker information (Pong-Wong *et al.*, 2001). Marker information allows us to discriminate among individuals of the same group while pedigree information allow us to distinguish whether two alleles from two individuals are identical in state or identical by descent. The effectiveness of the procedure will depend on the informativeness and density of the marker map. It should be noted that the exclusive use of markers to compute relationships is an unsolved problem as the inference of the expected (pedigree-based) matrix from the observed (marker-based) matrix has been proven to be difficult, as least for complex pedigrees (Toro *et al.*, 2002).

The rate of inbreeding determines levels of neutral genetic variability and the use of $\mathbf{A}_p$ in the optimisation tools ensures maintenance of ΔF and genetic variation at the desired level for loci unlinked to those under selection but not for selected loci or loci linked to them (Hill and Zhang, in this publication). Marker-based **A** can be also constructed at specific loci to evaluate the loss of variability at linked loci. This is illustrated in Figure 3, where optimised selection, aimed to restrict ΔF to 1% per generation, was applied using $\mathbf{A}_p$ and the trait under selection was genetically controlled by a single biallelic QTL and polygenes. A marker-based **A** was computed for each of four markers linked with the QTL and located at different distances away from the QTL (d= 5, 10,

20 and 40 cM). Inbreeding at the QTL position (d= 0 cM) was also evaluated. The optimisation tool was able to restrict ΔF computed from the pedigree at the desired level but not the ΔF at specific positions linked to the QTL. As expected, the closer the marker to the QTL the higher the inefficiency of the optimised method for restricting inbreeding.

Figure 3.
Inbreeding coefficient at different positions linked to the selected QTL when applying optimised selection using the pedigree-based **A** matrix.

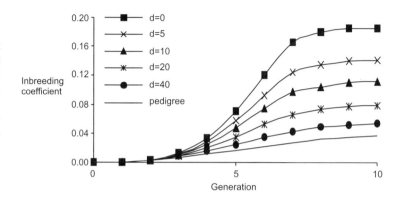

The control of diversity could in principle be achieved by combining all the available information (markers and pedigree), appropriately weighted, in the optimisation approach (equation 1). Thus it is necessary to reconsider constraints applied in the optimisation tools to restrict ΔF and loss of variation at specific positions, genome regions or across the whole genome. Conceptually, different restrictions could be applied to different **A** matrices. Also, in multiple trait scenarios, the same marker-based **A** may not be desirable for different traits and therefore, it is necessary to develop computationally efficient techniques to handle genetic evaluation of multiple traits when using molecular information and to optimise this type of schemes.

Eliminating deleterious identified alleles such as scrapie alleles

One of the clearest examples of the application of molecular information in livestock breeding is selection based on genotypes for the PrP locus to reduce and eventually eliminate occurrence of TSEs (scrapie and risk of BSE) in sheep. National Scrapie Plans (NSP) have been established in several European countries with this purpose. In the UK, the plan is based on genotyping rams at the PrP locus, and on breeding only from rams of specific (i.e less susceptible) genotypes. Whilst selective breeding for particular genotypes could speed up the control and eradication of scrapie, it is essential that selection for resistance to scrapie is seen in the wider context of overall genetic improvement programmes. Attempts to quickly achieve a higher frequency of scrapie-resistant alleles may well prove detrimental if they result in the loss of other desirable attributes (i.e. production traits of economic importance and health, welfare-

and adaptation-related traits). In particular, in some rare and traditionally farmed sheep breeds, pursuing the strategy proposed in the NSP could lead to important losses in genetic variability. Concerns over these issues appear to be constraining involvement of some breeds and sheep breeders in the NSP.

Existing optimisation methods for maximising genetic progress while restricting the loss of genetic variability have only dealt with scenarios where both the identified gene(s) and polygenes affect the quantitative trait. These methods could be extended to optimise contributions when selection is applied in favour of desirable alleles (scrapie resistance) that may have no effect on the quantitative trait or may even be negatively associated with it. In this case, the objective would be to fix the favourable gene without sacrificing large responses to selection for other important traits. An optimum balance between selection against scrapie and selection for other valuable traits would be thus achieved.

Conclusions

Methods are available for managing genetic resources in commercial breeding populations. The methods have the important property that by incorporating the management of diversity not only is gain not reduced, but also is generally improved when compared at the same ΔF. Optimised selection methods have proved to be succesful for restricting ΔF and still allow increased gains when molecular information on QTLs affecting the selected trait, or on markers linked to them, is incorporated. Therefore commercial populations can only benefit from using such tools for managing diversity.

Acknowledgements

Funding from the Biotechnology and Biological Sciences Research Council (BBSRC), the Scottish Executive Environment and Rural Affairs Department (SEERAD) and the Department for Environment, Food and Rural Affairs (Defra) is acknowledged. BV acknowledges financial support from the Secretaría de Estado de Educación y Universidades (Ministerio de Educación, Cultura y Deporte, Spain). Figure 1 is reproduced by kind permission of Cambridge University Press.

References

Avendaño, S., Villanueva, B. and Woolliams, J.A. 2003a. Expected increases in genetic merit from using optimised contributions in two livestock populations of beef cattle and sheep. *Journal of Animal Science* 81:2964-2975.

Avendaño, S., Visscher, P.M. and Villanueva, B. 2002. Potential benefits of using identified major genes in two trait Breeding goals under truncation and optimal selection. *Proceedings of the 7th World Congress on Genetics Applied to Livestock Production*. CD-ROM communication no. 23-03.

Avendaño, S., Woolliams, J.A. and Villanueva, B. 2004. Mendelian sampling terms as a selective advantage in optimum breeding schemes with restrictions on the rate of inbreeding. *Genetical Research, Cambridge* 83: 1-10.

Avendaño, S., Woolliams, J.A. and Villanueva, B. 2003b. Predicting genetic gain when rates of inbreeding are constrained to pre-defined values. *Proceedings of the British Society of Animal Science*, 26-28 March, University of York, York, p. 46.

Borrow, H.M. 1993. The effects of inbreeding in beef cattle. *Animal Breeding Abstracts* 61: 737-751.

Brisbane, J.R. and Gibson, J.P. 1995. Balancing selection response and rate of inbreeding by including genetic relationships in selection decisions. *Theoretical and Applied Genetics* 91: 421-431.

Caballero, A. and Santiago, E. 1998. Survival rates of major genes in selection programmes. *Proceedings of the 6th World Congress on Genetics Applied to Livestock Production*, University of New England, Armidale, Australia, Vol. 26, pp. 5-12.

Caballero, A., Santiago, E. and Toro, M.A. 1996. Systems of mating to reduce inbreeding in selected populations. *Animal Science* 62: 431-442.

Christensen, K., Fredholm, M., Wintero, A.K., Jorgensen, J.N. and Andersen, S. 1996. Joint effect of 21 marker loci and effect of realized inbreeding on growth in pigs. *Animal Science* 62: 541-546.

Dekkers, J.C.M. and van Arendonk, J.A.M. 1998. Optimizing selection for quantitative traits with information on an identified locus in outbred populations. *Genetical Research, Cambridge* 71: 257-275.

Fernando, R.L. and Grossman M. 1989. Marker assisted selection using best linear unbiased prediction. *Genetics, Selection, Evolution* 21: 467-477.

Fernández, J. and Caballero, A. 2001. Accumulation of deleterious mutations and equalization of parental contributions in the conservation of genetic resources. *Heredity* 86: 480-488.

Fernández, A., Toro, M.A., Rodríguez, C. and Silió, L. 2002. Genetic analysis of fluctuating asymmetry for teat number in Iberian pigs. *Proceedings of the 7th World Congress on Genetics Applied to Livestock Production*. CD-ROM communication no. 15-08.

Frankham, R., Ballou, J.D. and Briscoe, D.A. 2002. *Introduction to Conservation Genetics*. Cambridge University Press, Cambridge, UK.

Gibson, J.P. 1994. Short term gain at the expense of long term response with selection on identified loci. *Proceedings of the 5th World Congress on Genetics Applied to Livestock Production*, University of Guelph, Guelph, Canada, Vol. 21, pp. 201-204.

Grundy, B., Villanueva, B. and Woolliams, J.A. 1998. Dynamic selection procedures for constrained inbreeding and their consequences for pedigree development. *Genetical Research, Cambridge* 72: 159-168.

Hall, S.J.G. and Hall, J.G. 1988. Inbreeding and population dynamics of the Chillingham cattle (*Bos taurus*). *Journal of Zoology, London* 216: 479-493.

Kinghorn, B.P., Meszaros, S.A. and Vagg, R.D. 2002. Dynamic tactical decisions systems for animal breeding. *Proceedings of the 7th World Congress on Genetics Applied to Livestock Production*. CD-ROM communication no. 23-07.

Larzul, C., Manfredi, E. and Elsen, J.M. 1997. Potential gain from including major gene information in breeding value estimation. *Genetics, Selection, Evolution* 29: 161-184.

Lynch, M. and Walsh, B. 1998. Genetics and analysis of quantitative traits. Sinauer Associates, Sunderland.

Manfredi, E., Barbieri, M., Fournet, F. and Eisen J.M. 1998. A dynamic deterministic model to evaluate breeding strategies under mixed inheritance. *Genetics, Selection, Evolution* 30: 127-148.

Meuwissen, T.H.E. 1997. Maximizing the response of selection with a predefined rate of inbreeding. *Journal of Animal Science* 75: 934-940.

Meuwissen, T.H.E. and Woolliams, J.A. 1994a. Response versus risk in breeding programmes. *Proceedings of the 5th World Congress on Genetics Applied to Livestock Production*, University of Guelph, Guelph, Canada, Vol. 18, pp. 236-243.

Meuwissen, T.H.E. and Woolliams, J.A. 1994b. Effective sizes of livestock populations to prevent a decline in fitness. *Theoretical and Applied Genetics* 89: 1019-1026.

Nejati-Javaremi, A., Smith, C. and Gibson, J.P. 1997. Effect of total allelic relationship on accuracy of evaluation and response to selection. *Journal of Animal Science* 75: 1738-1745.

Nicholas, F.W. 1987. *Veterinary Genetics*. Oxford University Press, New York, USA.

Pong-Wong, R., George, A.W., Woolliams, J.A., Haley, C.S. 2001. A simple and rapid method for calculating identity-by-descent matrices using multiple markers. *Genetics, Selection, Evolution* 33: 453-471.

Pong-Wong, R. and Woolliams, J.A. 1998. Response to mass selection when an identified major gene is segregating, *Genetics, Selection, Evolution* 30: 313-337.

Saccheri, I., Kuussaari, M., Kankare, M., Vikman, P, Fortelius, W. and Hanski, I. 1998. Inbreeding and extinction in a butterfly metapopulation. *Nature* 392: 491-494.

Schoen, D.J., David, J.L. and Bataillon, T.M. 1998. Deleterious mutation accumulation and the regeneration of genetic resources. *Proceedings of the National Academy of Science, USA.* 95: 394-399.

Simm, G., Oldham, J.D. and Coffey, M.P. 2001. Dairy cows in the future. In: *Fertility in the High-Producing Dairy Cow.* Edited by M.G. Diskin. British Society of Animal Science Occasional Publication 26(1): 1-18.

Sonesson, A.K. and Meuwissen, T.H.E. 2000. Mating schemes for optimum contribution selection with constrained rates of inbreeding. *Genetics, Selection, Evolution* 32: 231-248

Toro, M., Barragán, C., Ovilo, C., Rodrigañez, J., Rogriguez, C. and Silió, L. 2002. Estimation of coancestry in Iberian pigs using molecular markers. *Conservation Genetics* 3: 309-320.

Toro, M. and Nieto, B. 1984. A simple method for increasing the response to artificial selection. *Genetical Research* 69: 145-158.

Toro, M. and Perez-Enciso, M. 1990. Optimization of selection response under restricted inbreeding. *Genetics, Selection, Evolution* 22: 93-107.

Villanueva, B., Dekkers, J.C.M., Woolliams, J.A. and Settar, P. 2002a. Maximising genetic gain with QTL information and control of inbreeding. *Proceedings of the 7th World Congress on Genetics Applied to Livestock Production.* CD-ROM communication no. 22-18.

Villanueva, B., Pong-Wong, R., Grundy, B. and Woolliams, J.A. 1999. Potential benefit from using an identified major gene and BLUP evaluation with truncation and optimal selection. *Genetics, Selection, Evolution* 31: 115-133.

Villanueva, B., Pong-Wong, R. and Woolliams, J.A. 2002b. Marker assisted selection with optimised contributions of the candidates to selection. *Genetics, Selection, Evolution* 34: 679-703.

Villanueva, B. and Woolliams, J.A. 1997. Optimization of breeding programmes under index selection and constrained inbreeding, *Genetical Research, Cambridge* 69: 145-158.

Villanueva, B., Woolliams, J.A. and Simm, G. 1994. Strategies for controlling rates of inbreeding in MOET nucleus schemes for beef cattle. *Genetics, Selection, Evolution* 26: 517-535.

Visscher, P.M., Smith, D. and Hall, S.J.G. 2001. A viable herd of genetically uniform cattle. *Nature* 409: 303.

Weigel, K.A. and Lin, S.W. 2002. Controlling inbreeding by constraining the average relationship between parents of young bulls entering AI progeny test programs. *Journal of Dairy Science* 85: 2376-2383.

Weller, J.I. 2001. *Quantitative trait loci analysis in animals.* CAB International, Wallingford, UK.

Woolliams, J.A., Pong-Wong, R. and Villanueva, B. 2002. Strategic optimisation of short- and long-term gain and inbreeding in MAS and non-MAS schemes. *Proceedings of the 7th World Congress on Genetics Applied to Livestock Production.* CD-ROM communication no. 23-02.

Woolliams, J.A. and Thompson, R. 1994. A theory of genetic contributions. *Proceedings of the 5th World Congress on Genetics Applied to Livestock Production*, vol.19, University of Guelph, Canada, pp.127-134.

Wray, N.R. and Goddard, M.E. 1994. Increasing long-term response to selection. *Genetics, Selection, Evolution* 26: 431-451.

Wray, N.R. and Hill, W.G. 1989. Asymptotic rates of response from index selection. *Animal Production* 49: 217-227.

Appendix

Optimal contributions of selection candidates for minimising the rate of inbreeding with a constraint on genetic gain

The problem of minimising $\mathbf{c}^T\mathbf{A}\mathbf{c}$ subject to the constraints $\mathbf{c}^T\mathbf{g} \geq K$ and $\mathbf{Q}^T\mathbf{c} = \frac{1}{2}\mathbf{1}$ can be solved by using Lagrangian multipliers. The optimal selection decisions at generation t can be obtained by minimising the following objective function

$$H_t = \mathbf{c}_t^T\mathbf{A}_t\mathbf{c}_t - \lambda_0(\mathbf{c}_t^T\mathbf{g}_t - K_t) - (\mathbf{c}_t^T\mathbf{Q}_t - \frac{1}{2}\mathbf{1}^T)\boldsymbol{\lambda}$$

where $\mathbf{c}_t$ is the vector of solutions (i.e. contributions or proportions of total offspring to be left by each candidate born at generation t), $\mathbf{A}$ is the additive genetic relationship matrix of candidates, $\mathbf{g}$ is the vector of BLUP-EBVs, K_t is the desired ΔG (i.e. K_t = average breeding value at t − 1 plus desired ΔG), $\mathbf{Q}_t$ is a known incidence matrix for sex, $\mathbf{1}$ is a vector of ones of order 2 and λ_0 (scalar) and $\boldsymbol{\lambda}$ (a vector of order 2) are Lagrangian multipliers. The second term in the right hand side ensures that the constraint on genetic response is met and the third term ensures that half of the contributions come from males and half from females.

Taking the first derivative of H_t and equating it to zero gives

$$2\mathbf{A}_t\mathbf{c}_t - \lambda_0\mathbf{g}_t - \mathbf{Q}_t\boldsymbol{\lambda} = 0$$

and solving for $\mathbf{c}_t$:

$$\mathbf{c}_t = \mathbf{A}_t^{-1}(\mathbf{Q}_t\boldsymbol{\lambda} + \lambda_0\mathbf{g}_t)/2 \qquad \text{[A1]}$$

Given the constraint $\mathbf{Q}_t^T\mathbf{c}_t = \frac{1}{2}\mathbf{1}^T$, substituting $\mathbf{c}_t$ by its value in [A1] and solving for $\boldsymbol{\lambda}$ gives

$$\boldsymbol{\lambda} = (\mathbf{Q}_t^T\mathbf{A}_t^{-1}\mathbf{Q}_t)^{-1}(\mathbf{1} - \lambda_0\mathbf{Q}_t^T\mathbf{A}_t^{-1}\mathbf{g}_t) \qquad \text{[A2]}$$

and given the constraint $\mathbf{c}_t^T\mathbf{g}_t = K_t$, substituting $\mathbf{c}_t$ by its value in [A1] and solving for λ_0 gives

$$\lambda_0 = 2\,(K_t - \frac{1}{2}\,\mathbf{g}_t^T\mathbf{A}_t^{-1}\mathbf{Q}_t\boldsymbol{\lambda})\,/\,\mathbf{g}_t^T\mathbf{A}_t^{-1}\mathbf{g}_t$$

Substituting the value of $\boldsymbol{\lambda}$ by its value in [A2]:

$$\lambda_0 = 2\,[K_t - \frac{1}{2}\,\mathbf{g}_t^T\mathbf{A}_t^{-1}\mathbf{Q}_t\,(\mathbf{Q}_t^T\mathbf{A}_t^{-1}\mathbf{Q}_t)^{-1}\,\mathbf{1}]\,/\,\mathbf{g}_t^T\mathbf{A}_t^{-1}\,[\mathbf{1} - \mathbf{Q}_t\,(\mathbf{Q}_t^T\mathbf{A}_t^{-1}\mathbf{Q}_t)^{-1}]\,\mathbf{g}_t$$

Finally, the optimal contributions (vector $\mathbf{c}_t$) are obtained by substituting the values of λ_0 and $\boldsymbol{\lambda}$ in [A1].

10

The value of genome mapping for the genetic conservation of cattle

J.L. Williams

Department of Genomics and Bioinformatics, Roslin Institute (Edinburgh), Roslin, Midlothian, EH25 9PS, UK

Abstract

Molecular markers can be used to explore the diversity present in livestock populations. In cattle the diversity among breeds, revealed using molecular markers, is greater than the within breed diversity. Therefore both at the phenotypic and genetic level breeds form discrete populations, which could be used to conserve maximum diversity in the species. The best way to conserve breeds is to ensure their commercial utility; therefore selection of breeds for commercially advantageous phenotypes should be encouraged. Gene mapping studies suggest that, even for complex traits, there may be very few genes involved in controlling variation in the phenotype. Therefore selection for a particular trait does not necessarily affect the genetic base of the population, other than at the genes under selection. This seems to be the situation in the Hereford breed, where the phenotype has changed considerably over the past 50 years, while blood group data suggests that the genetic base of the population has not been greatly affected. Genome mapping approaches allow first the chromosomal location and ultimately the genes controlling traits to be identified. This information provides DNA markers for favourable alleles at trait genes that can be used in selective breeding programmes to improve breeds for a range of traits. Work on double muscling in Belgian Blue cattle has shown that a single gene, myostatin, can be responsible for an extreme phenotype, so selection for double muscling potentially only affects diversity around this gene. However, the examination of the phenotypes associated with this gene in other breeds suggests that genes in addition to myostatin are involved in the development of the extreme phenotype. Thus information on single genes is too narrow to be of value in making conservation decisions. With the current state of knowledge genetic information can aid the choice of individuals to use for breeding and for conservation of diversity, but the information should be used with caution and in conjunction with additional information, such as pedigree, phenotype or function data.

Introduction

There are over 1,200 million cattle worldwide that provide a source of food, motive power and clothing. The EU alone consumed 7.26M tonnes of beef and around 1.2 M tonnes of liquid milk in 2001, which was produced from a dairy herd of over 20M cows (source EC DG-Agriculture "Prospects for Markets 2001-2008"). Cattle were first domesticated about 12,000 years ago and both the archaeological (Grigson, 1989) and molecular evidence (Loftus et al., 1994 and Bradley et al., 1996) suggest that this occurred in the near east and that domesticated cattle then spread to Africa and Europe. Traditionally, breeding was carried out at a local level, often using a limited number of shared bulls. Selection for individuals with particular characteristics suited to local environments, needs and preferences led to the emergence of distinct breeds with characteristic phenotypes. In 1993 there were 783 cattle breeds worldwide (Loftus and Scherf, 1993). Although the definition of a breed is often vague, it is a useful concept that may help to conserve genetic diversity.

With the introduction of artificial insemination (AI) in the 1950s it became possible for bulls with desirable characteristics to be used more widely. Although AI has been used in both beef and dairy breeds, in the latter it has been used more extensively for selective breeding to improve easily measured production traits. Breed improvement has been further enhanced by the development of statistical methods to devise breeding strategies that maximise genetic gain, with a focus on traits that are readily measured (e.g. Phillipson et al., 1994, Visscher et al., 1992). Consequently, there has been dramatic progress in milk yield and the quantities of meat produced from the 'improved' breeds. However, unless great care is taken in choosing the individuals for breeding, the unfortunate consequence of intensive selection is the reduction of genetic diversity within the selected breeds, as only a limited number of the superior individuals are used as breeding stock. The temptation to replace traditional stock with the improved breeds, as a route to increased productivity, is short sighted and potentially results in an even greater loss of diversity, as diversity between breeds is greater than within breeds. In addition, as the local breeds have often become adapted to local environments, e.g. are tolerant of temperature extremes, drought or disease challenge, replacement of the local breeds with improved breeds not only looses diversity *per se,* but also potentially looses functionally important alleles, or combinations of alleles. In West Africa the local White Fulani cattle have been crossed with Holstein-Friesian cattle to improve milk production (Mbap and Ngree, 1995). The first generation cross-bred individuals are indeed both productive and resistant to local diseases. However, in many places further breeding has been focussed on productivity with the F1 cattle back-crossed to Holsteins, with the resulting cattle no longer disease resistant (Dr O.A. Adebambo, pers.

comm). Indeed the introduction of the Holstein has meant that pure individuals of the local bred are rare, and therefore there is the risk that the genetic resistance to disease they have established will be lost.

The challenge for breeders is to produce cattle that are of the desired type, dairy, beef, dual purpose or suited as draught animals, that are both productive and matched to their environments. As the future requirements from cattle production are unknown, e.g. in terms of product quality or the management regimes that will be adopted, it is important that the broadest genetic base is maintained in cattle populations throughout the world as a resource for future selection. In Europe in recent years we have seen the demand for leaner meat, lower fat milk and the use of more extensive, environmentally friendly production systems. Animals selected for production in intensive, high energy systems are unlikely to thrive under extensive management. Therefore even in the relatively friendly production environments of the west, access to genetically different types of cattle remains important. In harsher environments animal production faces even greater challenges. It is important that breeds suited to environments found in developed countries are not simply transplanted to developing countries. As with the example of the White Fulani cattle given above, this may replace breeds that have the potential for improvement in the local environment and impose the need for expensive feed and veterinary treatments that are simply unavailable, both physically and economically.

The genetic composition of individuals is broadly unknown, and selection programmes invariably rely on breeding from animals with superior characteristics for a limited number of traits. Results emerging from gene mapping studies suggest that relatively few genes control the majority of variation in production traits, while other visible traits, such as differences in colour, pattern or the presence of horns are controlled by single genes. Therefore, for genes that are involved in traits that are not under selection, which represents the majority of the genome, it is chance whether unique alleles are maintained or lost from populations. In livestock species the concept of breeds as separate populations that are genetically different allows the development of a strategy of maintaining maximum diversity through the conservation of breeds (Hall and Bradley, 1995). The loss of breeds, and hence erosion of the genetic diversity available may restrict the ability of producers to meet future agricultural requirements (Barker et al., 1993). Between 1993 and 1995 112 of the 783 cattle breeds worldwide were at risk of extinction (Loftus and Scherf, 1993, Scherf, 1995).

Definitions of breeds

Cattle were domesticated in the near east around 12000 years ago,

and since that time they have been distributed across the world (Loftus
et al., 1994). The physical appearance of cattle in different locations
was initially dependent on the cattle available in the area followed by
local breeding decisions. During the 18th century selection on traits
such as colour, pattern and horns resulted in the development of
distinctive breeds. This prompted the formation of breed societies in
the early 19th century that ensured that breed characteristics were fixed
(Wright, 1977, Willham, 1987), for instance the registration of Hereford
cattle with the breed society was conditional on animals having pure
white faces and white socks. The creation of these (partially) closed
populations and specific selection strategies led to breeds that are
distinctive not only at the phenotypic but also at the genetic level.

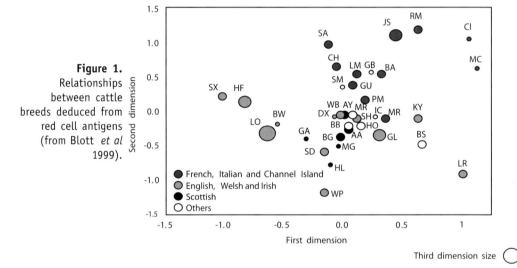

Figure 1.
Relationships
between cattle
breeds deduced from
red cell antigens
(from Blott *et al*
1999).

The genetic distinctiveness of breeds is also observed at the molecular
level, which allows development and relationships between breeds to
be studied. Blood group markers have been used extensively over the
past 40 years for verifying pedigrees recorded in herd books, to ensure
that the purity of breeds is maintained. Analysis of these data has
shown that breeds have a distinctive genetic composition with minimal
overlap (see Blott et al., 1998 and Figure 1). Using a diverse range of
genetic markers, such as the blood groups or DNA markers, individuals
can be assigned to particular breeds. The origins and relationships
between breeds can also be deduced from genetic markers, allowing
breeds with common origins to be clustered together based on similarities
in allele frequencies. Blott et al. (1999) found from an analysis of the
blood group data of European breeds that there were genetic similarities
between breeds that broadly reflect geographic distributions. In Figure
1, which represents genetic diversity in 3 dimensional space, the French,
Italian and Channel Island breeds clustered together (at the top right),

the English Welsh and Irish breeds are generally in the lower part of the figure, while Scottish breeds cluster in the centre of the space.

Relationships between genotypes and phenotypes

The intense selection pressure applied in recent years has resulted in changes in the phenotypes associated with breeds. The Hereford breed, which was small, stocky, slow-growing and suited to rough pastures up to the 1960s, is now a taller and larger breed (Blott *et al.*, 1998). Nevertheless, analysis of the gross genetic composition of the breed, using the blood groups, shows that the modern Hereford shows greater genetic similarity to the historic Herefords than to other breeds, which are phenotypically similar (see Figure 2). Therefore it should be noted that a random survey of genetic markers does not provide information on the phenotypes of the individuals being studied.

Figure 2.
The relationship between Hereford cattle between 1960 and the present day from various Hereford populations and other cattle breeds based on comparisons of red cell antigens (from Blott *et al.*, 1998). Numbers indicate branch length: lower numbers indicate higher similarity.

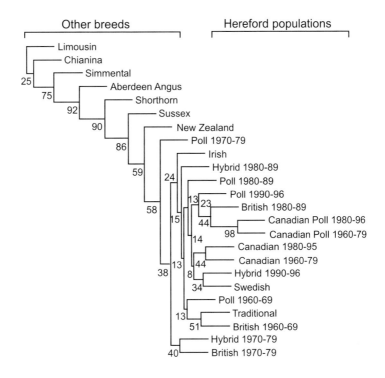

Use of DNA based markers

Recent work has focussed on using DNA markers to assess diversity of breeds and define their origins. The advantage of DNA markers over blood groups is that more loci can potentially be used, hence increasing the accuracy of the analysis. A further advantage is that markers can be selected to examine either the genome as a whole, or to focus on specific regions. Different types of DNA marker can be used to address different questions. Restriction Fragment Length Polymorphism (RFLP) markers are generally within genes and can be used to monitor the effects of selection on particular loci, for example, the impact of selecting cattle for milk production has affected the frequencies and distributions of alleles for the casein genes, which are a major constituent of milk (Freyer *et al.*, 1999). Variations in the mitochondrial DNA, which are inherited via oocytes and not sperm, provide markers that allow the maternal lineage to be investigated (Loftus *et al.*, 1994). A large number of micro-satellites are now available, whose genomic locations have been defined by genetic mapping, and which are available to define diversity across the entire genome (MacHugh *et al.*, 1994, Moazami-Gourdazi *et al.*, 1997). Although these DNA markers have been widely used to define breeds and explore the bovine genome, in general they do not provide information on the phenotypes of individuals.

As resources for *in situ* conservation are limited, it is necessary to choose between the breeds and individuals within breeds that can be conserved. The characterisation of genetic diversity within and between breeds has been achieved using a limited number of microsatellite markers. This approach provides some information on the distinctiveness of breeds, and hence is one criterion on which to prioritise breeds for conservation. However, extreme care must be exercised in using this information. An example of the genetic relationships between breeds derived from microsatellite data is presented in Figure 3. If we were to use this data to identify breeds that appear to be most genetically distant, as a way of preserving the maximum genetic diversity, much important information will be lost. In this example we would most likely select two beef breeds, eg Hereford and Charolais as the most distinct. The variations between breeds observed using genetic markers have most likely occurred as a result of genetic drift and in many cases do not reflect any functionally significant difference. In addition, many cattle populations have undergone several expansions and contractions in numbers, and it is known that when populations go through such genetic bottlenecks the genetic distance between breeds increases (Takezaki and Nei, 1996), although the range of alleles at functionally important loci may not differ appreciably. Thus, genetic distance information, based on few randomly selected markers, does not provide us with information on important phenotypic variations, e.g. dairy or beef type, or on specific genetic variations in the breeds controlling distinctive phenotypes such as the double muscling allele in the Belgian Blue, the milk quality of the Channel Island breeds or the dwarf gene of the Dexter. The current genetic information provides only a small part of the picture that can be used to identifying important genetic diversity in populations and should be used in conjunction with other information, such as pedigrees, breeding history and selection policies, where this information is available.

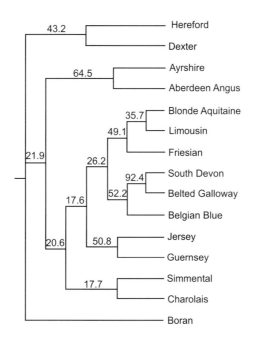

Figure 3.
Illustrative phylogentic tree based on DNA micro-satellite markers. Relationships are derived from Reynolds genetic distances (adapted from Usha, 1995).

Risks associated with intense selection

From the dual-purpose Friesian cattle of the 1950s and 60s, intense selection for milk production has led to the Holstein breed yielding 3 to 4 times as much milk. This dramatic increase in yield has been largely achieved by progeny testing, then using the bulls whose daughters are above average in the test for breeding by AI. Sons of these bulls are then progeny tested and again the best used for breeding. This 'truncation selection' approach can lead to a very small number of bulls being used, many of which are closely related. Whilst impressive improvements in easily measured production traits have been achieved by this approach, most notably in milk yield and more recently in fat and protein composition, the approach has also led to a decline in the genetic base of selected populations. Despite there being hundreds of thousands of Holstein Friesian cattle worldwide, the effective population size in some counties has been estimated to be as low as 20 or 30 [Dr M. Lund, DIAS Denmark, pers. comm.]. Thus, the sires selected for breeding subsequent generations are likely to be related. As the level of inbreeding increases, the risk of problems from recessive defects is also increased. In the early 1990s a few Holstein calves were observed that grew poorly and invariably died within the first 6 months of life. These animals were found to be homozygous for a mutation in the gene coding for the ß-sub-unit of the CD18 molecule that is involved in immune function, allowing neutrophils to access site of infection.

Animals homozygous for this mutation are affected by Bovine Leukocyte Adhesion Deficiency, or BLAD (Shuster *et al.*, 1992). This mutation was found at high frequency in Holstein populations world-wide: 15% among bulls in the USA in 1992 (Shuster *et al.*, 1992), 10% in young bulls in Germany in 1995 and over 16% in Japan in 1996 (Nagahata *et al.*, 1997). By studying the pedigrees of affected and carrier animals the source of the mutation was tracked to a single bull, Osbornedale Ivanhoe, that had very high genetic merit and so was used very widely. Subsequently his sons were also used widely spreading the defect through a large part of the Holstein population (Shuster *et al.*, 1992).

A simple DNA test for BLAD carriers has been developed (e.g. Mirck *et al.*, 1995) and this allows the occurrence of the defective allele in Holsteins to be monitored and managed. Although the eradication of such defects is desirable, in the case of BLAD carriers are so wide spread among the top ranking animals that management of the mutation has initially focussed on avoiding carrier – carrier matings, and preferentially selecting non-carriers as breeding sires so that the impact of the defective allele is initially minimised and then the allele eradicated.

The challenge for genome mapping is to localise, then identify the genes controlling particular traits. Knowledge of the genes controlling particular traits, and favourable alleles at those genes, will allow selection to be carried out on a wider cross section of the population and allow breeding plans to be designed to bring together the best combination of alleles, possibly from diverse genetic backgrounds or from different breeds. However, considerable work will be required before sufficient knowledge is available to make this possible.

Genome Mapping - the story so far

A considerable amount of work has been carried over the past 10 or so years to produce genetic and physical maps of the bovine genome. In the first instance these maps were composed predominantly of anonymous markers (Barendse *et al.*, 1997, Georges *et al.*, 1995, Bishop *et al.*, 1994), but more recently genes, and expressed sequence tags (ESTs), have been added (eg Band *et al.*, 2000). The anonymous markers were mainly microsatellite loci that show high levels of polymorphism in most populations. This type of marker usually has several alleles, often at fairly even frequencies. Microsatellite markers are therefore particularly appropriate for linkage mapping, where it is necessary to identify individual chromosomes and track their segregation through generations in families. At present there are around 2000 microsatellite markers on publicly available bovine genetic maps (see *www.marc.usda.gov/ genome/genome.html*, *www.cgd.csiro.au* and Figure 4). These genetic maps can be used to select markers spanning the entire genome to use in studies to localise regions of chromosomes harbouring genes controlling traits of interest.

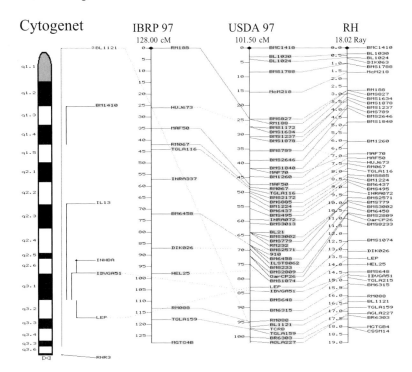

Figure 4. Examples of genomic maps of bovine chromosome 4. The physical map of positions of markers on the chromosome (Cytogenetic map) is compared with published linkage maps (IBRP97 and USDA 97) and with a map constructed from a radiation hybrid panel (RH) (adapted from Williams *et al.*, 2002).

Most production traits are under the control of several genes, which have varying effects on the trait and are generally referred to as Quantitative Trait Loci (QTL). Mapping these QTL requires families in which the trait is segregating and highly polymorphic DNA markers distributed throughout the genome to track the inheritance of chromosomal regions through generations in families. By correlating measurements on the traits and the inheritance of the markers, and from knowledge of the positions of the markers on the genomic maps (e.g. Figure 4), the locations of trait genes can be obtained.

Two approaches have been adopted for mapping trait genes. The first makes use of national cattle populations to localise putative QTL for easily measured traits, such as milk yield and composition (e.g. Georges *et al.*, 1995, Spelman, 1996). However, more complex traits, such as feed conversion, product 'quality', disease resistance or fertility traits are more difficult to quantify and are not systematically recorded. Thus, information on those traits that are currently most difficult to select for, and hence would most benefit from knowledge of the genes involved, is not available from existing commercial herds and a directed approach is required. The second approach involves specifically breeding herds of cattle from founders selected so that the traits to be studied are segregating within families in the herd. This approach has been successfully used to map QTLs for meat quality (e.g. Casas *et al.*, 2001), twinning rates (Kappes *et al.*, 2000) and disease resistance (Teale *et al.*, 1999), and is being used at Roslin Institute to localise QTL for a wide range of commercially important traits, such as growth, conformation, product quality, immune function and fertility (Williams, 2002).

Knowledge of the loci controlling individual traits will allow the direct selection for favourable alleles at these loci. In the first instance this can be achieved through marker-assisted selection, using markers linked to the gene involved in the control of the trait, which have been identified from linkage mapping studies. However, ultimately, knowledge of the gene itself and the allelic variation within that gene will allow more efficient use of the information to develop selection strategies. Once the genes have been localised to a region on a chromosome a number of approaches can be used for identifying the gene itself. These approaches include cloning and sequencing the region to identify the genes present. However, in most cases the precision with which QTL can be localised by linkage mapping makes the amount of sequencing required to cover the whole QTL region prohibitively large. The most popular and successful approach so far has been to identify 'positional candidate' genes, i.e. genes mapped to particular chromosomal regions that are known to have a biological function that putatively affects the trait.

The search for positional candidate genes is helped by knowledge that extended chromosome segments are conserved between different species in terms of the genes present (Hayes, 1995; Solinas-Toldo *et al.*, 1995; Chowdhary *et al.*, 1996). Using a limited number of genes it is possible to align regions of conserved synteny between cattle and other species where more detailed information is available, e.g. from man where the draft DNA sequence is available. Examination of the equivalent regions across species, together with known functions of the genes in controlling phenotypes in other species, will provide candidate genes which can then be tested to see if they are involved in the trait of interest. It is therefore important to have sufficient information to be able to align the regions of conserved synteny in genomes between species. Radiation hybrid mapping methods have been adopted to map genes and derive detailed comparative maps across species. This approach relies on random fragmentation of the donor genome by radiation, then the fusion of the irradiated cells with recipient hamster cells that then retain some fragments of the donor genome (see McCarthy, 1996). The presence or absence of donor DNA fragments containing the markers in a panel of radiation hybrid cells is then used to estimate the relationship between the markers to build the RH map as markers that are close together are more likely to be retained in the same cells. Several RH maps have now been constructed for cattle (Band *et al.*, 2000, Williams *et al.*, 2002, Figure 3) that allow the genetic mapping information to be correlated with the order of genes and hence comparative mapping information across species.

Early successes

Considerable success has been reported in localising QTL for a wide range of traits to relatively broad chromosomal regions (e.g. Kühn *et al.*, 1999, Stone *et al.*, 1999), and two notable successes have identified the major genes involved in increased muscling (Grobet *et al.*, 1997, McPherron and Lee, 1997) and milk production (Grisart *et al.*, 2002).

The Myostatin story

Some cattle breeds have individuals that show a pronounced level of muscling, caused by both muscular hyperplasia (increase in muscle fibre number) and hypertrophy (increase in muscle fibre size), this phenotype is referred to as 'double-muscling'. In commercial terms, the increase in lean muscle yield in double muscled cattle is balanced by increased dystocia (larger calves) and associated calving difficulty.

The gene involved in double muscling was first localised to bovine chromosome 2 in Belgian Blue and Spanish Austurian cattle by a classical linkage mapping approach (Charlier *et al.*, 1995; Dunner *et al.*, 1997). However, candidate genes at the appropriate region on bovine chromosome 2 proved not to be involved in the phenotype, until work in mice discovered a member of the TGFß gene family, a growth-determining factor (GDF8), which was expressed exclusively in muscle and predominantly in skeletal muscle (McPherron *et al.*, 1997). Mice in which expression of GDF8 gene was ablated showed an increased size, with the homozygotes 2-3 times bigger than litter-mates because of increased muscle mass but with no increase in fat or connective tissue (McPherron *et al.*, 1997). The GDF8 gene has been shown to specifically affect skeletal muscle development in mice by regulating early muscle fibre formation and hence has been named myostatin.

The myostatin gene was mapped to the appropriate position on bovine chromosome 2 and so was sequenced from normal and double muscled cattle. Different mutations affecting the mature protein were found which were associated with the double muscling phenotype in Belgian Blue (Grobet *et al.*, 1997) and Piedmontase cattle (McPherron and Lee, 1997). Subsequent studies have shown a myostatin allelic series, with loss-of-function mutations invariably associated with the double muscling phenotype in different cattle breeds (Grobet *et al.*, 1997).

Given the profound effect of the myostatin gene on muscularity, a survey of beef breeds was undertaken to see if variations in this gene could be identified that would increase muscular size without causing the extreme double muscled phenotype and associated calving problems. Instead, the same allele that is found in the Belgian Blue was discovered in several breeds in the UK: the South Devon breed has the allele at about 40% frequency (Smith *et al.*, 2000) and it is at low frequencies in the Aberdeen Angus, British Longhorn and Highland breeds (J. Williams, unpub). The phenotype associated with this allele is highly variable in the South Devon (Weiner *et al.*, 2002), and likely to be very mild in the Highland. Although a deletion in the myostatin gene accounts for the double muscled phenotype in eg the Belgian Blue breed, the effect of this gene in other breeds is variable, which suggests that other genes interact with myostatin to modify the phenotype.

Although the genetic control of phenotypic variation is being explored and increasing numbers of the loci involved a range of traits reported, this information is incomplete. The example of myostatin shows that genes that have a major effects on the phenotype in one genetic background, may have no effect, or indeed a different effect when associated with a different genetic background. It is therefore clear that we need considerably more information on genes and gene function than is currently available, and perhaps need to consider the preservations of *combinations* of genetic diversity at the animal level, rather than just preserving the maximum number of alleles conserved at particular loci, independent of their genetic background.

Future perspectives

Given limited resources available for conservation, it is impossible to conserve everything. Therefore there is a need to prioritise the individuals for preservation. Decisions can be made on several criteria, one of which is the diversity observed at the DNA level. However, we need to be aware that this information is currently very limited and has to be used in conjunction with other information, e.g. on phenotypes and utility of individuals. Using molecular markers, breeds form reasonably distinct groups and so preservation of breeds is a reasonable starting point for the conservation of diversity, assuming the prioritisation of breeds is done using all available information. As QTL mapping projects are completed more and more QTLs for various traits will be localised on chromosomes. This information will pave the way to identifying variations in the genes that control the differences between individuals. This information can be added to the choice making processes for prioritising breeds or individuals for conservation. A project sequencing of the bovine genome started in 2003 and as a by-product of the sequencing large numbers of single nucleotide polymorphisms (SNPs) will be found that will cover the entire genome. These SNPs can be used as genetic markers, and methods are becoming available that will allow the high throughput typing of large numbers of SNPs on large numbers of individuals. The cost of using large numbers of SNPs to characterise individuals on a genome-wide bases is still too high to consider in conservations programmes. However, as costs decrease SNPs will potentially allow increasing proportions of the genome to be scanned and used to define the genetic variation within and between populations. The new technologies that are becoming available, e.g. large scale gene expression studies using micro-arrays, together with genome sequence data, will allow better characterisation of genes and gene function and better definition of the genetic variation that is functionally significant and hence important to conserve. Therefore at some point in the future it may be possible to define accurately the genetic make-up of individuals and use this information in selection decisions, for breeding and conservation.

References

Band, M.R., Larson, J.H., Redeiz, M., Green, C.A., Heyen, D.W., Donovan, J., Windish,.R., Steining, C., Mahyuddin, P., Womack, J.E., and Lewin, H.A. 2000. An ordered map of the cattle and human genomes. *Genome Research* 10: 1359-1368.

Barendse, W., Vaiman, D., Kemp, S.J., Sugimoto, Y., Armitage, S.M., Williams, J.L., *et al.* 1997. A medium-density genetic linkage map of the bovine genome. *Mammalian Genome* 8: 21-28.

Barker, J.S.F., Bradley, D.G., Fries, R., Hill, W.G., Nei, W. and Wayne, R.K. 1993. An integrated global programme to establish the genetic relationships among the breeds of each domestic species. *Animal production and health division* Food and Agriculture Organisation of the United Nations (FAO), Rome, Italy.

Bishop, M.D., Kappes, S.M., Keele, J.W., Stone, R.T., Sunden, S.L.F., Hawkins, G,A., Toldo, S.S., Fries, R., Grosz, M.D., Yoo, J.Y., Beattie, C.W. 1994. A genetic linkage map for cattle. *Genetics* 136: 619-639.

Blott, S.C., Williams, J.L., and Haley, C.S. 1998. Genetic variation within the Hereford breed of cattle. *Animal Genetics* 29: 202-211.

Blott, S.C., Williams, J.L. and Haley, C.S. 1999. Discriminating among between cattle breeds using genetic markers. *Heredity* 6: 613-619.

Bradley, D.G., MacHugh, D.E., Cunningham, P. and Loftus, R.T. 1996. Mitochondrial diversity and origins of African and European cattle. *Proceedings of the National Academy of Sciences, USA.* 93: 5131-5135.

Casas, E., Stone, R.T., Keele, J,W., Shackelford, S.D., Kappes, S.M., and Koohmaraie, M. 2001. A comprehensive search for quantitative trait loci affecting growth and carcass composition of cattle segregating alternative forms of the myostatin gene. *Journal of Animal Science* 79: 854-860.

Charlier, C., Coppieters, W., Farnir, F., Grobet, L., Leroy, P.L., Michaux, C., Mni, M., Schwers, A., Vanmanshoven, P., Hanset, R. and Georges, M.1995. The Mh Gene Causing Double-Muscling in Cattle Maps to Bovine Chromosome-2. *Mammalian Genome* 6: 788-792.

Chowdhary, B.P., Fronicke, L., Gustavson, I., and Scherthan, H. 1996. Comparative analysis of the cattle and human genomes: detection of ZOO-FISH and gene mapping based chromosomal homologies. *Mammalian Genome* 7: 297-302.

Dunner, S., Charlier, C., Farnr,F., Brouwers, B., Canon, J. and Georges, M. 1997. Towards interbreed IBD fine mapping of the mh locus: Double-muscling in the Asturiana de los Valles breed involves the same locus as in the Belgian Blue cattle breed. *Mammalian Genome* 8: 430-435.

Freyer, G., Liu, Z., Erhardt, G. and Panicke, L. 1999. Casein polymorphism and relation between milk production traits. *Journal of Animal Breeding and Genetics* 116: 87-97.

Georges, M., Nielsen, D., Mackinnon, M., Mishra, A., Okimoto, R., Pasquino, A. T., Sargeant, L. Sorensen, A., Steele, M. R., Zhao, X., Womack, J., and Hoeschele, I. 1995. Mapping quantitative trait loci controlling milk production in dairy cattle by exploiting progeny testing. *Genetics* 139: 907-920.

Grigson, C. (1989) Size and sex: the evidence for domestication of cattle in the near east. In: *The Beginnings of Agriculture*. Edited by Milles, A., Williams, D., and Gardener, G. BAR International Series 496. British Archaeological Reports, Oxford, UK. pp. 77-109.

Grisart, B., Coppieters, W., Farnir, F., Karim, L., Ford, C., Berzi, P., Cambisano, N., Mni, M., Reid, S., Simon, P., Spelman. R., Georges, M., and Snell, R. 2002. Positional candidate cloning of a QTL in dairy cattle: Identification of a missense mutation in the bovine DGAT1 gene with major effect on milk yield and composition. *Genome Research* 12: 222-231.

Grobet. L., Martin, L. J. R., Poncelet, D., Pirottin, D., Brouwers, B., Riquet, J., Schoeberlein, A., Dunner, S., Ménissier, F., Massabanda, J., Fries, R., Hanset, R., and Georges, M. 1997. A deletion in the bovine myostatin gene causes the double-muscled phenotype in cattle. *Nature Genetics* 17: 71-4.

Hall, S.J.G., and Bradley, D.G. 1995. Conserving livestock breed diversity. *Trends in Ecology and Evolution* 10: 267-270.

Hayes, H., 1995. Chromosome painting with human chromosome-specific DNA libraries reveals the extent and distribution of conserved segments in bovine chromosomes. *Cytogenetics and Cell Genetics* 71: 168-174.

Kappes, S.M., Bennett, G.L., Keele, J.W., Echternkamp, S.E., Gregory. K.E., Thallman, R.M. 2000.Initial results of genomic scans for ovulation rate in a cattle population selected for increased twinning rate. *Journal of Animal Science* 78: 3053-3059.

Kühn, C. H., Freyer, G., Weikard, R., Goldammer, T., and Schwerin, M. 1999. Detection of QTL for milk production traits in cattle by application of a specifically developed marker map of BTA6. *Animal Genetics* 30: 333-340.

Loftus, R.T., MacHugh, D.E., Bradley, D.G., Sharp, P.M. and Cunningham, P. 1994. Evidence for two separate domestications of cattle. *Proceedings of the National Academy of Sciences, USA.* 91: 2757-2761.

Loftus, R.T. ands Scherf, B. (eds) 1993. World watch list for domestic animal diversity. Food and Agriculture Organisation of the United Nations, Rome, Italy.

MacHugh, D.E., Loftus, R.T., Bradley, D.G., Sharp, P.M. and Cunningham, E.P. 1994. Microsatellite DNA variation within and among European cattle breeds. *Proceedings of the Royal Society of London B* 256: 25-31.

Mbap, S.T., and Ngere, L.O. 1995. Upgrading of White Fulani Cattle in Vom Using Friesian Bulls. *Tropical Agriculture* 72: 152-157.

McCarthy, L.C. 1996. Whole genome radiation hybrid mapping. *Trends in Genetics* 12: 491-493.

McPherron, A.C., Lawler, A.M., and Lee, S.J. 1997. Regulation of skeletal muscle mass in mice by a new TGF-beta superfamily member. *Nature* 387: 83-90.

McPherron , A. C., and Lee, S-J. 1997. Double-muscling in cattle due to mutations in the myostatin gene. *Proceedings of the National Academy of Sciences, USA.* 94: 12457-12461.

Mirck, M.H., vonBannissehtwijsmuller, T., Timmermansbesselink. W.J.H., vanLuijk, J.H.L., and Buntjer, J.B. 1995. Optimization of the pcr test for the mutation causing bovine leukocyte adhesion deficiency. *Cellular and Molecular Biology* 41: 695-698.

Moazami-Goudazi, K., Laloe, D., Furet, J.P. and Grosclaude, F. 1997. Analysis of genetic relationships between ten cattle breeds with seventeen microsatellites. *Animal Genetics* 28: 338-345.

Nagahata, H., Miura, T., Tagaki, K., Ohtake, M., Noda, H., Yasuda, T. and Nioka, K. 1997. Bovine Leukocyte Adhesion Deficiency (BLAD) in Holstein-Friesian cattle in Japan. *Journal of Veterinary Medical Science* 59: 233-238.

Philipsson, J., Banos, G. and Arnason, T. 1994. Present and future uses of selection index methodology in dairy cattle. *Journal of Dairy Science* 77: 3252-3261.

Scherf, B. 1995. World Watch List for Domesticated Animal Diversity 2nd edition. Edited by B. Scherf. Food and Agriculture Organisation of the United Nations (FAO), Rome.

Shuster, D.E., Kehrli, M.E., Akermann, M.R. and Gilbert, R.O. 1992. Identification and prevalence of a genetic defect that causes leukocyte adhesion deficiency in Holstein Cattle. *Proceedings of the National Academy of Sciences, USA.* 89: 9225-2229.

Smith, J. A., Lewis, A. M., Wiener, P. and Williams, J L. 2000. Genetic variation in the bovine myostatin gene in UK beef cattle: Allele frequencies and haplotype analysis in the South Devon. *Animal Genetics* 31: 306-309.

Solinas-Toldo, S., Lenguaer, C., and Fries, R. 1995. Comparative genome mapping of human and cattle. *Genomics* 27: 489-496.

Spelman, R. J., Woppieters, W., Karim, L., van Arendonk, J. A. M., and Bovenhuis H. 1996. Quantitative trait loci analysis for five milk production traits on chromosome *six* in the Dutch Holstein-Friesian population. *Genetics* 144: 1799-1808.

Stone, R.T., Keele, J.W., Shackelford, S.D., Kappes, S.M. and Koohmaraie, M. 1999. A primary screen of the bovine genome for quantitative trait loci affecting carcass and growth traits. *Journal of Animal Science* 77: 1379-1384.

Takezaki, N. and Nei, M. 1996. Genetic distances and reconstruction of phylogenetic trees from micro-satellite DNA. *Genetics* 144: 389-399.

Tammen, I., Klippert, H., Kuczka, A., Treviranus, A., Pohlenz, J., Stober, M., Simon, D., and Harlizius, B. 1996. An improved DNA test for bovine leucocyte adhesion deficiency. *Research in Veterinary Science* 60: 218-221.

Teale, A., Agaba, M., Clapcott, S., Gelhaus, A., Haley, C., Hanotte, O., Horstmann, R., Iraqi, F., Kemp, S., Nilsson, P., Schwerin, M., Sekikawa, K., Soller, M., Sugimoto, Y., and Womack. J. 1999. Resistance to trypanosomosis: of markers, genes and mechanisms. *Archives of Animal Breeding* 42: 36-41.

Usha, A.P. 1995. PhD Thesis, University of Edinburgh.

Visscher, P.M., Hill, W.G. and Thompson, R. 1992. Univariate and Multivariate Parameter Estimates for Milk-Production Traits Using an Animal-Model .2. Efficiency of Selection When Using Simplified Covariance-Structures. *Genetics, Selection and Evolution* 24: 431-447.

Wiener, P., Smith, J.A., Lewis A.M. Woolliams, J.A. and Williams, J.L. 2002. Muscle-related traits in cattle: The role of the myostatin gene in the South Devon breed. *Genetics, Selection and Evolution* 34: 221-232.

Willham, R.L. 1987. Taking Stock. Iowa State University, Ames, USA.

Williams J.L. 2002. Advances in molecular genetics applied to animal breeding. *British Cattle Breeders Club Digest* 57: 12-13.

Williams, J.L., Eggen, A., Ferretti, L., Farr, C., Gautier, G., Amati, G., Ball, G., Caramori, T., Critcher, R., Costa, S., Hextall, P., Hills, D., Jeulin, A., Kiguwa, S.L., Ross, O., Smith, A.L., Saunier, K.L.,

Urquhart, B.G.D., and Waddington D. 2002. A Bovine Whole Genome Radiation Hybrid Panel and Outline Map. *Mammalian Genome* 13: 469-474.

Wright, S. 1977. *Evolution and the Genetics of Populations*, Vol 3. University of Chicago Press, Chicago, USA.

11

Conservation genetics of UK livestock: from molecules to management

M.W. Bruford

School of Biosciences, Cardiff University, PO Box 915, Cathays Park, Cardiff, CF10 3TL, UK

Abstract

Analysis of molecular genetic diversity in livestock potentially allows for rational management of genetic resources experiencing the serious pressures now facing the livestock sector. The potentially damaging effects of genetic erosion are an ongoing threat, both through loss of breeding stock during the 2001 FMD crisis and potentially as a result of the ongoing National Scrapie Plan. These factors and an increasing focus through the Food and Agriculture Organisation of the United Nations (FAO) on the conservation of animal genetic resources force us to consider seriously how to measure, monitor and conserve diversity throughout the genomes of livestock. Currently debated ways to optimally conserve livestock diversity, particularly the 'Weitzman Approach', may fail to take into account the significance of within-breed genetic diversity and its structuring, and apply relatively simplistic models to predict the probability of extinction for breeds over defined periods of time under certain management scenarios. In this paper I argue, using examples from our work and that of others, that within-breed diversity, in particular, should not be ignored when conserving livestock diversity, since breeds may be genetically structured at a variety of scales and there is little evidence for a convincing relationship between effective population size and genetic diversity within rare UK breeds. Furthermore, until we understand the population genetic forces that shape diversity in breeds in more detail, using raw indices of genetic variation or distances to rank or prioritise breeds in terms of some notional threat of extinction has questionable conservation value.

Molecular approaches to livestock genetic resource management

Increasingly, and in particular since the beginning of the 1990's, molecular data have played an important role in the characterisation of genetic diversity at the level of the individual, population, breed and species, in both wild and domestic animals and plants (Frankham *et al.*, 2002). In domestic livestock the applications are diverse. Some of the most interesting recent work has enabled a more detailed understanding of the initial processes that gave rise to modern domestic 'species' (e.g. Kadwell *et al.*, 2001; Luikart *et al.*, 2001; Troy *et al.*, 2001). These domestications have been shown in some cases to be quite complex, involving several putatively independent events and genetically distinct populations (MacHugh and Bradley, 2001 and references therein). The DNA sequence most favoured in domestication studies has been the control region of the mitochondrial genome (see below) since it is well suited to phylogeographic studies. Recent analyses of genetic diversity at the breed and population level however have tended to use more variable autosomal genetic markers and, in particular, microsatellites. Such studies have been applied to questions such as the history of African cattle pastoralism (Hanotte *et al.*, 2002) and genetic differentiation among European and Middle Eastern breeds and have been applied either in geographically distinct or commercial populations (e.g. MacHugh *et al.*, 1998; Loftus *et al.*, 1999; Diez-Tascón *et al.*, 2000). Recently, more applied DNA-based genetic management tools have started to be used on a large scale, such as the screening of prion protein (PrP) variants in sheep (Arnold *et al.*, 2002), since the possession of certain 3 amino-acid variant genotypes confers different levels of susceptibility or resistance to scrapie in the carriers (e.g. Bossers *et al.*, 1997). The application of such technologies and the potential impact of their results on the management of genetic diversity within breeds using pedigree data or DNA fingerprinting methods such as multi-locus microsatellite genotyping to maximise genetic diversity (if so desired) are extremely important challenges for geneticists and breed managers alike. Paternity testing in cattle has been a commercial reality for ten years and such typing services are assisting in breed management on a regular basis (e.g. Vankan and Faddy, 1999). Microsatellite marker systems, which are highly variable, individual-specific, and are easily automated, also now offer one of several approaches to the application of fork-to-farm diagnostic traceability for meat products (e.g. Cunningham and Meghen, 2001).

Threats to livestock resources from genetic erosion

The phenomenon of 'genetic erosion' in both wild and domestic populations refers to the anthropogenic alteration of gene-pools originally constituted, in the case of wild species, through a combination of natural selection, gene-flow, genetic drift/inbreeding and mutation, and in domestic species through a combination of the aforementioned factors plus (and in many cases, far more importantly) controlled reproduction and artificial selection. In livestock, genetic erosion can take a number of forms but includes, at its most drastic, the extinction of entire breeds, populations or genetic lineages, but may also take the form, for example, of genetic 'pollution', more commonly referred to as introgression or hybridisation with another species, breed or differentiated population. The latter phenomenon is potentially of extreme importance due to its invidious nature in many breeds (often brought about through the process of genetic improvement, e.g. Hanslik *et al.*, 2000) but also because it is often undocumented, unlike breed extinction, which thanks to the efforts of a number of organisations is now becoming well understood and ameliorable. It remains, therefore, the processes which occur within breeds which pose the greatest challenge to the genetic manager. In rare breeds and those with fragmented distributions and small populations, demographic isolation may result in genetic drift (random fluctuations in variation due to stochastic demographic processes) and/or inbreeding (mating among related animals). Loss of diversity within the genomes of individuals and the populations they belong to, can result in genetically depauperate breeding stock, possibly suffering from the effects of inbreeding depression which may in turn reduce their reproductive potential or capacity for future selection and adaptation, whether artificial or natural in origin. The largely anecdotal evidence for such phenomena in the past has perhaps resulted in insufficient attention being paid to within population management for diversity both in natural and domestic populations, but recent data, especially from wild species, has reinforced the fact that such phenomena can be real and can have catastrophic consequences (e.g. Madsen *et al.*, 1999; Keller *et al.*, 2001 although see Visscher *et al.*, 2001). Taken together with the ongoing threats posed by recent epizootics within livestock populations and the sometimes inaccurate pedigrees (e.g. Visscher *et al.*, 2002) and management systems which can confound even the best efforts of breed managers and societies, it becomes clear that the maintenance and judicious management of genetic variation within breeds is a challenging task ahead.

Molecular markers

The two markers chosen for discussion here, mitochondrial DNA (especially the control region) and microsatellites are highlighted primarily because of their recent large-scale utilisation, not because other markers are of less potential importance. Indeed some of the most interesting studies of diversity within and among breeds have used blood typing data to evaluate the differentiation within the global population of Hereford cattle (Blott *et al.*, 1998a,b) and other protein polymorphisms, variation in the major histocompatibility complex (MHC) and other gene sequences and a whole suite of other methods and molecules have an important role to play. Indeed it seems highly likely that SNP (Single Nucleotide Polymorphism) technology will eventually supplant all the above thanks to their abundant nature, simple pattern of variation and easy automation (e.g. Heaton *et al.*, 2002). However it is undoubtedly the case that mitochondrial and microsatellite DNA markers have figured largely in many of the recent studies of livestock diversity and will therefore be the focus here.

Mitochondrial DNA

Mitochondrial (mt)DNA has a variety of characteristics which lend its application particularly to phylogenetic analysis. It is a closed circular molecule of approximately 16,000 base pairs found only in mitochondrion and has been used since the early 1980's not only to examine the evolutionary relationships (and produce gene 'trees') among species but also to measure and partition genetic diversity within taxa. MtDNA is inherited usually clonally as a single copy (through the maternal line) and thus does not undergo recombination. This means that new genetic variation almost exclusively accumulates through single point mutations, and the rate of accumulation of these changes throughout its 37 genes is well understood in mammals. On average, mitochondrial DNA evolves at a rate approximately an order of magnitude faster than nuclear DNA, but certain regions evolve even faster. The control region (which contains the displacement loop, or d-loop and the origin of replication) evolves the most rapidly of all and has been used extensively in studies of livestock diversity. MtDNA is also convenient to analyse, since it is found in hundreds to thousands of copies per cell which means that DNA can be retrieved through PCR amplification from minute quantities DNA template, such as that found in non-invasively sampled biological material, museum specimens and even some archaeozoological remains. Highly variable elements such as the control region have been used to identify different evolutionary groups within modern livestock (see above), although the interpretation of some of these data has been questioned and, in particular, the use of molecular 'clocks' to estimate times of evolutionary divergence among domestic lineages and even to time domestication events has proven controversial and almost always imprecise on an archaeological timescale (Luikart *et al.*, 2001; Vila *et al.*, 1997). This is because the high evolutionary rate and polymorphism of the control region makes it susceptible to phylogenetic sampling bias, primarily due to demographically mediated lineage sorting. 'Phylogeographic' analysis of mtDNA, which explores the distribution of evolutionary lineages in space to infer recent patterns of movement, gene exchange and population expansion or decline has been especially illuminating in livestock species (e.g. Troy *et al.*, 2001; Luikart *et al.*, 2001). The major disadvantage of mtDNA analysis in livestock stems from its maternal pattern of inheritance, since much gene-flow is male mediated – a classic example of which is the almost complete absence of *Bos indicus* mtDNA in African cattle, despite the high rates of introgression of zebu cattle into taurine populations on that continent (MacHugh *et al.*, 1997).

Microsatellites

Microsatellites are inherited in a profoundly different way to mtDNA, with different evolutionary mechanisms implicated in generating their variation and they are found predominantly in the nuclear genome – these markers are thus applied to quite different problems. Their importance to livestock diversity was recognised early on (MacHugh *et al.*, 1994) and the FAO, through DAD-IS has attempted to standardise the use of these markers to characterise genetic variation within and among breeds. Microsatellites are found in tens of thousands of localities in the genomes of most vertebrates and comprise short repetitive sequences such a poly (CA) stretches. Variation, which can reach spectacular levels, but which commonly sees single locus microsatellites displaying 10 alleles per locus and 70% heterozygosity, is generated by the gain or loss of one or more of the (for example) CA repeat units through a process of slipped strand mispairing during replication (Bruford and Wayne 1993). Their ubiquitous distribution in the genomes of eukaryotes and high information content makes them ideal markers for measuring indices of genetic variation, genome mapping, population genetics, parentage and relatedness analysis. More recently, in combination with genealogical modelling, Bayesian likelihood approaches and more computationally elaborate simulations (e.g. Beaumont and Bruford 1999; Chikhi *et al.*, 2001), microsatellites have become extremely powerful tools to recover the signatures of past demographic events, including introgression, drift, inbreeding and changes in population size. Their high polymorphism makes them extremely sensitive to changes in effective population size and they can therefore accumulate major changes in allele frequencies very rapidly. This property has both advantages and disadvantages, since it means, for example, that microsatellites can tell us very quickly whether a population has been through a recent demographic bottleneck or indeed an expansion (e.g. Luikart and Cornuet 1998; Beaumont 1999). However, this sensitivity to changes in allele frequency means that the type of analytical methods traditionally employed for calculating genetic distances to infer relationships within and among breeds (and which is still recommended by the FAO's DAD-IS) are poorly suited to recovering meaningful and statistically robust estimators of genetic relationships using microsatellites for many breeds and commercial populations of livestock (e.g. MacHugh *et al.*, 1994, 1998). Genealogical analysis, which does not summarise or transform the allelic configuration of an individual or population as a distance value is potentially much more powerful (e.g. Beaumont, 1999).

Conservation genetics – from wild species paradigms to breeds?

Molecular data have been used directly or indirectly to make management decisions in endangered wild populations since the 1980's (e.g. Ryder, 1986). Using these data remains a major challenge facing biodiversity managers today – including those protecting domestic livestock, and much of the ongoing debate focuses on which units of biological diversity should form the core basis of conservation planning. Whereas biological species traditionally occupied this role in wildlife conservation, increasing evidence points to the fact that this approach often poorly estimates the amount of diversity necessary for the conservation of biological units with future evolutionary potential. The rise of phylogenetics in conservation decision making (see Kraaijeveld, 2000; Vogler and DeSalle, 1994) has even led to proposals to abandon taxon concepts in their entirety and raises serious questions about the universality of even the idea of a species, subspecies and breed and its meaning in conservation (Moritz, 1994a,b; Crandall *et al.*, 2000, Hendry *et al.*, 2000, Avise and Walker, 2000). Moritz (1994b) defined two levels of conservation unit which could be recognised using standard molecular markers: the **Evolutionary Significant Unit** (ESU) – defined as being *reciprocally monophyletic for mitochondrial DNA sequences and having significant allele frequency differences for nuclear DNA* (such as microsatellites). Reciprocal monophyly is a condition achieved once populations have become sufficiently demographically isolated that they no longer share mtDNA alleles, either due to divergence under drift or due to mutation. The second level he described was the **Management Unit** (MU) – defined as having *significant allele frequency differences at both mitochondrial and nuclear DNA regardless of the level of evolutionary distinctiveness of the alleles*. These definitions have served as useful guides in subsequent studies, but have come under significant criticism due to their inability to take into account demographic scenarios that could potentially mimic the evolutionary processes giving rise to these patterns and because they ignore adaptive divergence (Barratt *et al.*, 1999; Crandall *et al.*, 2000).

Once a biological unit for conservation has been identified, the thorny question of prioritisation arises, and here the debate becomes even more complex. The relative importance of preserving distinctive units (e.g. phenotypically, behaviourally, genetically) – also known as *taxic diversity* (e.g. Vane Wight *et al.*, 1991) as opposed to evolutionarily active groups which demonstrate evidence of ongoing diversification - also known as *evolutionary fronts* (e.g. Erwin, 1991) is another factor for consideration in assigning conservation priorities. At its most extreme, an advocate might argue that prioritising taxic diversity conserves as much of our evolutionary heritage as possible, whereas a counter-argument is that prioritising evolutionary fronts at least ensures that extant diversity has the potential for future adaptation in a rapidly changing world. This is a particularly important debate in breed conservation and prioritisation. Should we protect those populations with unique adaptations, regardless of their within-breed diversity and long-term viability (livestock populations may survive with low genetic diversity if their genetic load has already been purged by selection e.g. Visscher et al., 2001) or should we instead target those breeds which possess large amounts of diversity (either partitioned within or between populations and/or subtypes) since these have the greatest possibility for adaptation to future environmental and market conditions?

My thesis in this paper is that we are faced with a major problem in understanding the nature of genetic resources in livestock species and breeds, and that unless we resolve these issues first, the piecemeal application of management criteria for livestock is naïve and potentially dangerous – why? Firstly, there is every reason to believe that livestock evolution has happened in a unique (with respect to evolution under natural selection) and variable fashion since man first domesticated animals. In the UK we are faced with a wide variety of breeds, some of which may be comparatively ancient (having their origins long before the arrival of modern agricultural improvement and which may have been extensively farmed over long periods of time) but many of which are more recent and may have arrived with the major changes in agricultural technology and science and are thus likely to have originated less than 100 generations ago. During such a short time period significant amounts of genetic diversity and differentiation are unlikely to have accumulated through mutation, even if effective population sizes are high. Novel phenotypic mutations do occasionally arise (such as the callipyge gene in sheep, e.g. Freking et al., 1998) but mutation is nevertheless unlikely to predominate as an evolutionary force in domestic breeds. On the contrary, demographic fluctuations, such as those that may be imposed by disease outbreaks, inbreeding, low effective population size (e.g. in males) and population isolation are likely to be much more profound moulding influences on breed diversity. These processes may be concurrent, recurrent and may indeed be ongoing in the current agricultural climate. But they are inadequately incorporated in nearly all population and phylogenetic models for understanding, comparing and prioritising genetic diversity.

Other problems may also confound simple interpretation of population genetic data in modern breeds. The role of introgression in livestock has already been referred to. Where this is documented and has been to some extent quantified, it can be simply accounted for and several reliable allele frequency-based methods are available to calculate the parental proportions in the admixed population (e.g. Bertorelle and Excoffier, 1998). However, one rarely has the luxury of being able to measure the admixture proportions in the direct descendent population and in livestock breeds genetic drift may confound any attempts to get a admixture precise estimate (see Figure 1).

Figure 1.
Here two hypothetical parental populations are admixed to form a third 'hybrid' population, similar to an event such as an improvement exercise. However subsequent genetic drift may have profound effects on all three population's allele frequencies, such that the event can become rapidly obscured (cartoon courtesy of Lounès Chikhi).

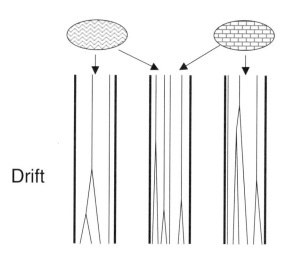

In this case, methods need to be developed which take this scenario (among others) into account when estimating admixture proportions of any parental breed into its 'offspring'. We have attempted to do this using Bayesian likelihood analysis of genealogies simulated using the Markov Chain Monte Carlo process (Chikhi *et al.*, 2001). However, although the method takes into account drift since admixture, sampling effects and uncertainty (i.e. a lack of knowledge) of the ancestral allele frequencies, a more realistic scenario of ongoing or recurrent admixture events is still under development and is extremely computationally intensive. Nonetheless, simple allele frequency-based analysis of the modern populations of these three breeds could produce an extremely misleading result. Other remaining issues include the relative rates of evolution of neutral DNA markers such as microsatellites and the mitochondrial control region in comparison to the often spectacular rates of phenotypic change brought about by artificial selection. Unlike in wild populations, where a combination of the application of a molecular clock and the fossil record allow relatively robust correlations between species age and genetic divergence (e.g. Irwin *et al.*, 1991), it seems unlikely that the highly variable tempo and mode of breed evolution will enable such a correlation to prove robust. Indeed it is the contention of this paper that livestock breed evolution and genetic structure is likely to violate so many of the assumptions under which most population genetic and genetic distance estimators were defined that their application is fraught with difficulties and must be treated with extreme caution. Finally, since mtDNA is maternally inherited and male gene-flow is most common, very few livestock populations are likely to be defined as ESU's regardless of their evolutionary distinctiveness because of the requirement for reciprocal monophyly with mtDNA (c.f. MacHugh et al., 1997). Further, since, as I will describe later, many livestock populations have been shown to, and are indeed expected to display highly significant allele frequency differences at microsatellite (nuclear) loci, this renders their distinctiveness essentially equivalent demographically and thus these data are uninformative in a prioritisation context. Thus it seems likely that the standard ESU and MU designations simply do not apply to livestock breeds and either need to be modified or disregarded.

Current approaches to livestock diversity and prioritisation

Recently, following a seminal paper by Weitzman (1992) attention in the livestock and economics community has focused on prioritisation approaches for livestock breed conservation. Weitzman's concept resolves around a 'diversity function' $D(S)$ – which is the diversity encompassed by a set of breeds (for any character, although this is usually genetic in origin and recently has become viewed as primarily molecular diversity). To become a priority for conservation funding any breed that is to be added to an initial list should, under the Weitzman scenario, add new diversity elements – i.e they must add to 'S' to produce 'S+' to justify incorporation. This additional diversity may be summarised by pairwise genetic distance/dissimilarity values (see above) to the existing breeds in the set and may be represented in a cluster analysis as part of a dendrogram. Breeds that possess the highest overall genetic distance to the remainder of the set (and which should be located at the tips of the most distinct branches of the dendrogram) would be incorporated in conservation programmes, whereas those having low genetic distance would be of lower priority. Thus, this approach sensibly, favours the most parsimonious set to encompass the broadest diversity.

The first problem with this approach is that genetic distance data, as we will see, can often be misleading in livestock where they are demographically (i.e. genetic drift/founder effect/inbreeding/admixture) mediated and contain little or no evolutionary (i.e. biodiversity relevant) signal – the very element which genetic distance estimators are designed to recover. Furthermore, except at the very deepest (and economically least helpful) evolutionary levels, dendrograms of breeds based on some method to group units on their genetic distance are poorly statistically supported (see below) and may, at worst, represent random agglomerations.

In the Weitzman approach, breeds possess a *marginal diversity* value, which is represented by the change in the overall diversity of the set if the probability of extinction $P(E)$ of the breed is changed by $+/-1$ unit.

One current model (Simianer *et al.*, 2003) proposes ΔF (an increase in the inbreeding coefficient) as being proportional to $P(E)$. The ΔF value may be linked to a breed's effective population size (N_e) by $\Delta F = 1/2N_e$. An increase in conservation budget is assumed to have the desirable property of causing a reduction in ΔF, hence $P(E)$ under either additive (incremental) or multiplicative models. The end point of such analysis is to multiply the marginal diversity value by the change in extinction probability of the breed being analysed (calculated as above) to produce the relative increase in the expected diversity of the whole set of breeds under study. The calculation is then applied to each breed in turn to enable prioritisation of those breeds which rank highest or which increase the total diversity by a certain threshold value.

The second problem with this approach is that calculating a $P(E)$ value based on ΔF summarises just one component of the extinction risk that may be applied to a breed if it becomes small in size. The expected increase in probability of extinction of small populations as a function of inbreeding is a matter of considerable debate (e.g. Keller and Waller, 2002). Indeed, it is now accepted in the conservation literature that small populations are considerably more likely to become extinct due to demographic factors in the short term than they are to go extinct because of genetic causes (e.g. Frankham *et al.*, 2002), in which case estimates of the population census size N_c are of more significance for extinction than the effective population size (N_e). Furthermore, some populations that have withstood small effective size in isolation for long periods may have purged any genetic load that would contribute to the deleterious consequences of only having a small number of breeders (Groombridge *et al.*, 2000, Visscher *et al.*, 2001). As an alternative, an integrated approach to assessing extinction risk has become popular in conservation biology over the last ten years, which usually involves stochastic simulation of populations utilising a combination of genetic, demographic and extrinsic parameters in a flexible model. Programs such as VORTEX or RAMAS (see Frankham *et al.*, 2002) have proved extremely useful in coping with the wide variety of intrinsic and extrinsic factors affecting extinction probabilities through the more formal process of Population Viability Analysis (PVA – Coulson *et al.*, 2001). While such approaches have yet to be applied extensively to livestock, they potentially offer a more sensitive and flexible way of deriving extinction probabilities for the future.

The third problem with this approach is that it effectively ignores the partitioning of genetic variation within livestock breeds and treats each breed as a single, equivalent unit (Toro *et al.*, in this publication). Livestock breeds, ranging from those cosmopolitan breeds now found the world over, to rare breeds found in isolated populations, are likely to show different levels of genetic differentiation at geographically different scales. This phenomenon has been demonstrated in cosmopolitan breeds, for example in the Hereford cattle (Blott *et al.*, 1998a) and in rare breeds, for example between the mainland and island flocks of Soay sheep (Byrne *et al.*, in press). Genetic differentiation can sometimes be obvious phenotypically, (e.g. coat colour, body size, resistance to diseases such as scrapie) but, more importantly, may have been overlooked or may be completely unsuspected. It is necessary therefore, that prior to any breed being included in a prioritisation scheme, within breed diversity and divergence is both measured and included as a criterion in prioritisation. Such an approach has never been as potentially important as it is now, with schemes such as the National Scrapie Plan having the potential to homogenise currently diverse gene pools or create new genetic structure within breeds. Breeds with large geographic distributions and known phenotypic differentiation should ideally be targeted for such surveys prior to any large-scale management.

How applicable are these approaches to livestock? Examples

To assess the scale of genetic partitioning among sheep breeds, a recent study from our laboratory (Byrne *et al.*, in press) focused on within- and among-breed genetic diversity in a phenotypically and geographically diverse set of European and Near-Eastern native breeds. The objectives were to determine whether there is a relationship between breed population size and genetic diversity and to investigate the level of genetic cohesiveness and differentiation between regional types on a large geographical scale. The study included five breeds classified as rare or endangered (Table 1). Genetic diversity within and among breeds was measured using twenty microsatellite loci (found on 15 chromosomes). Interestingly, extremely high genetic diversity was found at the individual level, including 297 alleles in 275 individuals, 69 of which are found in only one breed (Awassi). Expected heterozygosity ranged between 0.55 and 0.77 per breed and there was significant inbreeding in all breeds (mean $F_{IS} = 0.11$). Consequently, genetic differentiation between breeds was also high with a mean F_{ST} of 0.18, and all estimates were significantly different from 0 (often at the $P < 0.01$ level).

Breeds were highly discrete when analysed using factorial correspondence analysis or neighbour joining, but no consistent relationships among any breeds could be statistically supported using bootstrap resampling (Figure 2). In general, clustering algorithms have proved poor at recovering the evolutionary (or demographic) relationships among breeds since summarising high levels of allelic differentiation as frequency differences among discrete demographic units using highly variable markers such as microsatellites, has proved uninformative. It is possible that reconstructing demographic relationships among breeds will always be a technically demanding

161

Table 1. Breeds included in the study – type, status and current population sizes.

Breed	Origin of sample[1]	Breed type[2]	Breed status[3] (year data collected)	No. breeding females[4]
Awassi	Israel	Near-East Fat-tailed	Not at risk (1991)	1778639*
Chios	Greece and Cyprus	European Fat-tailed	Not at risk (1986)	16000
Merino	Spain	Fine wool	Endangered -maintained (1994)	788
Andalucian Churra	Spain	Churra	Endangered (1997)	100-1000
Racka	Hungary	Zackel	Not at risk (1998)	2705
Skudde	Germany	Heath	Endangered (1997)	658
Bizet	France	Hill	Not at risk (1996)	9000
Soay	Hirta and Mainland UK	Northern short-tailed	Endangered (1999)	689**
Icelandic	Iceland	Northern short-tailed	Not at risk (1994)	473000
Mouflon	Massif Central France	Feral/wild	Rare (1989)	5655***

[1] The origin given here is the geographic origin of the samples included in this study. Some populations may originate from another location, e.g. Chios from the island of Chios, Mouflon from Corsica; [2] Breed types are described in the text; [3] The breed status follows FAO criteria (www.fao.org/dad-is). Mouflon data is from Shackleton (1997), and is IUCN status; [4] Number of breeding females was used for estimates of population size because this figure was available for all domestic breeds. *Awassi - figures were not available for Israel, this number is from Jordan. **Soay figures are for mainland population, feral population is approximately 1500. ***Mouflon refers to total population size on mainland France, Corsican population is less than 1000 animals and is listed as Vulnerable

task, but utilising approaches which allow examination of the data at a number of factors or dimensions (such as using Principle Components, Factorial Correspondence or Multi Dimensional Scaling) at least maximises the opportunity to discover meaningful variance components at the breed level. In this case, nearly all individuals clustered with others from the same breed, with the exception of all Andalucian Churra individuals, which grouped within the Merino cluster (the only other Iberian breed in the data set). There was no consistent relationship between within-breed genetic diversity and population size. The Churra-Merino grouping was the only geographic pattern that could be recovered but is suggestive that such patterns may occur within regions.

The impact of past demography on levels of genetic diversity, regardless of present-day rarity was clear from the results for the Soay and Awassi (the least and most variable breeds in the study). The Soay is a rare breed that originates from a small island where populations can never have been large. The effect of drift and natural selection caused by the harsh environment and population cycling are strongly implicated in this loss of diversity. This was confirmed by a large F_{ST} value observed between the two Soay flocks (island and mainland). In contrast, the Awassi originates from the Near East, close to one of the centres of sheep domestication, and is a widespread and numerous breed, which harbours a relatively large amount of genetic diversity. Loftus *et al.*, (1999) found a comparable result in cattle where the most variable breeds also originated from the Near or Middle East, and the least variable from Europe and West Africa. In a wider context, then, the geographic position of a breed may be at least as important in determining its genetic variation as its current census or effective size. In an expanded analysis including UK sheep, we have not found a significant correlation between female population census size and expected heterozygosity (Byrne *et al.*, in prep).

The consequences of these data for devising rational conservation policies are therefore important and should take into account the fact that (i) there is a high level of genetic divergence between breeds even if they are geographically proximate, (ii) this level may be equivalently elevated among many breeds (at least in the present study), (iii) even a common breed may be under great risk of loss of diversity due to breeding practices. One consequence of the first two points is that when a breed is lost, an entire cohesive genetic cluster, which may even be considered a recent separate evolutionary lineage that has diverged from the base of the phenogram (Figure 2), may be lost. However, since data are normally generated using microsatellite markers, which have been found to be predominantly free from the effects of selection, it is reasonable to assume that the microsatellites results correspond to a large extent with neutral variation. It is therefore possible that differences in genes involved in adaptive traits, such as disease resistance and adaptation to particular environments, may be larger. A consequence of the third point is that conservation policies cannot focus simply on the rarity status of a breed but should also take into account the effects of modern breeding practices, which represent a threat to many breeds. For instance, the choice of a limited number of males to sire large populations when taken together with the effects of large imports from a limited number of countries could lead, even in demographically expanding breeds, to a paradoxical and extreme genetic impoverishment. Therefore, defining priorities (i.e. which breeds to conserve) will be highly problematic using current genetic management approaches (see Moritz *et al.*, 1994; Barratt *et al.*, 1999) and urgently requires new conservation priority setting criteria to be established for potential implementation.

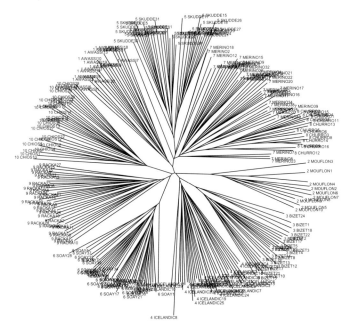

Figure 2.
Neighbour-joining phenogram of allele sharing distances among all individuals in all breeds studied. Bootstrap values for relationships among breed clusters are not shown since none exceeded 50% (from Byrne *et al.*, in press).

To assess the scale and partitioning within breeds in the absence of confounding problems such as incomplete pedigrees and introgression, we recently studied genetic diversity within the island population of Jersey cattle, an ubiquitous and successful cattle breed that originates from the UK Channel Island of Jersey (Chikhi *et al.*, unpublished results). While the breed has been exported extensively, no imports have taken place to the island since 1789. This has recently led to a concern regarding possible loss of genetic diversity and increased inbreeding. This large-scale genetic analysis used only samples from the island. Two hundred and twenty three cattle from all parishes except one were genotyped for 12 microsatellite loci. The average number of alleles per locus and expected heterozygosity were found to be unexpectedly high

(n_A=4, H_e=0.64) and broadly comparable to observed values in a number of geographically proximate continental breeds. Only breeds that have been improved and possess elevated diversity due to recent introgression are clearly more variable than the Jersey. Limited, but significant amounts of genetic differentiation were found between parishes (F_{ST}=0.013) and even between farms (F_{ST}=0.035).

This high diversity and low structure was confirmed by the application of novel statistical approaches. For example, a Bayesian partition analysis (Dawson and Belkhir, 2001) showed that the most probable value for the number of possible hidden partitions in the sample, was 1 (p~0.98). Thus there seems to be sufficient background gene flow across the island to overcome local genetic drift. Overall, the observed level of genetic diversity in island Jersey cattle and its distribution does not require augmentation by imports from other countries. A careful choice of bulls from the island should allow the adequate maintenance of both diversity and performance.

Perspective

ESU/MU designations appear to be essentially irrelevant in livestock, since under current definitions all breeds are likely to be MU's, and since the designation of ESU's depends on monophyly of mtDNA. Under current definitions, very few breeds are likely to be ESUs, and those which are monophyletic will be do due to chance lineage sorting as opposed to new mutations. The Weitzman approach may hold promise for some breeds and in some parts of the world, but in the UK and Europe several key assumptions may be violated, for example there is no clear correlation between genetic diversity and population size. Further, other approaches are needed to assess P(E) and diversity within breeds is currently effectively ignored. Thus currently, although breeds appear to represent the most sensible 'management unit', much more attention needs to be paid to complete within-breed sampling.

It is reasonable to suppose that breeds have often disappeared and been created throughout human history and are constantly being reshaped by a combination of human-mediated selection, natural selection, and genetic drift. The consequence of attempting to conserve these unique evolutionary trajectories is that the many breeds, which are likely to contain specific and unusual alleles and gene combinations, may be worthy of protection in their own right. The challenge for domestic breed conservation will be devising a method of combining genetic diversity data with measures of phenotypic, economic and cultural significance such that rational prioritisation of these unique genetic resources prevails.

Acknowledgements

The examples described here were carried out in the Conservation Genetics Group of the Institute of Zoology and the Biodiversity and Ecological Processes Group, School of Biosciences, Cardiff University. In particular, Kate Byrne, Saffron Townsend, Lawrence Alderson, Lounès Chikhi, Benoît Goossens, Anna Gomercic, Robert Cruickshank, Mark Beaumont, Jan de Ruiter, Trinidad Perez made major contributions to the sampling (KB, ST, LA, JR), laboratory analysis (KB, ST, LC, BG, AG, RC) and theoretical development (LC, MB). I thank Defra for funding and other support (especially John Caygill, Emma Hennessey and Elizabeth Henson), the Rare Breeds Survival Trust and Mark Hughes of Liverpool University for providing samples.

References

Arnold, M., Meek, C., Webb, C.R., Hoinville, L.J. 2002. Assessing the efficacy of a ram-genotyping programme to reduce susceptibility to scrapie in Great Britain. *Preventive Veterinary Medicine* 56: 227-249.

Avise, J.C., Walker, D. 2000. Abandon all species concepts? *Conservation Genetics* 1: 77-80.

Barratt, E.M., Gurnell, J., Malarky, G., Deaville, R. and Bruford, M.W. 1999. Genetic Structure of fragmented populations of red squirrel (*Sciurus vulgaris*) in Britain. *Molecular Ecolology* S12: 55-65.

Beaumont, M.A. 1999. Detecting population and expansion using microsatellites. *Genetics* 153: 2013-2029.

Beaumont, M.A. and Bruford, M.W. 1999. Microsatellites in conservation genetics. In: *Microsatellites: Evolution and Applications.* Edited by DB Goldstein and C Schlötterer, Oxford University Press, Oxford, UK. pp. 165-182.

Bertorelle, G. and Excoffier, L. 1998. Inferring admixture proportions from molecular data. *Molecular Biology and Evolution* 15: 1298-1311.

Blott, S.C., Williams, J.L. and Haley, C.S. 1998a. Genetic variation within the Hereford breed of cattle. *Animal Genetics* 29: 202–211.

Blott, S.C., Williams, J.L. and Haley, C.S. 1998b. Genetic relationship among European cattle breeds. *Animal Genetics* 29: 273–282.

Bossers, A., Belt, P.B.G.M., Raymond, G.J., Caughley, B., De Vries, R. and Smits, M.A. 1997. Scrapie susceptibility-linked polymorphisms modulate the in vitro conversion of sheep prion protein to protease resistant forms. *Proceedings of the National Academy of Sciences, USA.* 94: 4931-4936.

Bruford, M.W. and Wayne, R.K. 1993. Microsatellites and their application to population genetic studies. *Current Opinion in*

Genetics and Development 3: 939-943.

Byrne, K., Chikhi, L., Townsend, S.J., Cruickshank, R.H., Alderson, G.L.H. and Bruford, M.W. 2004. Extreme genetic diversity within and among European sheep types and its implications for breed conservation. *Molecular Ecology*. In press.

Chikhi, L., Bruford, M.W. and Beaumont, M.A. 2001. Estimation of Admixture Proportions: A likelihood-based approach using Markov Chain Monte Carlo. *Genetics* 158: 1347-1362.

Coulson, T., Mace, G.M., Hudson, E. and Possingham, H. 2001. The use and abuse of population viability analysis. *Trends in Ecology and Evolution* 16: 219-221.

Crandall, K.A., Bininda-Emonds, O.R.P., Mace, G.M., Wayne, R.K. 2000. Considering evolutionary processes in conservation biology. *Trends in Ecology and Evolution* 15: 290-295.

Cunningham, E.P., Meghen, C.M. 2001. Biological identification systems: genetic markers. *Revue scientifique et technique* 20: 491-499.

Dawson, K. and Belkhir, K. 2001. A Bayesian approach to the identification of panmictic populations and the assignment of individuals. *Genetic Research* 78: 59-77.

Diez-Tascón, C., Littlejohn, R.P., Almeida, P.A.R. and Crawford, A.M. 2000. Genetic variation within the Merino sheep breed: analysis of closely related populations using microsatellites. *Animal Genetics*. 31: 243–251.

Erwin, T.L. 1991. An evolutionary basis for conservation strategies. *Science* 253: 750-752.

Frankham, R.J., Ballou, J.D. and Briscoe, D.A. 2002. *Introduction to Conservation Genetics*. Cambridge University Press, Cambridge, UK.

Freking B.A., Keele, J.W., Beattie, C.W., Kappes, S.M., Smith, T.P., Sonstegard, T.S., Nielsen, M.K., Leymaster, K.A. 1998. Evaluation of the ovine callipyge locus: I. Relative chromosomal position and gene action. *Journal of Animal Science* 76: 2062-2071.

Groombridge, J.J., Jones, C.G., Bruford, M.W. and Nichols, R.A. 2000. Ghost alleles in the Mauritius kestrel. *Nature* 403: 616.

Hanotte, O., Bradley, D.G., Ochieng, J.W., Verjee, Y., Hill, E.W. and Rege, J.E.O. 2002. African pastoralism: genetic imprints of origins and migrations. *Science* 296: 336-339.

Hanslik, S., Harr, B., Brem, G., Schlötterer, C. 2000. Microsatellite analysis reveals substantial genetic differentiation between contemporary New World and Old World Holstein Friesian populations. *Animal Genetics* 31: 31-38.

Heaton, M.P., Harhay, G.P., Bennett, G.L., Stone, R.T., Grosse, W.M., Casas, E., Keele, J.W., Smith, T.P., Chitko-McKown, C.G., Laegreid, W.W. 2002. Selection and use of SNP markers for animal identification and paternity analysis in US beef cattle. *Mammalian Genome* 13: 272-281.

Hendry, A.P., Vamosi, S.M., Latham, S.J., Heilbuth, J.C., Day, T. 2000. Questioning species realities. *Conservation Genetics* 1: 67-76.

Irwin. D.M., Kocher, T.D. and Wilson, A.C. 1991. Evolution of the cytochrome-b gene of mammals. *Journal of Molecular Evolution* 32: 128-144.

Kadwel,l M., Fernandez, M., Stanley, H.F., Wheeler, J.C., Rosadio, R. and Bruford, M.W. 2001. Genetic analysis reveals the wild ancestors of the llama and alpaca. *Proceedings of the Royal Society of London* B 268: 2575-2584.

Keller, L,F., Jeffery, K.J., Arcese, P., Beaumont, M.A., Hochachka, W.M., Smith, J.N.M. and Bruford, M.W. 2001. Immigration and the Ephemerality of a Natural Population Bottleneck: Evidence from Molecular Markers. *Proceedings of the Royal Society of London* B 268: 1387-1394.

Keller, L.F., and D.M. Waller. 2002. Inbreeding effects in wild populations. *Trends in Ecology and Evolution*. 17:230-241.

Kraaijeveld, K. 2000. The phylogenetic species concept and its place in modern evolutionary thinking. *Ardea* 88: 265-267.

Loftus, R.T., Ertugrul, O., Habra, A.H., El-Barody, M.A., MacHugh, D.E., Park, S.D.E. and Bradley, D.G. 1999. A microsatellite survey of cattle from a centre of origin: the Near East. *Molecular Ecology* 8: 2015–2022.

Luikart, G., Gielly, L., Excoffier, L., Vigne, J-D., Bouvet, J. and Taberlet, P. 2001. Multiple maternal origins and weak phylogeographic structure in domestic goats. *Proceedings of the National Academy of Sciences USA*. 98: 5927-5932.

Luikart, G. and Cornuet, J.M. 1998. Empirical evaluation of a test for identifying recently bottlenecked populations from allele frequency data. *Conservation Biology* 12: 228-237.

MacHugh, D.E., Loftus, R.T., Bradley, D.G., Sharp, P.M. and Cunningham, P. 1994. Microsatellite DNA variation within and among European cattle breeds. *Proceedings of the Royal Society of London* B 256: 25–31.

MacHugh, D.E., Shriver, M.D., Loftus, R.T., Cunningham, P. and Bradley, D.G. 1997. Microsatellite DNA variation and the evolution, domestication and phylogeography of taurine and Zebu cattle (*Bos taurus* and *Bos indicus*) *Genetics* 146: 1071-1086.

MacHugh, D.E., Loftus, R.T., Cunningham, P. and Bradley, D.G. 1998. Genetic structure of seven European cattle breeds assessed using 20 microsatellite markers. *Animal Genetics* 29: 333–340.

MacHugh, D.E. and Bradley, D.G. 2001. Livestock domestic origins: goats buck the trend. *Proceedings of the National Academy of Sciences USA*. 98: 5382-5384.

Madsen, T., R. Shine, M. Olsson, and H. Wittzell. 1999. Restoration of an inbred adder population. *Nature* 402:34-35.

Moritz, C. 1994a. Applications of mitochondrial DNA analysis in conservation - a critical review. *Molecular Ecology* 3: 401-411.

Moritz, C. 1994b. Defining evolutionary significant units for conservation. *Trends in Ecology and Evolution* 9: 373-375.

Ryder, O.A. 1986. Species conservation and systematics: the dilemma of subspecies. *Trends in Ecology and Evolution* 1: 9-10.

Simianer, H., Marti, S.B., Gibson, J., Hanotte, O. and Rege, J.E.O. 2003. An approach to the optimal allocation of conservation funds to minimize loss of genetic diversity in livestock breeds. *Environment and Development Economics* 45: 377-392.

Troy, C.S., MacHugh, D.E., Bailey, J.F., Magee, D.A., Loftus, R.T., Cunningham, P., Chamberlain, A.T., Sykes, B.C. and Bradley, D.G. 2001. Genetic evidence for near-east origins of European cattle. *Nature* 410: 1088-1091.

Vane Wright, R.I., Humphries, C.J., Williams, P.H. 1991. What to protect – systematics and the agony of choice. *Biological Conservation* 55: 235-254.

Vankan, D.M. and Faddy, M.J. 1999. Estimations of the efficacy and reliability of paternity assignments from DNA microsatellite analysis of multiple-sire matings. *Animal Genetics* 30: 355-361.

Vilà, C., Savolainen, P., Maldonado, J.E., Amorim, I.R., Rice, J.E., Honeycutt, R.L., Crandall, K.A., Lundberg, J. and Wayne, R.K. 1997. Multiple and ancient origins of the domestic dog. *Science* 276: 1687-1689.

Visscher, P.M., Smith, D., Hall, S.J., Williams, J.L. 2001. A viable herd of genetically uniform cattle. *Nature* 410: 36.

Visscher, P.M., Woolliams, J.A., Smith, D., Williams, J.L. 2002. Estimation of pedigree errors in the UK dairy population using microsatellite markers and the impact on selection. *Journal of Dairy Science* 85: 2368-2375.

Vogler, A.P., DeSalle, R. 1994. Diagnosing units of conservation management. *Conservation Biology* 8: 354-363.

Weitzman, S. 1992. On diversity. *Quarterly Journal of Economcis* 107: 363-405.

12

Role of reproductive biotechnologies: global perspective, current methods and success rates

M. Thibier[1], P. Humblot[2] and B. Guerin[2]
[1]*Conseil Général Vétérinaire, Ministère de l'Agriculture, de l'Alimentation, de la Pêche et des Affaires Rurales, Paris,France;*
[2]*UNCEIA, Services Techniques, 13 Rue Jouët , 94703 Maisons Alfort, France*

Abstract

Artificial Insemination (AI), the first generation of Reproductive Biotechnologies (RB), is widely used in cattle with more than 110 million females, accounting for 20% of the total global population of breeding females, inseminated annually. There is still great potential for expansion, but further increases will depend on improved conception rates on commercial premises. AI is also used in sheep and goats but to a much lesser extent. AI in pigs has become increasing popular in recent years (close to 50%) with further expansion (around 10% or more) expected in the next decade. The major challenge for AI in the years to come is to reduce the cost of the offspring that are produced. Irrespective of the species, the key element determining this cost is the percentage of inseminations that result in viable offspring. In vivo collected embryo transfer has also been widely used across the world but to a much lesser extent than AI. In cattle around 500,000 embryos are transferred annually, generating an active international exchange of germplasm. Technical limitations associated with variability in superovulatory responses between individuals will limit further uptake of this technology. In contrast, in vitro embryo production, particularly in the bovine, has greater potential for expansion due to its ability to (1) generate a large number of offspring, (2) reduce costs of embryo production, (3) facilitate adoption of nuclear transfer and transgenesis and (4) regulate and minimise potential hazards associated with disease transmission. A remaining factor limiting uptake of this technology is the achievement of satisfactory pregnancy rates following the transfer of cryopreserved embryos, although recent developments suggest these problems can be overcome. Nuclear transfer and transgenesis present numerous opportunities for genetic conservation, but are still very much at the experimental stage. Genetically modified animals may only be tolerated by society if they lead to benefits in human health or to the environment. Emphasis should be directed to disease resistance in relation to both animal and zoonotic diseases.

Introduction

Of the Reproductive Biotechnologies (RB), artificial insemination (AI) has played a pioneering role both experimentally and through its uptake by industry between the last two world wars in the former USSR, Europe and North America. However, it is with the major discovery by J Rostand in the late 1930s (Rostand, 1941, 1946) that, with the addition of 'glycerin' (now called glycerol) frog gametes could survive freezing, that the reproductive biotechnologies have really emerged as a technology. There have been many advances in reproductive biotechnologies since then. American, British and French scientists, in particular, have made major contributions over the last 50 years to developing new techniques and applying them to the farmers' benefit. Four generations of such technologies are now recognized: (1) artificial insemination, (2) *in vivo* embryo collection and transfer, (3) *in vitro* embryo production and transfer and (4) nuclear transfer and transgenesis. Before reviewing some of the main features of these four technologies, it should be recognised that as one generation adds to the former, so both financial and technical investments increase, limiting the application of each successive generation. Furthermore, it is most relevant to note that major societal changes have occurred during the last five decades that require any intervention with food producing animals to be assessed, not only in terms of benefits to producers, but also in terms of benefits to consumers and society as a whole. Clearly, at least in developing countries, the consumer is interested in getting food at an affordable price, but quality and safety are also paramount. The highly urbanised society of the developed world has limited, if any, knowledge of agricultural production systems, but expresses strong views on issues related to animal welfare and ecological sustainability.

Globalization represents a further significant change to the environment in which our livestock industries and associated reproductive technologies must operate in future. It is beyond the scope of this review to consider that factor in detail, but it is clear that the international movement of people, capital and goods has dramatically increased in recent years. This has many consequences at the political, economical and social level, but also in terms of the inherent responsibilities of our industry and our technologies. The aim of this paper is to briefly review the global perspective of each of the 4 generations of reproductive biotechnologies.

The first generation of reproductive biotechnologies: artificial insemination

Insemination by artificial means has now been developed in many species of mammals, birds and insects. Here we will concentrate on livestock.

Distribution of artificial insemination

Recent worldwide surveys of the implementation of AI in livestock around the world by Thibier and Wagner (2000, 2002) have shown, not surprisingly, that AI is most widely used in cattle and buffaloes. Almost all bovine inseminations are carried out using frozen/thawed semen and more than half of the 260 millions doses or so are produced in Europe (Table 1). There are several reasons for this, one being the mode of implementing the breeding programmes and, in particular, storing semen from sires under progeny testing before their breeding value is determined. The dairy breeds of cattle account for three-quarters of the total number of prepared AI semen doses around the world. This is also of significance, particularly in terms of genetic progress within various breeds. It is also of interest to note that there is an active international trade of semen, facilitating the transport of germplasm at minimal cost with minimal risk of disease transmission. Although the number of doses imported and exported (Table 2) do not quite match for several reasons, the order of magnitude of such a trade is considerable at around 20 million doses. The total number of first service inseminations is reported in Table 3. It is interesting to note that the largest number of inseminations is in the Far East, in particular China, India and Pakistan. Europe and North America lag far behind. The impact of AI is of the order of 20% with extremes of 61% in Europe and 1.7% in Africa. This indicates that AI is very widely used, but that there is still huge potential to increase its use further. As briefly discussed below, any significant further increase will rely on technical and economic factors.

Artificial insemination is also widely used in the other farm animal species (Table 4), although to a lesser extent than in the bovine, depending on species. In sheep, over 7 million AI doses were produced in 1998 with just over 3 million inseminations. Corresponding figures for goats are 0.7 million doses produced with 0.5 million inseminations. Almost 65 million fresh semen doses were produced in pigs with over 40 million inseminations. A complementary survey by Weitze (2000) estimated that 24 million sows were inseminated in the 29 largest pig producing countries, i.e. close to 50% of the 50 million sows recorded in those countries (range: 10% in Brazil to 85% in the Netherlands and Spain). Weitze also predicts a further significant increase, as opposed to what is forecast in other species such as sheep and goat. In the next 10 years, such an increase would account for an additional 1.5 million sows in Europe, 1.8 million in North America, 0.9 million in South America and roughly 3 million in the Asia- Pacific region, i.e. a total of 7 million more sows.

Table 1.
The regional distribution of semen collection centres, bulls and semen doses produced in cattle and buffaloes (Thibier and Wagner, 2001).

Region	No. of collection centres	No. of semen banks	No. of bulls	No. of doses produced (x 1000)		
				Fresh	Frozen	Total (%)
Africa	18	161	646	55	1,484	1,540 (0.6%)
North America	69	73	9,627	0	43,270	43,270 (16.4%)
South America	71	138	530	0	5,917	5,917 (2.2%)
Far East	188	644	9,228	8,875	69,938	72,812 (27.5%)
Near East	17	124	268	31	2,559	2,590 (1.0%)
Europe	285	495	20,785	2,694	135,563	138,258 (52.3%)
TOTAL	648	1,635	41,084	11,656	252,733	264,390 (100%)

Note: Total number of doses is calculated separately from data given by each the unit, which explains the slight difference from the single addition of those figures above.

Table 2.
Semen doses of cattle imported and exported by region (Thibier and Wagner, 2001).

Region	No. doses imported	No. doses exported
Africa	568,589	2,360
North America	1,366,100	13,564,500
South America	5,318,595	120,650
Far East	1,467,388	658,700
Near East	402,640	5,120
Europe	4,884,999	5,010,510
TOTAL	14,008,311	19,361,840

Technology and fertility rates

Further increases in the use of AI will rely on both technical advances and the prevailing economics as illustrated in the developing world (Thibier, 1993). The success of AI, particularly in the developing world, will depend on interactions between each of the following factors: farming

Region	Total females of breeding age -A- (40% of total cattle and buffaloes)	Total first service AI - B - (% From total first AI's)	Impact ratio B/A x 100 %
Africa	51,577,000	870,892 (0.8%)	1.7
North America	45,206,000	11,203,880 (10.2%)	24.8
South America	124,460,000	1,366,678 (1.2%)	1.7
Far East	236,850,000	58,181,005 (52.7%)	24.6
Near East	23,433,000	1,068,991 (0.9%)	4.6
Europe	6,175,000	37,738,142 (34.2%)	61.1
Total	543,276,000	110,429,588 (100%)	20.3

Species	Sheep		Goats		Pigs	
	Fresh	Frozen	Fresh	Frozen	Fresh	Frozen
No. of doses produced within the country in 1998	6,686,594	435,552	318,410	385,554	64,974,650	15,900
No. of first inseminations in 1998	3,168,260	159,213	211,010	256,942	40,327,293	6,810

system, management, breed organisations, and added value to the end products. Focusing on the AI technology itself, however, recent developments such as long-term storage of sperm at chilled or room temperature (Evans and Maxwell, 2000) provides more flexibility and is less costly than deep freezing. Lower numbers of sperm per dose allows the production of much larger numbers of offspring from each ejaculate and each male (Rath, 2002; Morris and Allen, 2002), and sexed semen, which facilitates predetermined gender selection, are just some other examples of recent developments with this technology (see Holt, 2003).

The use of fresh semen in cattle is generally not very popular, although sometimes it is used on a large scale such as in New Zealand.

However in France some attempts have been made to use fresh semen for the purpose of reducing the number of sperm cells per dose and hence intensify the distribution of genes from outstanding sires. In one study, Gerard *et al.* (1998) collected semen from sires of very high genetic merit during a 6 month period (November to March), and conducted 110, 000 inseminations with inseminates containing 5 million sperm per dose. The mean non-return rate (between 60 and 90 days post AI) was 60.6% for fresh semen compared to 62.5% for frozen/thawed semen for the same sires (157,000 doses of frozen semen inseminated). There were marked differences between bulls in terms of their ability to sustain high fertility rates with fresh semen according to the day of insemination (Day 0 to Day 3). This is a good illustration of some of the problems encountered in the management of fertility: mediocre results (although often considered satisfactory by the industry), variation between sires, and difficulty to sustain good fertility in some individuals with fresh semen over 48h. A number of tentative approaches to increase fertility on the female side include aids for the detection of estrus and AI with early diagnosis of pregnancy. This is certainly appealing in theory but, unfortunately, often incompatible with the economics of most production systems.

The major challenges facing AI in the years to come are (1) to make AI more readily accessible, in particular to beef cattle under range-land conditions and (2), to reduce the cost of offspring produced. Irrespective of the species, one main factor determining cost is the percentage of inseminations that result in viable offspring. In dairy cows, calving rates after AI have continually decreased over the last fifteen years (Humblot, 2001). In a well-documented retrospective study, Chevallier and Humblot (1998) showed from a set of more than 150,000 bovine dairy females inseminated in the West of France between 1987 and 1997, that the non return rates (between 60 and 90 days) following first AI had declined from 60.0% to 53.4% with only a slight increase during the last two years (53.9% and 54.2%). One of the key issues for AI in ruminants is to better understand why this gradual decline between fertilization and calving, which results in a loss of about half of the conceptuses that had initiated development, occurs and how to correct this. As illustrated by Humblot (2000), there are four major sources of pregnancy failure that contribute to lower fertility rates: females not in estrus at time of insemination (NE), early embryonic mortality (EEM), late embryonic mortality (LEM) and abortion (A). The frequency of occurrence of the first three factors have been investigated in two different

M. Thibier et al.

environments: temperate (France) and sub tropical (La Réunion) (Humblot, 2001). High milk producing herds in France had, on average, 10% lower fertility than the mean of herds in this country (Table 5). The main cause of poor fertility was EEM, accounting for half of the losses (abortions excluded). In the La Reunion herds, pregnancy rates were also very low, even without taking into account the 5 to 8% of abortions that occur regularly. Early embryonic mortality corresponded to more than one third of the losses in itself. Late Embryonic Mortality does not have such a great influence on fertility (about half that of EEM).

Table 5.
Embryonic loss and pregnancy rates following first insemination in Holstein herds in two distinct environments (adapted from Humblot, 2001).

Events	France (temperate)	La Réunion (Sub Tropical)
No of females	1,395	847
Not in Estrus(*)	4.7%	7.8%
Early Embryonic Mortality	31.6%	36.8%
Late Embryonic Mortality	14.7%	14.0%
Pregnancies (**)	42.9%	25.6%

(*) Inseminated although confirmed not to be in estrus at time of insemination.
(**) Abortions not reported in the present study as the pregnancy rate was assessed by rectal palpation 60/90 days after AI.
NB: The total of each column does not add up to 100% because of certain cows that were not assigned to categories shown or for which data were missing.

Those studies and many others show the multi-factorial causes of reproductive failure. Classically there are genetic causes, and environmental and managerial causes. The latter are well recognised if not easily corrected. It has become clear that genetic selection over the last three decades has contributed to the reduction in fertility (Humblot and Denis, 1986; Boichard and Manfredi, 1995). In the study of Humblot, (2001) these genetic effects were specifically expressed as early embryonic mortality whereas the effect of selection on the frequency of late embryonic mortality was not significant. This is a complex problem and represents a major challenge to this first generation of reproductive biotechnologies in cattle.

The situation may not be quite as bad in other species although improvements in birth rates following AI, either with fresh or frozen/thawed semen, are still of primary concern. The fact that late embryonic mortality does not play a major role in gestation losses in cattle is not generally true for all ruminants. In goats, for example, the rate of early and late embryonic mortality is similar (11 to 30% for early losses and 9 to 42% for late losses; Humblot, 2000). With an incidence of around 20%, pseudo pregnancy is a common occurrence in this species. The factors that influence fertility, therefore, may differ from one species to another. Leboeuf *et al.*, (1998) have reported kidding rates of 65 to 69% in Saanen and Alpine breeds, respectively, following oestrous synchronisation and AI. Similar fertility rates can be obtained in sheep. For example, Paulenz *et al.*, (2000) reported rates between 56 and 63% using 100 or 200 x 10^6 sperm cells of fresh semen after vaginal or cervical insemination. These results are frequently considered satisfactory by industry but, in fact, they contribute to the increase in cost of offspring conceived by AI and, unless this is improved, there is little hope for further increase of AI in such species.

The situation to which each professional should tend to is obviously that of the pig. In a recent field study, Eriksson *et al.*, (2002) reported fertility rates (farrowing rates and litter size) between frozen/thawed (F/T) semen and natural mating/fresh semen (NM/FS). The frozen semen in question was exported from Sweden. These authors showed that from several hundreds of sows so inseminated, the farrowing rates were 72% (F/T) and 81% (NM/FS). Live piglets born were identical in the two groups. The farrowing rates reported for fresh and frozen/thawed semen are interesting because they allow strict comparisons but they are not exceptional as rates of over 90% for fresh semen have been recorded in the field (Table 6).

Artificial insemination has the most favourable cost/benefit ratio of the reproductive biotechnologies and should be the first procedure to be considered by farmers in order to improve any given herd or flock on a medium or long-term basis. Apart from pigs, the future increase in the number of females inseminated will be moderate and will rely on the ability of scientists and industry to generate significantly higher fertility rates following first insemination. Sexed semen has the potential to dramatically modify the market for AI, but it still has to prove its efficacy, consistency and cost effectiveness. It is, however, unlikely that it will contribute to any increase in fertility rates.

Table 6. Pregnancy and farrowing rates in pigs after artificial insemination with fresh or frozen thawed semen.

Semen	Pregnancy rate	Farrowing rate	References
Fresh	67-94% (60days)	90-95%	Althouse *et al.*, 1998 Juonala *et al.*, 1998
Frozen		70-80 %	Thilmant, 2001

The second generation of reproductive biotechnologies: *in vivo* derived embryos

Global distribution of *in vivo derived embryos*

It is now over a quarter of a century since the second generation of RB became operational in the field. Due to the relatively high cost of obtaining offspring (around €700, for cattle in the European Union), its global uptake has been restricted mainly to cattle. As shown by the annual survey of the IETS Data Retrieval Committee, around 500,000 bovine embryos are transferred annually across the world. In 2001, the number transferred was slightly less due, in part, to a slight reduction of activity in some parts of the world, notably in Europe, but also due to the lack of data recorded that year for some Asian countries (Table 7). Nevertheless, more than 100,000 cattle were flushed during that year. It is difficult to record precisely breed distributions across the world due to certain semantic difficulties in the definition of beef, dual purpose and draught breeds. However, from countries where this information is available, it seems that *in vivo*-derived embryos are recovered from all breeds, with a somewhat similar proportion from dairy and beef breeds. As can be seen from Table 7, in 2001 almost half of the 450,000 bovine embryos transferred (Thibier, 2002), were in North America. Europe accounted for less than one quarter of transfers and most of the remaining activity was shared between South America, Asia (notably Japan and Korea) and to a lesser extent Oceania. International trade is quite active, although hard to report accurately due to the fact that some countries, particularly in Europe, are reluctant to publish their figures. Such figures are available in North America and around 30,000 bovine *in vivo*-derived embryos have been exported from this region. Regular records are also collected for small ruminants revealing that some tens of thousands of sheep embryos are collected and transferred each year. In goats, several thousand *in vivo*-derived embryos are transferred domestically or internationally. Several hundred embryos from cervids are also subject to *in vivo* collection and transfer in some parts of the world (Canada, New Zealand for example). There is also some activity going on in horses and more surprisingly in pigs, although mostly for experimental purposes. However as shown in Table 8, close to 100,000 *in vivo*-derived embryos were collected and transferred compared to almost one thousand sow recipients.

Technological limitations and future perspective

This second generation of RB became popular during the 1970s as soon as it became feasible to safely flush embryos from donors and transfer them to recipients. The attraction was being able to multiply the number of offspring per female donor. However, limitations in the number of good quality transferable embryos recovered per superovulatory session introduced a degree of realism. In cattle, many models predicted good financial returns provided that 10 or more offspring were born from a single donor session. However, such targets have not been realised. The mean number of transferable embryos in cattle is around 5.5 (Humblot, 2001), with some reports from the Canadian national survey suggesting a mean of 6.4 (Thibier, 2002).

Furthermore, these mean values are associated with a high degree of variability. On average, between 20 to 25% of donor animals will fail to respond to the superovulatory and embryo recovery regimen and deliver no offspring.

Table 7.
The recovery and transfer of *in vivo* derived bovine embryos in 2001 (Thibier, 2002).

Continents	Flushes	Transferable embryos	Number of transferred embryos		
			Fresh	Frozen	Total
Africa	929	5,218	2,284	2,142	4,426 (1 %)
N. America	55,981	315,628	110,619	111,082	221,701 (49 %)
S. America	11,007	53,610	47,655	10,034	57,689 (12.8 %)
Asia	10,440	80,521	14,703	39,574	54,277 (12.0 %)
Europe	19,594	109,698	44,890	49,713	98,000 (21.6 %)
Oceania	3,340	15,402	7,927	8,523	16,450 (3.6 %)
Total	101,291	580,077	228,078	221,068	452,546 (**)

Table 8.
Embryo transfer in pigs during 2001 (Thibier, 2002).

Continents	Flushes	Transferable embryos	Number of recipient females
Canada	111	2,120 (18 frozen)	
Korea	105	1,156	210
In vitro produced		1,200	15
Clone and transgenic		67,750	166
Europe		6,125(*)	245
	253(**)	3,702	57
Taiwan		504	21
USA			25
Total	469	82,557	739

(*) from AETE statistics
(**) in addition to the AETE statistics.

To make matters worse, it is very difficult to accurately predict which of these females are likely to respond. Humblot (2001) reported that the standard deviation of the response, expressed in terms of numbers of transferable embryos per session, is equal to the mean response. As shown in the IETS data retrieval report (Thibier, 2002), in vivo-derived embryos are transferred either fresh or frozen/thawed in roughly equal proportions. Pregnancy rates are in the order of 60% for fresh and 50% for frozen/thawed embryos (Nibart and Humblot, 1997). Some improvements in pregnancy rates following the transfer of frozen/thawed bovine embryos have been reported which, at around 60%, are similar to those for fresh embryos. The Canadians report pregnancy rates of 60.8% and 59.7% respectively for fresh and frozen embryos. One positive point developed during the last decade, is the development of non-surgical, transcervical methods of embryo transfer that are both quicker and cheaper. In a series of 8,042 such transfers using frozen/thawed bovine embryos in France, Humblot (2001) reported a pregnancy rate of 48.3%. Pregnancy rates as high as 60% have also been reported (Mapletoft, personal communication). This development has facilitated the development of other 'sister technologies,' including embryo splitting, embryo sexing and, of course, in vitro embryo production. Splitting and sexing requires equipment and expertise to micro-manipulate embryos, and dedicated facilities. Associated costs of production are, therefore, high, so marginalising these applications. In addition these manipulations make deep freezing of such embryos more difficult. Pregnancy rates are not as high as for non-manipulated embryos, so that many practitioners and farmers prefer to use fresh embryos with all the constraints that this brings. Nevertheless, in 2001, the Canadians reported pregnancy rates of 59.3% following the transfer of 2,511 sexed, fresh embryos. In addition, more than 300 embryos were split in that year resulting in a pregnancy rate of 52.7% (Mapletoft, personal communication).

This technology has greatly facilitated the development of the next generation of RB. This second generation will last for decades due to its inclusion in breeding programmes based on multiple ovulation and embryo transfer (MOET), the preservation of germplasm, and the national and international movement of germplasm. It is unlikely however, that it will develop very much due to limitations in the superovulatory response and, hence, limited numbers of offspring per donor. In addition, the newer generations of RB will now rely more on in vitro produced embryos. Its future is further limited due to the somewhat high investment, both in equipment and expertise, required. With pregnancy rates at 90 days of 60% and calving rates of 50% or so, it is unlikely that this generation of RB will find much wider favour. This is also a problem shared by AI, particularly in monotocous species. Late embryo and early fetal losses are a major problem to be solved in future.

The third generation of reproductive biotechnologies: *in vitro* embryo production and transfer

Global distribution of in vitro *embryo production*

The transfer of *in vitro*-produced embryos became operative for special purposes in cattle and, to a lesser extent, in other species some ten years ago. More than 30,000 bovine *in vitro*-produced embryos were transferred worldwide in 2001 (Table 9). This figure is likely to be an underestimate, being lower than that recorded in previous years due to lack of data from some countries in Europe, Oceania and North America. The level of activity is still much less than for *in vivo*-derived embryos. The reasons for this lack of activity are many but include the level of investment, particularly in laboratories and expertise required. The difficulty of freezing such embryos (see below) has also constituted an obstacle to further development. It is noteworthy, however, that half of the *in vitro*-produced embryos that are transferred have been cryopreserved. *In vitro* production of embryos has also been recorded in small ruminants, including sheep, goats and also in some cervidae (see review by Thibier and Guérin, 2000). However, apart from specific cases, it is not practiced on a wide commercial basis.

	Transferable embryos		Embryos transferred	
	Collected	Fresh	Frozen	Total
Africa	800	10	156	166 ↑
Asia	92,706	9,216	9,465	18,681↑
N.America(*)	n.d	498	78	576 ↓
S.America	668	199	202	401 ↓
Europe	13,031	4,482	4,830	9,312 ↓
Oceania	2,000	974	150	1,124 ≅
Total	109,205	15,379	14,881	30,260 ↓

Table 9. The number of bovine *in vitro*-produced embryos transferred in 2001 (from Thibier, 2002).

(*) only one country from this region has reported those figures

Development and future

This third generation of RB has great potential for development. The basic source of biological material, that is oocytes, are in unlimited supply if collected at the abattoir. Its flexibility also allows one to collect oocytes from cattle by ultrasound guided transvaginal follicular aspiration (so-called 'Ovum Pick Up'; OPU) or, in small ruminants, by laparoscopy. Moreover, such transvaginal aspirations can be repeated at short intervals (two to three sessions per week) over a prolonged period of time and at various physiological phases such as during gestation. Theoretically this should allow a large number of female gametes to be fertilized *in vitro* and overcomes the limitations described above for *in vivo*-derived embryos. In addition, the embryos so fertilized can be used for other technologies such as bisection, sexing, for the determinations of genetic markers (typing), nuclear transfer and transgenesis (see below).

In his report to the Association Européenne de Transfert Embryonnaire (AETE) on the status of *in vitro* produced bovine embryos throughout Europe, Heyman, (2002) indicated that out of 13,000 embryos produced, 10,000 (76%) were derived by OPU compared to just 3,000 from abattoir-derived oocytes. From 1,138 donors in 4,479 OPU sessions (i.e. 4 OPU sessions per donor), 10,065 blastocysts were produced (equivalent to 8.8 embryos per session). Results obtained in France show that on experimental farms, 17.9 oocytes were collected per OPU session from which 13.0 were inseminated and 3.5 embryos transferred (n = 104 sessions). On commercial farms, the results were respectively 10.5, 8.3 and 2.8 (Humblot, 2001). Pregnancy rates from fresh *in vitro*-produced embryos were 54.5 and 46.4% for experimental and commercial farms respectively (Table 10).

Table 10.
Pregnancy rates of fresh and frozen *in vitro*-produced bovine embryos in France.

Embryos	Early pregnancy rate (Day 35)	Pregnancy rate (Day 90)	Reference
Fresh			
Experimental farm	57%	52.8%	Guyader-Joly et al., 2000
		54.5%	Marquant-LeGuienne et al., 2001
Commercial farm		46.3%	Marquant-LeGuienne et al., 2001
Frozen (Selected grade 1)	69.8%	51%	Marquant-LeGuienne et al., 2001

A major difficulty encountered during the last ten years is the deep freezing of *in vitro*-produced embryos that behave differently during this process compared to their *in vivo* counterparts. In Italy many frozen/thawed *in vitro*-produced zygotes are still transferred to sheep for culture. Pregnancy rates are then similar to those of fresh embryos at around 50% or more (Galli, personal communication). However, such techniques are costly and time consuming. Recent progress in embryo culture technology and freezing have resulted in improved pregnancy rates (from approximately 30 to 50%; Marquant LeGuienne *et al.*, 2001). Should this progress be confirmed and this problem solved, then this technology certainly has a bright future. At present it is estimated that the number of calves produced per donor using *in vitro* techniques is about 3 times higher than that for *in vivo*-derived embryos. There are two points of concern at this stage. The first is that it is still difficult to establish a sufficiently large and regular supply of *in vitro*-produced embryos. The second refers to animal health hazards, in particular when oocytes are derived from ovaries collected post-mortem from the abattoir. There are, however, guidelines designed to overcome these difficulties as laid down by the International Embryo Transfer Society (IETS) and further implemented as international recommendations in the Office International des Epizooties (OIE) International Animal Health Code. This particular point will be treated elsewhere at this symposium. Provided the techniques can be made more robust, there is no reason to doubt that such a technique has a great future. Once the technology is accepted and widely used, the cost will decrease considerably. The laboratories, if farmer-led so as to control costs, will have specialists to handle such embryos. With high quality technical input, results should be good. Under such conditions, it should form a most valuable biotechnology to improve the genetics of a herd or flock without the limiting factors referred to above for *in vivo*-collected embryos. It will also be a valuable technology to preserve biodiversity and a safe, well-monitored, and easily controlled technique to minimise risks of disease transmission.

The fourth generation of reproductive biotechnologies: nuclear transfer and transgenesis

Cloning and nuclear transfer

Early methods of cloning by embryo bisection have been in existence since the early 1980s. As a technique for the production of clones, embryo bisection is less powerful than nuclear transfer (NT), although it does result in the generation of true clones as opposed to genomic copies. With NT, genetic material from a donor cell (karyoplast) is transferred to a suitable recipient cell from which the nuclear or genomic material has been removed (Campbell *et al.*, 2002). The donor cells may be of various types including fetal fibroblasts, fetal germ cells and a large range of adult-derived populations such as mammary gland, oviductal epithelial and ovarian follicle cumulus cells, and dermal fibroblasts. The use of blastomeres as donor cells has resulted in the creation of several thousand animals by NT over the last decade, but only hundreds of somatic cell-derived clones have been obtained so far. It is interesting to note that the only 'official' record to date of such clones consists of a letter from the Holstein Association of the USA sent to the chairman of the Health and Safety Advisory Committee of the IETS. It gave the following figures: the correct number of animals created by embryo splitting is 2,336; the correct number of animals created by NT using embryonic cells (blastomeres) is 171, and the correct number of animals created by somatic cell NT is 45 (Robertson, personal communication, October 2002). Clearly according to Renard (1999), the technology has not yet reached the level of expertise necessary for widespread application in the field. Success rates are still very low with, for example, the percentage of offspring born from reconstituted embryos ranging from 0.4 to 2.7% in cattle and sheep. Under 'optimal conditions' as many 10 to 15% of the reconstituted blastocysts transferred can complete gestation (Renard, 1999). In addition, several abnormalities of the young during the first two to three first months of life have been reported. Many technical points still have to be resolved although significant progress has been made over the last two to three years.

Nuclear transfer has many applications both from an experimental and a practical standpoint. In the commercial world, advantage can be taken from its ability to produce and store multiple copies of any individual. This has implications for preservation of endangered species, domestic or wild, from which artificial insemination or in vivo embryo collection may be difficult. This was first demonstrated by Wells et al., (1999), who cloned the last surviving cow of the Enderby Island breed. Another important application of this technology is in the creation and propagation of genetically modified animals.

Transgenic animals

This final and most recent RB to be considered involves the genetic modification of animals. Technically, it should be recognised that, at present, the overall efficiency of this technique is very low, with only 2 to 3% of the pro-nuclear injected eggs giving rise to transgenic offspring (Denning et al., 2002). These authors have also shown that gene targeting by homologous recombination will be the method of choice in the future, however difficult it may be in livestock at present.

Transgenesis certainly has great potential, but a cautionary approach in this area of research should be exercised so that society can appreciate the benefits whilst understanding the risks (Thibier, 1999). Briefly, there are three main applications for this technology: including biopharming, xenotransplantation and alterations in animal growth and metabolism. The two former goals are very specific and will not be discussed here any further. Regarding the third objective, and in agreement with Houdebine (1999), we consider that genetic modification to increase disease resistance should be considered a priority area of research with both economical and societal benefits. This would relate to animal diseases proper but also to diseases that are zoonotic and hence of some hazard to the human population. Genetically modified animals that minimise pollution within the environment might also be worthy of pursuit. In that regard, the creation of pigs with a phytase gene facilitating the digestion of organically bound phosphorus is worthy of mention. It will be important to assess the impact of this project both in environmental terms and also in terms of public acceptability. Other transgenic animals producing food products with direct nutritional and health benefits for the human population may also find public acceptance. Apart from that, we see little potential in the animal industry for transgenic animals where the modification only results in improved growth and body composition, i.e where there are no clear benefits to the human population.

Conclusion

The field of research, as applied to Reproductive Biotechnologies, has been very dynamic during the last 50 years. Artificial insemination will remain, in the years to come, the major technology to offer to farmers throughout the world. It has the most advantageous cost/benefit ratio. To further increase its impact, however, will require increased accessibility in rangeland situations and in extensive areas, particularly in the developing world. Mostly, efforts are required to significantly improve parturition rates in ruminants. Half or so of the pregnancies that are initiated in these species are lost and this holds true for the two more recent generations of RB, i.e. *in vivo*-derived and *in vitro*-produced embryos. This loss contributes to increased costs and is a barrier to wider access. The potential of sexed semen, once the various current limitations have been solved, will modify the current use of artificial insemination but it will not improve accessibility of the service to breeders nor will it contribute to improvements in parturition rates. The *in vitro* production of embryos, particularly in ruminants, has the brightest future, because of its numerous advantages in terms of numbers of offspring that can be generated per donor, and because it forms an essential step in nuclear transfer and transgenic technology. These latter two technologies will certainly develop in the next ten years, but with little impact on livestock production. Finally, it should be noted that all these technologies with their own respective advantages, provide the means to salvage gender, species or breeds threatened by extinction.

References

Althouse, G.C., Wilson, M.E., Kuster, C. and Parsey, M. 1998. Characterization of lower temperature strorage limitations of fresh extended porcine semen. *Theriogenology* 50: 535-543.

Boichard, D, and Manfredim E. 1995. Analyse génétique du taux de conception en population Holstein. *Elevage et Insémination* 269: 1-11.

Campbell, K.H.S., Alberio, R., Lee, J-H. and Ritchie, W.A. 2002. Nuclear transfer in practice. *Cloning and Stem Cells* 3: 201-208.

Chevallier, A. and Humblot, P. 1998. Evolution of non return rates after AI: effects of controlling the interval between calving and first AI on fertility results. *Rencontres Recherches sur les Ruminants* 5: 75-77. (Institut Technique de l'Elevage, Paris, France).

Denning, C., Dickinson, P., Burl,S., Wylie, D., Fletcher, J. and Clark, A.J. 2002. Gene targeting in primary fetal fibroblasts from sheep and pig. *Cloning and Stem Cells* 3: 221-231.

Eriksson, B.M., Petersson, H., Rodriguez-Martinez, H. 2002. Field fertility with exported boar semen frozen in the new FlatPack Container. *Theriogenology* 58: 1065-1079.

Evans, G. and Maxwell, W.M.C. 2000. Current status of embryo transfer and artificial insemination technology in small ruminants. In: *Proceedings, Reproduction in Small Ruminants*. Sandnes, Norway, 30 June-1 July 2000. Edited by I Engeland and C Steel. pp. 54-59.

Gerard, O., Goiset, C., Boffety, F. and Humblot, P. 1998. Utilisation de la semence fraîche pour optimiser la diffusion du progrès génétique dans l'espèce bovine. *Rencontres Recherches sur les Ruminants* 5: 45-48. (Institut Technique de l'Elevage, Paris, France).

Guyader-Joly, C., Durand, M., Morel, A., Ponchon, S., Marquant LeGuienne, B. Guérin, B. and Humblot, P. 2000. Source of variation in blastocysts production in a commercial ovum pick up *in vitro* embryo production program in dairy cattle. *Theriogenology* 53: 355 (Abstract).

Heyman, Y. 2002. European embryo transfer statistics. In: *Proceedimgs of the 18th meeting of the Association Européenne de Transfert Embryonnaire* (AETE), pp. 75-79.

Houdebine, L.M. 1999. La transgénèse animale. In: *Colloque scientifique de l'AFSSA: Biotechnologies de la reproduction animale et sécurité sanitaire des aliments*, Paris 29 Septembre 1999, Edited by M Thibier. AFFSA Publications, Maisons Alfort, France. pp. 29-33.

Humblot, P. and Denis, J.B.D. 1986. Sire effect on cow fertility and embryonic mortality in the Montbeliarde breed. *Livestock Production Science* 14: 139-148.

Humblot, P. 2000. The frequency and variation of embryonic mortality and the use of pregnancy specific proteins to monitor pregnancy failure in ruminants. In: Proceedings of the 3rd ESDAR meeting. *Reproduction in Domestic Ruminants* Suppl. 6: 19-27.

Humblot, P. 2001. Use of pregnancy specific proteins and progesterone assays to monitor pregnancy and determine the timing, frequencies and sources of embryonic mortality in ruminants. *Theriogenology* 56: 1417-1433.

Humblot, P. 2001. Les biotechnologies de la reproduction: répondre aux choix de demain. *Bulletin Technique de l'Insémination Artificelle* 100: 38-42.

Juonala, T., Lintukangas, K., Nurtilla, T. and Andersson, M. 1998. Relationship between semen quality and fertility in 106 AI boars. *Reproduction in Domestic Animals* 33: 155-158.

Leboeuf, B., Brice, G., Baril, G., Broqua, C., Bonne, J.L., Humblot, P. and Terqui, M. 1998. Importance of female selection to improve fertility after goat AI. *Rencontres Recherches sur les Ruminants* 5: 71-74.

Marquant-LeGuienne, B., Guyader-Joly, C., Ponchon, S., Delalleau, N., Florin, B., Ede, P., Ponsart, P., Guérin, B. and Humblot, P. 2001. Results of *in vitro* production in a commercial ovum pick up program. *Theriogenology* 55: 433 (Abstract).

Morris, L.H.A. and Allen, W.R. 2002. An overview of low dose insemination in the Mare. *Reproduction in Domestic Animals* 37: 206-210.

Nibart, M. and Humblot, P. 1997. Pregnancy rates following direct transfer of glycerol glucose or ethylene glucose cryopreserved bovine embryos. *Theriogenology* 47: 371 (Abstract).

Paulenz, H., Adnoy, T. and Fossen, O.H. 2000. Comparison of vaginal and cervical insemination with fresh semen in Norwegian ewes using two different sperm numbers. In: Proceedings, Reproduction in Small Ruminants. Sandnes, Norway, 30 June-1 July 2000. Edited by I Engeland and C Steel. p. 85.

Rath, D. 2002 Low dose insemination in the sow. A review. *Reproduction in Domestic Animals* 37:201-205.

Renard, J.P. 1999. Production de mammifères domestiques par clonage. In: Colloque scientifique de l'AFSSA: Biotechnologies de la reproduction animale et sécurité sanitaire des aliments, Paris 29 Septembre 1999, Edited by M Thibier. AFFSA Publiccations, Maisons Alfort, France. pp. 23-27.

Rostand, J. 1941. Congélation prolongée des cellules fécondantes. In: *La Vie*, Edited by J. Rostand and A. Tétry. 1962. Larrousse, Paris. p. 144.

Rostand, J. 1946. Glycérine et résistance du sperme aux basses températures. *C.R. Acad. Sci.* 22: 1524-1525.

Thibier, M. 1987. L'Insémination Artificielle dans l'espèce bovine, moyen privilégié" d'améliorer l'efficacité de la reproduction. Manuel pour l'Eleveur de Bovins. 9: 7-18. ITEB Publications, Paris, France.

Thibier, M. 1993. Analyse critique des services d'insémination artificielle dans les pays en voie de développement. In : Amélioration génétique des Bovins en Afrique de l'Ouest. Etudes FAO - Production et Santé Animales N° 110: 91-106 ; FAO Ed. Rome, Italy.

Thibier, M. 1999. Les Biotechnologies de la reproduction et la sécurité sanitaire des aliments. In: Colloque scientifique de l'AFSSA: Biotechnologies de la reproduction animale et sécurité sanitaire des aliments, Paris 29 Septembre 1999, Edited by M Thibier. AFFSA Publications, Maisons Alfort, France. pp. 7-8.

Thibier, M. 2002. A contrasted year from the world activity of the animal embryo transfer industry – A report from the IETS Data Retrieval Committee. *IETS Newsletter* 20(4) (in press).

Thibier, M. and Guérin, B. 2000. Embryo transfer in small ruminants, the method of choice for health control in germplasm exchanges. *Livestock Production Science* 62: 253-270.

Thibier, M. and Wagner, H.G. 2001. World Statistics for artificial insemination in cattle. *Livestock Production Science* 74: 203-212.

Thilmant, P. 2001. Congélation du sperme de verrat en paillettes fines de 0.25 ml. *Journées de la Recherche Porcine de France* 33: 151-156.

Wagner, H.G. and Thibier, M. 2000. World statistics for artificial insemination in small ruminants and swine. In: *Proceedings of the 14th International Congress on Animal Reproduction*, July 2000, Stockholm. Abstracts book, vol 2, n° 15:3, p. 77.

Weitze, K.F. 2000. Update on the worldwide application of swine AI. *Boar Semen Preservation IV*, 141-145.

Wells, D.N., Missica, P.M., Forsyth, J.T., Beg, M.C., Lange, J.M., Tervit, H.R. and Vivanco, W.H., 1999. The use of adult somatic cell nuclear transfer to preserve the last surviving cow of the Enderby Island cattle breed. *Theriogenonology* 51: 217 (Abstract).

13

Role of new and current methods in semen technology for genetic resource conservation

W.V. Holt[1] and P.F. Watson[2]

[1]Institute of Zoology, Zoological Society of London, Regent's Park, London, NW1 4RY, UK

[2]Basic Veterinary Sciences, Royal Veterinary College, Royal College Street, London NW1 0TU, UK

Abstract

The establishment of repositories of frozen semen, for the conservation of agricultural genetic resources, is not a simple matter of collecting and freezing semen in the hope that one day it will be suitable for use in an artificial insemination procedure. Important genetic issues need to be considered; for example, how many samples should be stored and from how many individuals? Aside from these, many biological and logistic issues must be considered. Cryopreservation technology does not work equally well in all species, often because of anatomical differences in the female reproductive tract leading to significant variability in the number of spermatozoa needed in order to achieve an acceptable conception rate. Moreover, spermatozoa from different species are not equally susceptible to cryoinjury. However, it is also emerging that semen samples from individuals within a species are also of different quality; several studies have revealed that these differences reflect the quality of DNA within the spermatozoon itself and also the efficacy of biochemical functions, including metabolic and signalling systems, within individual cells. As new possibilities to select spermatozoa for insemination arise, especially the use of flow-sorting for gender selection, these issues may become more significant. In this article we interpret the way in which some of this new information may impact upon the practical implementation of genetic resource conservation.

Introduction

Deciding upon a suitable policy for conserving genetic resources from the numerous agricultural breeds that have developed over many hundreds of years through specialised domestication requires an initial fundamental decision. Is the purpose to conserve 'genes' or 'genotypes?'

In the former case, the best policy is simple. Because there is no need to distinguish between breeds, the objective is merely to capture as many genes as possible irrespective of breed. However, in reality this approach is probably doomed to failure. Sets of genes have co-adapted over many years and it would be difficult to mix and match them from a simple menu of properties because they would undoubtedly depend on each other for function. This automatically leads to the view that the objective of conserving 'genotypes' is the more desirable option. Here, a genotype is equivalent to a breed, which has evolved through selection and agricultural practices to suit its environment and purpose. Conserving the diversity of extant breeds must be considered a massive task, where resource limitation would be a major factor. Storage of frozen semen in banks that would be available for the foreseeable future will require long-term funding, good quality-control practices, and an expectation that the frozen semen samples would remain fertile during storage for many years to come. None of these expectations can be guaranteed especially the non-technical, but nevertheless imperative, hope that the semen bank will be maintained in good condition for decades and more.

There is a widespread perception among the lay public and many science funding bodies that semen freezing, storage and use for artificial insemination (AI) is a widely applicable, routine and reliable technique. Naturally, it follows from this view that there is little to be gained by supporting research in the topic. However, the recent surge of interest in conserving genetic resources from diverse breeds of agricultural species underlines something that has always been clear to anyone with an interest in this area. AI has been tuned to work efficiently in a few species, including cattle, sheep and pigs. However, it suffers from a number of limitations. For example, sheep AI with frozen semen has a poor success rate unless performed using a surgically invasive method and pig AI performed with frozen semen is only efficient when insemination timing with respect to ovulation is managed very intensively.

In 1994 the Meat and Livestock Commission of the United Kingdom convened a meeting to discuss possible future technical improvements and potential research directions for increasing the efficiency of sheep artificial insemination (AI) methods. Their interest in this area was stimulated by a concern that laparoscopic insemination, which is the most effective AI method currently available for sheep, involves minor surgery and may eventually become unacceptable to public opinion at large. Despite the general agreement that something should be done, non-surgical methods have shown no significant improvement since 1994. This is partly because little or no extra funding was forthcoming to support the research. However, it is also because the sheep cervix presents a physical barrier to the insemination pipette, and the spermatozoa may never reach the egg. AI in cattle is more successful because it is easy to inseminate directly into the uterus, and a relatively small number of functional spermatozoa ($< 10 \times 10^6$) are sufficient to produce an acceptable conception rate (50-60%). Fortunately for the cattle industry, this requirement for low numbers of spermatozoa offsets the substantial degree of damage induced by cryopreservation. Calculations based on the relative conception rates achieved with fresh and frozen bull spermatozoa demonstrate that 90% of frozen/thawed spermatozoa are incapable of either fertilising an egg or supporting embryonic development (Shannon, 1978; Vishwanath *et al.* 1996).

With these practical limitations in mind, this review will attempt to examine some of the exciting new developments in semen technology that have occurred over the last 10 years or thereabouts. Some have revolutionised human clinical infertility treatment, and may therefore reinforce the view that semen technologies no longer present any problems. However, some new technologies have been designed with the agricultural sector in mind, and these merit serious consideration when planning to establish new repositories of frozen semen. Semen assessment has also moved forward considerably over the last few years, both with the introduction of new techniques for testing well known parameters such as sperm motility, but also with the realisation that completely novel assays, such as the assessment of sperm DNA integrity, are now possible. We will briefly describe these tests and examine their significance for agricultural species.

Some recent developments in semen technology

Bearing in mind that storing genetic resources in the form of spermatozoa is pointless unless the samples can be used at some time in the future for breeding, it is clear that improvements in the success of cryopreservation technology are urgently required. It is nevertheless salutary to consider that AI in humans performs badly too. The current success rate in the UK is still only about 8% per cycle (Ford *et al.* 1997), even though the patients receive several inseminations per cycle, thus compensating for inaccurate synchronisation with ovulation. It seems, however, that semen cryopreservation technology *per se* may not be entirely to blame in the case of humans, because there is naturally a very high level of early embryonic loss (possibly as high as 80% (Edmonds *et al.* 1982; Racowsky, 2002). However, much of this may originate through defective sperm DNA, as will be discussed below.

Despite the poor performance of semen cryopreservation and AI in mammals, the last decade has seen the introduction of some major advances in semen technology. Semen sexing, separation of X- and Y-chromosome-bearing spermatozoa, is becoming widely used for gender selection in both farm animals and humans. New and more efficient insemination methods have been developed to maximise the use of the small numbers of spermatozoa that are available.

Many of the more novel and adventurous techniques for use with spermatozoa have been developed with the specific intention of treating human infertility. In particular, the development of intracytoplasmic sperm injection (ICSI), whereby single oocytes are injected with single spermatozoa, has revolutionised the thinking behind much infertility treatment. Cryopreservation of single spermatozoa for ICSI has now been achieved (Cohen *et al*. 1997). Moreover, injection of spermatids, instead of spermatozoa, into oocytes has also been undertaken in a few cases of testicular dysfunction in men, a practice which has stimulated a lively ethical debate (Aslam *et al*. 1998).

Another esoteric approach to the conservation of genetic resources could involve the banking of testicular cells and tissues from particular individuals. This technology is now routine in human clinical medicine, where male cancer patients often opt for storage of semen or a testicular biopsy prior to chemo- or radiotherapy. A question mark still hangs over the success with which such biopsies can be used to restore fertility, but it seems likely that the cells are able to recolonise the individual's own testis with germ cells after treatment and recovery. Interestingly, experimental studies have shown that these cells are also capable of colonising testes in heterologous species (Clouthier *et al*. 1996), but whether spermatozoa produced in this way are capable of fertilisation and supporting normal embryonic development is still unclear.

The clinical focus of human infertility treatment is to assist the individual couple, and both patients and clinicians are prepared to invest considerable effort, discomfort and money in obtaining a child of their own. It is notable that the use of anonymously donated semen as a means of overcoming infertility is now sharply in decline. However, the rationale for conserving genetic resources and breeding agricultural species is quite different; here the focus is on the conservation of populations, so making major investment in a single individual is less of a priority.

The last decade has also seen a number of important advances in semen assessment technology, with much of the incentive again coming from clinical practice. While assessment may not at first sight seem as important as storage and use for breeding, the resource constraints inevitably imposed upon any scheme to conserve large numbers of samples, soon forces quality control to become a priority issue. Sperm motility assessment, where the objective is to quantify the proportion of spermatozoa that are moving in relation to those that are not, has typically been performed subjectively by trained technicians. However, computer-assisted techniques now have the capacity to measure not only this relatively simple parameter, but can also make extensive quantitative measurements of the way in which each cell moves along its trajectory, e.g. velocity, track linearity, amplitude of side-to-side movement.

A number of new techniques are currently emerging whereby the DNA within the sperm head can be assessed for strand breaks and stability to denaturing treatments. These defects are believed to arise during spermatogenesis, especially if the testes have been exposed to toxic chemicals or to radiation. Interestingly, it appears that the success rate of ICSI in humans is negatively correlated with the extent of DNA damage (Morris et al. 2002). The exact relevance of this type of test to agricultural species has yet to be firmly established, but some authors have reported negative correlations between the success of AI in cattle and the stability of the chromatin-DNA complex in the sperm heads (Evenson, 1999). Evaluation of mitochondrial function by the use of fluorescent probes that bind to metabolically active mitochondria represents yet another dimension of sperm viability that is now being evaluated. Earlier versions of such fluorescent probes were insensitive to anything but severe changes in metabolic activity; however, newer generations of probes can respond to differences in membrane potential by changing their fluorescent emission properties. There appears to be a degree of correlation between motility and mitochondrial activity detected in this way, and this will undoubtedly be a fruitful area of research for the immediate future.

The importance of fluorescent probes as reporters of cell activity, not just in spermatozoa, has itself stimulated new measurement technologies that have provided novel insights into the biology of spermatozoa. Foremost among these at present is the use of flow cytometry, whereby spermatozoa are labelled with specific dyes and passed individually through laser beams that excite the probes, and past detectors that analyse the resultant emitted fluorescent light. These instruments have provided novel insights into the biology of ejaculates, through their ability to display the characteristics of many individual cells. One of the outstanding and most important outcomes of using flow cytometry has been the clear demonstration that spermatozoa within any semen sample display heterogeneity in virtually all of their attributes. Whether measuring intracellular calcium concentration, membrane fluidity or specific processes that lead to capacitation, it is clear that the spermatozoa differ from each other and tend to fall naturally into three or four categories or subpopulations. This result is in agreement with equivalent data from computer-assisted measurements of sperm motility, and leads to the unifying view that the relative sizes of these sperm subpopulations may be the key factor in determining the fertility potential of an individual animal or ejaculate.

Sperm subpopulations; their origin and significance

While Shannon and Vishwanath (Shannon, 1978; Vishwanath *et al.* 1996) calculated that frozen bovine semen samples contain approximately 10-fold fewer fertile spermatozoa than their non-cryopreserved counterparts, Harrison (1998) provided even more startling calculations showing that a fresh ejaculate contains vanishingly small proportions of spermatozoa with all of the attributes necessary to fertilise an oocyte *in vitro*. Using published data derived from porcine *in vitro* fertilisation studies he calculated that this figure ranges from only 0.1% to 1% of spermatozoa. He argued that this is because a fully competent spermatozoon has to fulfill every one of many rigorous requirements; i.e. (1) maintain the basal intracellular environment by controlling ionic fluxes, pH, ATP concentrations, etc. (2) express motility, undergo capacitation in response to the suitable extracellular environment, (3) attach to the zona pellucida, undergo the acrosome reaction and express hyperactivated motility, (4) fuse with the oocyte plasma membrane and initiate early development (5), and contain paternal DNA that supports full embryonic and foetal development.

This conclusion reveals that the majority of spermatozoa would never be able to fertilise an egg, and that the process of spermatogenesis is extremely inefficient, especially in mammals. Mammals appear to have evolved mechanisms that favour the massive over-production of spermatozoa in the expectation that a few fully functional cells will be produced. This contrasts with the situation in some other species, particularly insects, where spermatogenesis is a more economical process and most spermatozoa are capable of supporting fertilisation and embryogenesis. Based on theoretical calculations, it has also long been recognised that many spermatogenic cells do not complete their maturation into spermatozoa, otherwise there would be many times more spermatozoa produced than is really the case (Johnson *et al.* 1987). Apoptosis is therefore a critically important process during spermatogenesis, presumably removing many defective products of mitosis and meiosis; nevertheless, the evidence clearly shows that many partially defective spermatozoa manage to survive this level of selection and reach the ejaculate.

Several different aspects of sperm DNA integrity can now be measured fluorometrically and visualised by microscopy. The Sperm Chromatin Structure Assay (SCSA) uses acridine orange fluorescence to measure the susceptibility of sperm DNA to mild denaturation *in situ* (Evenson, 1999). In turn, this has been correlated with lifetime non-return (fertility) rates (Ostermeier *et al.* 2001).

Damage to the DNA itself can be assessed using the TUNEL (Terminal deoxynuceotidyl transferase-mediated dUTP-biotin 'nick' end labelling) assay or by the 'comet' assay. The TUNEL assay, which can produce a fluorescent signal proportional to the extent of DNA damage, has the advantage of being amenable to quantitation by flow cytometry (Sun *et al.* 1997). The comet assay involves immobilising the spermatozoa in agarose, decondensing their chromatin, and subjecting the preparation to single cell electrophoresis, whereupon the DNA migrates. The fragmented and intact DNA migrate at different rates, with the intact DNA remaining as a compact spot. When visualised using ethidium bromide fluorescence, the sperm heads resemble comets, whose area and length can be measured by image analysis. Recent data indicates that semen samples from infertile patients contain higher proportions of spermatozoa showing fragmented, TUNEL-positive, DNA than their fertile counterparts (Ramos *et al.* 2002), and that fragmentation is negatively correlated with embryonic survival after ICSI (Morris *et al.* 2002).

These findings support the view that some spermatozoa are simply destined to be incapable of fertilisation because a number of processes leading to their formation have all been defective in one way or another. Why they should exist at all is still a matter of speculation and a subject of considerable interest in evolutionary terms. The defective spermatozoa appear to be particularly frequent in the human, although their presence is well documented in animals. There is considerable evidence that the mammalian female reproductive tract has evolved selective mechanisms to reject these spermatozoa, only selecting the fully competent sperm for fertilisation (Gomendio at al. 1998; Katila, 2001). The few references cited above are sufficient to show that considerable progress has now been made in the search for suitable ways to achieve the same level of selection in the laboratory.

Not only can sperm subpopulations be recognised in terms of genetic integrity and motility, but also recent studies have shown that signalling pathways controlling motility in boar spermatozoa also differ in effectiveness. Boar spermatozoa are unusually sensitive to the presence of bicarbonate; in the absence of bicarbonate they are poorly motile and possess low intracellular concentrations of the signalling molecule cyclic AMP. In the presence of bicarbonate, sperm motility increases and cyclic AMP concentration is rapidly elevated; moreover the plasma membrane lipids change their conformation and become more fluid (Harrison *et al.* 1996). Holt and Harrison (2002) examined these motility responses of boar spermatozoa in more detail, studying the effects of a range of modulators that stimulated or inhibited key steps along the bicarbonate-responsive pathway. By the use of computer-assisted semen analysis (CASA) to measure the velocity and linearity of individually tracked spermatozoa it was found that many spermatozoa rapidly (<2 min) respond to bicarbonate, or to caffeine and papaverine (which are also stimulators of intracellular cyclic AMP concentration) by switching from an uncoordinated and non-directional mode of movement to rapid and progressive motion. However, significant proportions of spermatozoa in the same samples are intrinsically unable to respond to these effectors, something that is readily seen when the CASA data for individual spermatozoa are plotted (Figure 1). Furthermore, between-boar comparisons (Figure 1) reveal that different animals show different proportions of these responding and non-responding spermatozoa. The impact of such differences on between-individual fertility is still unknown but intuitively it seems reasonable that the animals with larger proportions of spermatozoa, which respond sensitively to their environment, are also likely to be the more fertile. At present we are unable to correlate the extent of DNA fragmentation in an individual spermatozoon with the motility of that same cell, or the biochemical efficacy of its signalling pathways. However, these insights lead towards the view that the different cohorts of spermatozoa may be the products of equivalent cohorts of spermatogenic cells, some carrying more genomic errors than others.

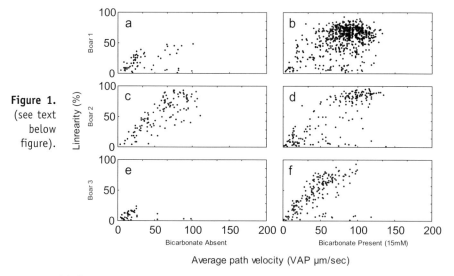

Figure 1.
(see text below figure).

This figure illustrates the motility responses of spermatozoa from three different boars to the addition of 15mM bicarbonate. (adjacent panels, e.g. **a** and **b**, represent spermatozoa from the same samples). Washed spermatozoa were pre-incubated for 10 min in a basal Tyrode's medium (at 39°C), whereupon 60 µl subsamples were taken for videorecording and CASA analysis using a Hobson Sperm Tracker. In these plots where each point represents the motion of an individual spermatozoon, two of the samples (Figure 1a and 1c) reveal the presence of forward progressive sperm (high linearity, high velocity), but the third (Figure 1e) does not. However, 2 minutes after the addition of bicarbonate to the sperm suspensions, considerable sperm motility activation was apparent, especially in the third boar (1e v 1f). Nevertheless, significant proportions of spermatozoa within each sample were unresponsive to bicarbonate (see Holt and Harrison, 2002).

Sperm morphology assessment has been performed for many years in andrology laboratories, but typically using preconceived and subjective judgements of what does, or does not, constitute a 'normal' sperm head. This approach has been improved by augmenting the subjective assessment with simple objective measures such as sperm head width and length, but opinion still plays a major role in this classification method. Modern methods for analysing sperm morphology are, however, beginning to indicate that the way in which DNA and chromatin are packaged during spermiogenesis influences the shape of the resultant sperm head.

Recently, two independent studies have used image analysis algorithms to describe sperm head shape in terms of Fourier harmonic amplitudes (Ostermeier *et al.* 2001; Thurston *et al.* 2001), with some interesting consequences. Because the resultant data is quantitative, it is amenable to multivariate pattern analysis and discriminant function analysis methods. Thurston *et al.* (2001) were able to distinguish three classes of boar spermatozoa within single ejaculates, which differed subtly from each other in head shape, being essentially tapered, rectangular and elongated. None of these differences would be considered abnormal and a human observer would find it very difficult to distinguish these forms by eye. Interestingly the frequencies of two of these subpopulations in individual ejaculates were reciprocally correlated with the susceptibility of that semen sample to cryopreservation-induced damage. In a similar vein, Ostermeier *et al.* (2001) assessed spermatozoa from several bulls using the Fourier technique and also identified subpopulations that were defined by sperm head shape, or more precisely, sperm nuclear shape. In this case the frequencies of sperm subpopulations were highly correlated with bull fertility measured by lifetime non-return rate, and the sperm shape identified as indicative of high fertility was rather elongated. As Ostermeier *et al.* (2001) had worked with sperm nuclear shape, they reasoned that the differences might be a reflection of the efficiency of chromatin packing during spermiogenesis. They supported this conjecture by showing a correlation between the SCSA assay for chromatin stability and sperm shape. If the fidelity of spermiogenesis affects all functional aspects of a resultant spermatozoon it would be no coincidence if the most fertile spermatozoa were also those most resistant to cryoinjury, hence explaining Thurston *et al.*'s (2001) data. These approaches to objective sperm morphology assessment are still in their infancy but seem to hold considerable promise.

Sperm sexing and low dose insemination techniques

The ability to preselect the sex of offspring born by AI or through IVF and embryo transfer has many major advantages for the agricultural industry. Not only is there often a gain in efficiency, but by avoiding the production of the undesired sex, issues of animal welfare associated with their fate can also be avoided.

Attempts to develop techniques for the separation of X- and Y-bearing chromosomes by density gradient centrifugation (Upreti *et al.* 1988) or differential migration through albumin gradients (Beernink *et al.* 1993) did not fulfil their early promise (Chen *et al.* 1997). Currently the only technique for the effective separation of male and female spermatozoa relies upon the use of flow sorting. The flow sorter is a modified flow cytometer, but the cells can be collected for use after sorting. To perform the flow sorting the spermatozoa must first be stained with a fluorescent DNA stain, and the separation system relies upon the detection of the small difference in DNA content between X- and Y-bearing sperm. The X-chromosome, being slightly larger, contains more DNA. The technique has been applied to several agricultural species, including cattle, sheep, pigs and horses, and has also been applied to humans. For reviews, see Abeydeera *et al.* 1998; Garner, 2001; Johnson, 2000; Seidel and Johnson, 1999.

The flow sorting technique still suffers from both practical and ethical drawbacks. The practical drawbacks are that because sorting is done serially, on a sperm by sperm basis, it is still relatively slow. Furthermore, the sorted spermatozoa, having been subjected to manipulations that involve exposure to high pressure and laser illumination, are rather fragile. This causes some problems with sperm cryopreservation and sperm storage, which is less effective than for non-sorted spermatozoa. Recently a novel method for semen cryopreservation has been developed (Arav *et al.* 2002). The technique differs from the traditional freezing protocol in that diluted semen is frozen in straws and tubes by physically sliding them across a temperature gradient. A freezing machine for this purpose has now been through several prototypes. Semen that has been frozen and thawed by this method is reportedly indistinguishable in terms of motility from semen that has never been frozen.

The slow sorting rate has presented a particular problem for AI in pigs, where it is normal practice to inseminate a relatively large number of spermatozoa (3×10^9) suspended in 70-80 ml of fluid, directly into the cervix. However, acceptable conception and farrowing rates can be achieved with about 100-fold fewer spermatozoa if they are deposited near to the utero-tubal junction (Krueger *et al.* 1999). Despite the complex anatomy of the pig cervix and uterine horns, a new fibreoptic endoscope has been developed which permits the non-surgical insemination of these minimal sperm numbers with relative ease (Martinez *et al.* 2001). This flexible catheter is 1.8 metres in length, and the latest versions have dispensed with the fibreoptic facility. Using this technique, Martinez *et al.* (2002) showed that insemination with 5 or 15×10^7 spermatozoa resulted in farrowing rates of 76.2% and 82.9% respectively, compared with 83.0% for controls that had been inseminated by the normal technique. While these results were not statistically different from the controls, reducing the number of spermatozoa still further, to 2.5×10^7, produced a significantly reduced farrowing rate (46.7%).

This technology is not only applicable for using sexed semen, but will also be of considerable benefit for any situation where sperm numbers or sperm quality are limiting factors. Use in conjunction with valuable cryopreserved semen is obviously a good option for using a genetic resource bank, particularly with pigs where frozen semen does not perform well after cryopreservation.

It is too early to know whether the ethical issues associated with the sperm sexing technique are a problem or not. Essentially, because the technique is entirely dependent upon the detection of nuclear DNA that has been stained with fluorochrome, a nagging doubt that this may lead to some unforeseen effects will remain unresolved for a number of years to come. Nevertheless, the agricultural industry has embraced this technology enthusiastically, and the technique is also being offered for human use in the USA, for both medical and social reasons.

Biosecurity

The risk of disease transmission associated with the translocation and use of gametes and embryos is generally lower than that associated with animal translocations. One of the motivations for introducing AI into the agricultural industry was to assist with disease control. However, some diseases can be transmitted via semen and the risks of causing widespread disease transmission are amplified if contaminated semen is used inadvertently. The area of disease risks and gamete transfer in relation to genetic resource banking has recently been reviewed (Kirkwood and Colenbrander, 2001).

There are two main types of risk involved, one arising if semen is collected from infected males and the other if a sample becomes infected by contact with other samples during storage. Obviously, good veterinary practice and adherence to animal health guidelines have protected AI centres for many years, although there are many debates about which precautions should be taken. Semen is routinely treated with a cocktail of antibiotics to guard against most bacteria. The issues surrounding inter-sample cross-contamination are, however, more insidious. Concern over the potential for transmission of viral infection between samples stored in liquid nitrogen has recently been expressed in the United Kingdom, particularly in relation to samples of human semen. Fears arose initially when transmission of hepatitis to a patient was attributed to stored bone marrow cells that had been contaminated when an infected sample was damaged within the same sample storage container. Following this case an independent review panel was convened to assess disease transmission risks between stored samples. The panel's recommendation that all straws of human semen should henceforth be stored in the vapour, rather than the liquid phase of the nitrogen container caused consternation in laboratories with semen storage facilities, and some clinics are now adopting this procedure.

This episode highlighted a problem with regard to semen storage that has hardly ever been addressed. In fact, there is virtually no data upon which to base an assessment, except for some experiments performed at The Royal Veterinary College, London (Russell *et al.*1997) to investigate the potential for viral leakage from sealed straws. Russell *et al.* (1997) could find very little evidence that viral leakage from inside the straw was possible, provided the straws were properly sealed. However, viral contamination of the external surface of straws could occur during filling if they were dipped into the semen-diluent mixture. Care to avoid this practice is therefore recommended, especially when dealing with samples whose disease status is unknown. Secondary protection against straw breakage may also be advisable in certain cases and so-called 'high security straws' with this extra containment are now available commercially. These precautions are considerably easier to implement with straws than with pellets. Here the semen is initially in contact with dry ice, and is then normally stored in a perforated tube, where direct contact with the liquid nitrogen itself occurs.

Concluding remarks

Establishing genetic resource banks for the storage of gametes has been practiced in the agricultural sector for many years. Commercial companies, which have invested considerable sums in the genetic improvement of their breeds and lines naturally, often conserve their own genetic resources 'in house'. Organisations dedicated to protecting specific breeds, whether rare or not, also wish to conserve the genetic diversity that is currently available in the expectation that one day they may need to retrieve their samples in case of disaster or to counter genetic drift. However, establishing a genetic resource bank is no trivial exercise and demands both extensive planning and sufficient resources. Many factors have to be considered when planning such genetic resource banks, and while issues are often context-specific, considerable guidance can be taken from the published literature. The authors of this paper recently published a book on genetic resource banking for wildlife conservation, and many of the planning issues are highly relevant and are discussed at length (Watson and Holt, 2001).

References

Abeydeera, L. R., Johnson, L. A., Welch, G. R., Wang, W. H., Boquest, A. C., Cantley, T. C., Rieke, A. and Day, B. N. 1998. Birth of piglets preselected for gender following *in vitro* fertilization of in vitro matured pig oocytes by X and Y chromosome bearing spermatozoa sorted by high-speed flow cytometry. *Theriogenology* 50: 981-988.

Arav, A., Yavin, S., Zeron, Y., Natan, D., Dekel, I. and Gacitua, H. 2002. New trends in gamete's cryopreservation. *Molecular and Cellular Endocrinology* 187: 77-81.

Aslam, I., Fishel, S., Green, S., Campbell, A., Garratt, I., McDermott, H., Dowell, K. and Thornton, S. 1998. Can we justify spermatid microinjection for severe male factor infertility? *Human Reproduction Update* 4: 213-222.

Beernink, F. J., Dmowski, W. P. and Ericsson, R. J. 1993. Sex preselection through albumin separation of sperm. *Fertility and Sterility* 59: 382-386.

Chen, M. J., Guu, H. F. and Ho, E. S. 1997. Efficiency of sex pre-selection of spermatozoa by albumin separation method evaluated by double-labelled fluorescence in-situ hybridization. *Human Reproduction* 12: 1920-1926.

Clouthier, D. E., Avarbock, M. R., Maika, S. D., Hammer, R. E. and Brinster, R. L. 1996. Rat spermatogenesis in mouse testis. *Nature* 381: 418-421.

Cohen, J., Garrisi, G. J., CongedoFerrara, T. A., Kieck, K. A., Schimmel, T. W. and Scott, R. T. 1997. Cryopreservation of single human spermatozoa. *Human Reproduction* 12: 994-100.

Edmonds, D. K., Lindsay, K. S., Miller, J. F., Williamson, E. and Wood, P. J. 1982. Early embryonic mortality in women. *Fertility and Sterility* 38: 447-453.

Evenson, D. P. 1999. Loss of livestock breeding efficiency due to uncompensable sperm nuclear defects. *Reproduction, Fertility and Development* 11: 1-15.

Ford, W. C. L., Mathur, R. S. and Hull, M. G. R. 1997. Intrauterine insemination: is it an effective treatment for male factor infertility. *Bailliere's Clinical Obstetrics and Gynaecology* 11: 1-20.

Garner, D. L. 2001. Sex-Sorting mammalian sperm: concept to application in animals. *Journal of Andrology* 22: 519-526.

Gomendio, M., Harcourt, A. H. and Roldan, E. R. S. 1998. Sperm competition in mammals. In: *Sperm competition and sexual selection*. Edited by Birkhead, T. R. and Moller, A. P. Academic Press Ltd, San Diego, London, Boston, pp.667-755.

Harrison, R. A. P. 1998. Sperm evaluation: what should we be testing?, In *Genetic diversity and conservation of animal genetic resources*. Ministry of Agriculture, Forestry and Fisheries, Japan., Tsukuba, Japan, pp. 135-154.

Harrison, R. A. P., Ashworth, P. J. C. and Miller, N. G. A. 1996. Bicarbonate/CO_2, an effector of capacitation, induces a rapid and reversible change in the lipid architecture of boar sperm plasma membrane. *Molecular Reproduction and Development* 45: 378-391.

Holt, W. V. and Harrison, R. A. 2002. Bicarbonate Stimulation of Boar Sperm Motility via a Protein Kinase A- Dependent Pathway: Between-Cell and Between-Ejaculate Differences Are Not Due to Deficiencies in Protein Kinase A Activation. *Journal of Andrology* 23: 557-565.

Johnson, L., Nguyen, H. B., Petty, C. S. and Neaves, W. B. 1987. Quantification of human spermatogenesis - germ-cell degeneration during spermatocytogenesis and meiosis in testes from younger and older adult men. *Biology of Reproduction* 37: 739-747.

Johnson, L. A. 2000. Sexing mammalian sperm for production of offspring: the state-of-the- art. *Animal Reproduction Science* 60-61: 93-107.

Katila, T. 2001. Sperm-uterine interactions: a review. *Animal Reproduction Science* 68: 267-272.

Kirkwood, J. K. and Colenbrander, B. 2001. Disease control measures for genetic resource banking. In: *Cryobanking the genetic resource; wildlife conservation for the future?* Edited by Watson, P. F. and Holt, W. V. Taylor and Francis, London, UK. pp.69-841.

Krueger, C., Rath, D. and Johnson, L. A. 1999. Low dose insemination in synchronized gilts. I52: 1363-1373.

Martinez, E. A., Vazquez, J. M., Roca, J., Lucas, X., Gil, M. A., Parrilla, I., Vazquez, J. L. and Day, B. N. 2001. Successful non-surgical deep intrauterine insemination with small numbers of spermatozoa in sows. *Reproduction* 122: 289-296.

Martinez, E. A., Vazquez, J. M., Roca, J., Lucas, X., Gil, M. A., Parrilla, I., Vazquez, J. L. and Day, B. N. 2002. Minimum number of spermatozoa required for normal fertility after deep intrauterine insemination in non-sedated sows. *Reproduction* 123: 163-170.

Morris, I. D., Ilott, S., Dixon, L. and Brison, D. R. 2002. The spectrum of DNA damage in human sperm assessed by single cell gel electrophoresis (Comet assay) and its relationship to fertilization and embryo development. *Human Reproduction* 17: 990-998.

Ostermeier, G. C., Sargeant, G. A., Yandell, B. S., Evenson, D. P. and Parrish, J. J. 2001. Relationship of bull fertility to sperm nuclear shape. *Journal of Andrology* 22: 595-603.

Racowsky, C. 2002. High rates of embryonic loss, yet high incidence of multiple births in human ART: is this paradoxical? *Theriogenology* 57: 87-96.

Ramos, L., Kleingeld, P., Meuleman, E., van Kooy, R., Kremer, J., Braat, D. and Wetzels, A. 2002. Assessment of DNA fragmentation of spermatozoa that were surgically retrieved from men with obstructive azoospermia. *Fertility and Sterility* 77: 233-237.

Russell, P. H., Lyaruu, V. H., Millar, J. D., Curry, M. R. and Watson, P. F. 1997. The potential transmission of infectious agents by semen packaging during storage for artificial insemination. *Animal Reproduction Science* 47: 337-342.

Seidel, G. E., Jr. and Johnson, L. A. 1999. Sexing mammalian sperm–

overview. *Theriogenology* 52: 1267-1272.

Shannon, P. 1978. Factors affecting semen preservation and conception rates in cattle. *Journal of Reproduction and Fertility* 54: 519-527.

Sun, J. G., Jurisicova, A. and Casper, R. F. 1997. Detection of deoxyribonucleic acid fragmentation in human sperm: correlation with fertilization in vitro. *Biology of Reproduction* 56: 602-607.

Thurston, L. M., Watson, P. F., Mileham, A. J. and Holt, W. V. 2001. Morphologically distinct sperm subpopulations defined by Fourier shape descriptors in fresh ejaculates correlate with variation in boar semen quality following cryopreservation. *Journal of Andrology* 22: 382-394.

Upreti, G. C., Riches, P. C. and Johnson, L. A. 1988. Attempted sexing of bovine spermatozoa by fractionation on a Percoll density gradient. *Gamete Research* 20: 83-92.

Vishwanath, R., Pitt, C. J. and Shannon, P. 1996. Sperm numbers, semen age and fertility in fresh and frozen bovine semen. *Proceedings of the New Zealand Society of Animal Production* 56: 31-34.

Watson, P. F. and Holt, W. V. 2001. (Editors) *Cryobanking the genetic resource; wildlife conservation for the future?* Taylor and Francis, London, UK.

14

Oocytes and assisted reproduction technology

H.M. Picton[1], M.A. Danfour[1] and H. Coulthard[2]
[1]Academic Unit of Paediatrics, Obstetrics and Gynaecology, University of Leeds, Leeds, LS2 9NS, UK
[2]Cattle Tech Ltd, Selby, YO8 9HL, UK

Abstract

A number of methods can now be used to store the female germ plasm from farm species. New assisted reproductive techniques such as ultrasound guided oocyte pickup, followed by the in vitro maturation of oocytes together with cryopreservation enable the collection and storage of germinal vesicle or metaphase II secondary oocytes and more practically embryos after in vitro fertilisation. Following freezing using equilibration or vitrification protocols, oocytes and embryos can be stored at liquid nitrogen temperatures for as long as required. Despite much research interest, the efficiency of secondary oocyte freezing is low and the developmental potential of stored oocytes is directly affected by the local cellular and hormonal environment during maturation, fertilisation and extended culture in vitro. An alternative strategy which avoids many of the technical difficulties associated with mature oocyte freezing may be to cryopreserve primordial oocytes in situ within ovarian cortex. This approach has the added advantage that it may also provide a means of conserving the oocytes of rare species and it can be used to bank cells obtained from postmortem tissue samples or species for which IVF protocols may not have been fully optimised. Although the methodology is still in its infancy, when ovarian cryopreservation is combined with autografting, xenografting, or follicle culture, ovarian tissue freezing has the potential to restore or extend the fertility of domestic animals so maximising their genetic potential.

Introduction

With the advent of assisted reproduction technology (ART) and an improved understanding of cryobiology, strategies have been developed which allow the collection and long-term storage of the sperm and embryos from many species. The options available for the preservation and storage of oocytes from farm animals are usually indirect and are frequently limited to the preparation and storage of embryos as these can be easily flushed from the oviducts after *in vivo* fertilisation, or they can be created *in vitro* after oocyte harvest and *in vitro* fertilisation (IVF). Embryos can easily be frozen and banked long-term for transfer at a later date (Trounson *et al.*, 1994). In contrast to the successful freezing of embryos, oocyte cryopreservation had proved far less reliable. Consequently, in recent years attention has focused on two areas: (i) improving the protocols for the freeze-storage of secondary oocyte and the associated technologies of *in vitro* maturation (IVM) and fertilisation from different species; and (ii) the low temperature banking of immature oocytes and ovarian tissue from different species. This review will focus therefore on the recent advances made in ART which will support preservation of female gametes (Figure 1).

Oocyte biology

The germ plasm of mammalian females consists of hundreds of thousands of primordial oocytes which are enclosed within primordial follicles. The full number of the primordial oocytes available for reproduction during adult life in mammalian species is laid down in the ovary early in fetal development. Thus the tiny primordial follicles represent the reserve from which all fertile oocytes will ultimately development. However, primordial oocytes are not fertile and they have to grow in follicles for several months before they become competent to undergo cytoplasmic maturation, resume nuclear maturation, undergo fertilisation and support early cleavage divisions of the embryo before the embryonic genome is activated. Throughout their growth phase there is direct contact between follicle cells and oocytes to facilitate the transfer of nutrients, metabolites, small peptides, growth factors and gases to and from the oocyte (Anderson and Albertini, 1976). During growth, oocytes increase in size and they acquire the full complement of cytoplasmic organelles such as endoplasmic reticulum, cortical granules, and most importantly mitochondria (Hyttel *et al.*, 1997; Picton *et al.*, 1998) and they sequentially accumulate the payload of proteins and RNAs which are required to support fertilization and early embryo development. Throughout growth, oocytes are arrested at the diplotene stage of the first meiotic prophase which is characterised by the presence of a Germinal Vesicle (GV) nucleus. Although both oocyte growth and follicle growth occur continuously from fetal life onwards, it is not until puberty that the oocyte and follicles receive the signals that they need to induce the final changes required to support the production of a fertile gamete.

Once puberty is reached, the preovulatory surge of the gonadotrophins Luteinising Hormone (LH) and Follicle Stimulating Hormone (FSH) in each reproductive cycle will trigger the resumption of meiosis in the

Figure 1.
Options available for
oocyte preservation.
GV- germinal vesicle;
MII-metaphase II;
IVF- *in vitro*
fertilisation; ICSI-
intracytoplasmic
sperm injection.

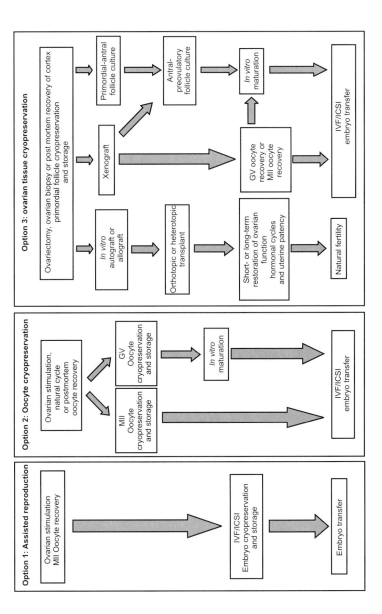

209

oocytes present within the preovulatory follicle(s). The surge levels of LH result in: (i) a decrease in intracellular cyclic adenosine monophosphate (camp) production by the follicular granulosa cells; (ii) the loss of gap junctional contacts between the granulosa cells and the oocyte; and (iii) the production of hyaluronic acid which leads to the mucification and expansion of the cumulus granulosa cells immediately surrounding the oocyte (Andriez *et al.*, 2000; Picton 2002). The shift in follicular steroidogenesis from predominately oestrogen to progesterone production during the process of follicular luteinisation, post surge, may also help to modulate the meiotic status of the oocyte and may actively promote the resumption of meiosis by stimulating the production of a GV breakdown-inducing signal within the oocyte. The resumption of meiosis is characterised by the breakdown of the GV nucleus, chromosome condensation, formation of the first meiotic spindle, expulsion of the first polar body and arrest in metaphase of the second meiotic division (metaphase II, (MII)). After ovulation and fertilisation of the oocyte within the oviduct, meiosis is completed and the second polar body is emitted.

Oocytes and assisted reproduction

Oocytes to be used for ART can be harvested from the ovaries by ultrasound guided aspiration techniques at either the GV stage of nuclear maturation, or the MII stage if they have been exposed to an endogenous or exogenous gonadotrophin surge *in vivo* (Picton 2002). Additionally GV oocyte can be recovered from postmortem tissue samples by needle aspiration. While mature oocytes can be used immediately for IVF and IVP, GV oocytes must first be matured for 24-48hrs *in vitro* (Danfour, 2001; Schramm and Bavister 1994; Trounson *et al.*, 1994; Wynn *et al.*, 1998) before fertilisation and embryo production is possible. Furthermore, in comparison with IVF oocytes, where the mature gamete is surrounded by expanded and mucified cumulus granulosa cells, GV oocytes collected for IVM are enclosed in tightly packed cumulus cells. This difference in the size and morphology of the cumulus-oocyte complex makes the recovery and identification of oocytes for IVM far more difficult than for IVF. The maturational changes which occur in the oocyte and surrounding cumulus cells during IVM must replicate those seen *in vivo*. Thus IVM oocytes will progress through GV breakdown and progress to MII following the extrusion of the first polar body.

In recent years the technology of IVM has improved markedly and oocyte IVM has become a routine method for the commercial production of embryos from genetically superior animals in species such as sheep and cattle and it is also used in humans (Trounson *et al.*, 1994, Picton 2002). Improvements in IVM, IVF and *in vitro* embryo production (IVP) systems have resulted in enhanced rates of embryo development and improved pregnancy rates following transfer of embryos derived from IVM oocytes. This means that when the culture conditions have been optimised, as with for example cattle oocytes, oocytes harvested from different sources can have similar maturation and developmental potential (Figure 2). However, oocytes from different species may differ significantly in their capacity to be matured and fertilised in *vitro*. Optimisation of the protocols for IVM, IVF and IVP is therefore required for each species if ARTs are to be used successfully.

Figure 2.
(See text below figure).

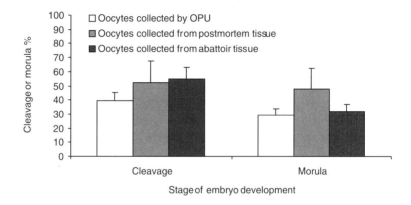

Comparison of *in vitro* production of early cleavage stage bovine embryos or morulae derived from: oocytes collected by ultrasound guided oocyte pickup (OPU); oocytes collected by ovarian dissection of postmortem tissue specimens immediately after slaughter for injury; and oocytes collected by needle aspiration from ovaries obtained from a local abattoir. All of the oocytes were matured *in vitro* in serum-free medium. There was no significant difference between the groups in cleavage rate or morula production (P> 0.05).

Irrespective of species, the success of IVM, IVF and IVP systems depends upon a number of factors including follicular maturity, the components of the culture environment, the presence or absence of cumulus cells, the capacity to induce cumulus expansion *in vitro* and the number of oocytes cultured together.

(i)   *Follicular maturity*

The data of Lonergan *et al.*, (1994) and Danfour (2001) clearly demonstrate that better quality bovine and ovine oocytes are derived from large follicles of >6mm and > 2.5mm in cows and sheep respectively. The improved oocyte maturation and developmental competence of oocytes harvested from larger follicles is due to a longer exposure to a favorable follicular microenvironment *in vivo*. This enables the oocytes to complete the storage of components, such as proteins, mRNAs, and any other molecules that will be used to sustain early embryonic development post fertilisation.

(ii)   *Components of the culture environment*

While suboptimal culture systems may compromise both the quantity and the developmental competence of oocytes and embryos developed *in vitro*, research on oocytes and embryo production will serve to increase our understanding of the mechanisms controlling oocyte maturation and embryo development and so help define the optimal culture and cellular environment. It is clear that the production of embryos *in vitro* exposes them to hazards not normally present *in vivo*. Numerous reports show that culture media components and culture conditions can affect the meiotic regulation of oocytes (for example: Gandhi *et al.*, 2000; Kito and Bavister, 1997; Thompson *et al.*, 1998).

While the majority of the literature indicates that the addition of serum to culture medium supports oocyte IVM, fertilisation and embryo development, other data also suggests that serum has adverse effects on embryo metabolism and development *in vitro*. When serum is added to the culture medium, it acts as a source of albumin, which balances the osmolality and scavenges potentially harmful molecules and metal ions that can act as a source of free oxygen radicals. Serum also contains many factors, such as serum proteins, amino acids, carbohydrates, trace elements, hormones, growth factors, and some as yet undefined components. Inclusion of serum in oocyte maturation medium may also prevent premature loss of cortical granules and zona hardening – an effect which is attributed to the serum component fetiun. The use of serum in IVM and IVF medium has been observed to increase the number of large lipid droplets in the cytoplasm of sheep embryos, which is thought to be due to impaired mitochondrial function (Danfour, 2001; Thompson *et al.*, 1995). Also, Walker *et al.*, (1996) reported that addition of serum to culture medium cause cytoplasmic fragmentation and premature blastulation of sheep embryos. This finding suggests either that there are highly active mitogenic factors present in the serum itself or that embryonic development may be influenced by altered metabolic or growth factor mediated responses. In both animals and humans, serum could be a source of transmission of pathogenic diseases and in particular viral diseases, such as HIV and Hepatitis B and C in humans, or bovine viral diarrhea virus in cattle (Meyling *et al.*, 1990).

Another major side effect of using serum in IVM culture media is the potential production of large offspring syndrome (LOS) in fetuses derived from IVP embryos produced by extended culture in the presence of serum (Thompson et al., 1995, Young and Fairburn, 2000). This condition is often associated with slightly extended gestation and an increase in post-natal mortality, due to a greater incidence of dystocia (Thompson et al., 1995; Walker et al., 1996; van Wagtendonk-de Leeuw et al., 2000, Young and Fairburn, 2000). Increasingly LOS of IVP embryos is thought to be associated with disruption of key imprinted genes (Young and Fairburn (2000). The link between LOS and serum-based culture media is further established by the fact that LOS is considerably reduced when serum is replaced with amino acids and albumin in the culture medium (van Wagtendonk-de Leeuw et al., 2000). While it is not yet clear precisely what components of serum cause LOS in sheep and cattle, factors at any stage of the sequential processes of maturation, fertilisation, and embryo culture may play a decisive role.

(iii)   *Cumulus cell mass and expansion*

The cellular environment in immediate proximity to the oocyte, namely the cumulus cells, appears to influence the ability of oocytes to undergo maturation, fertilisation and development to the blastocyst stage. Cumulus cells play a very important role during oocyte development, supplying nutrients, gases and small molecules such as cAMP and purines, and they mediate the effect of hormones on the cumulus oocyte complex. Furthermore, cumulus cells in many mammalian species are involved in the interaction between male and female gametes, guiding the sperm toward the oocyte, inducing capacitation, maintaining sperm mobility and viability and enhancing sperm penetration and *in vitro* fertilisation (Fukui, 1990). The factors known to be secreted by the cumulus cells to enhance cumulus expansion include heparin sulphate, chondroitin sulphate, and hyaluronic acid (Chen et al., 1993). In light of this evidence it is perhaps not surprising that optimal expansion of the cumulus mass appears to be essential for both normal ovulation (Chen et al., 1993) and subsequent fertilisation (Vanderhyden and Armstrong, 1989). Furthermore, when defined culture systems are used for IVM/IVF and IVP the degree of cumulus expansion during oocyte maturation *in vitro* appears to correlates with the cleavage rate of two cell mouse embryos (Chen et al., 1993), the cleavage potential of equine oocytes (Boyazoglu et al., 2000) and the blastocyst formation rate of bovine embryos (Danfour et al., 2002). Several other reports (reviewed by Danfour, 2001) demonstrate that the presence or absence of cumulus cells and the cumulus cell mass during maturation *in vitro* affect IVM potential and subsequent embryo cleavage and blastocyst rates.

(iv)   *Oocyte and embryo group culture*

The great majority of embryos derived from ART in large animals have been generated by the culture of multiple oocytes or embryos in open wells of culture media. Our own recent data (Danfour *et al.*, 2002) also show that acceptable blastocyst production rates can be obtained when oocytes have been cultured singly in microdrops of serum-free IVM media, when compared to the blastocyst rates produced from oocytes cultured in groups of 5 or 10 or more. Furthermore, the transfer of morulae or blastocyts derived from oocytes matured individually in serum-free medium has achieved a pregnancy rate of 45% (Danfour, 2001). This pregnancy rate is comparable to published rates of 42% which have been obtained from oocytes matured either in the presence of serum or in large groups (50 – 70 oocytes/group) (Numabe *et al.*, 2000 and van Wagtendonk-de Leeuw *et al.*, 2000).

Oocyte cryopreservation

While IVM and IVF embryo production systems are immensely useful as a means to exploit the reproductive potential of animals of high genetic merit, they are of reduced value in the preservation of the gene pool of rare or endangered species as the success rate of these approaches are highly species dependent. For this group of animals we have to look to the development and application of cryopreservation strategies which will support oocyte freezing rather than IVF and embryo production. At the present time two practices are used to freeze-store oocytes, each method has its own advantages and limitations. Isolated secondary oocytes can be stored at either the GV or MII stages of development or alternatively immature primordial oocytes can be cryopreserved. Additionally, there are 2 defined methods of cryopreservation, namely slow freezing /equilibration protocols or rapid freeze / vitrification protocols (Picton *et al.*, 2003b).

Secondary oocyte freezing

Although MII oocytes are routinely collected as part of ART practices from a number of species, and despite intense research interest, the results of MII and GV egg freezing from domestic species are disappointing. The low success rates of MII oocyte freezing can be attributed in part to the biological characteristics of the oocyte. Mature oocytes have a short fertile life span, have little capacity for repairing damage and have a low surface:volume ratio, all of which are characteristics which increase the likelihood of damage due to the formation of large ice crystals inside the cytoplasm during freezing and thawing. This damage can affect the meiotic spindles and so increase the chance of genetic abnormality and it can also disrupt the cellular cytoskeleton leading to changes in the organisation and trafficking of molecules and organelles. Consequently, mature oocyte freezing is frequently associated with low post-thaw viability, poor fertilisation rates and reduced developmental competence of the embryos. In contrast to freezing MII oocytes, banking of GV stage oocytes reduces the risk of cytogenetic errors. However, cytoskeleton damage may still occur and GV oocytes must undergo extended culture and IVM in order to reach MII and become fertile. Furthermore, if GV oocytes are to be successfully matured after cryopreservation then both the junctional contacts and the viability of the cumulus cells must also be retained post-thaw to support IVM. Although it has proved possible to obtain live births after the slow freezing and storage of GV and MII oocytes in laboratory species such as mice, the efficiency of secondary egg freezing is far lower for large animals and humans where only a limited number of births have been achieved (Picton *et al.*, 2003b). Recently some encouraging results have been obtained after the vitrification of oocytes but the data is inconsistent and the success of the methodology appears to be dependent on the stage of oocyte nuclear maturity. It is also highly species specific.

Ovarian tissue cryopreservation

Following advances in ovarian cryobiology and the development of strategies designed to preserve the fertility of young cancer patients, the radical new strategy of low temperature banking of immature oocytes *in situ* within the ovarian cortex can now be applied to the storage of eggs from humans, farm animals or endangered species (Picton *et al.*, 2001). This approach is possible because of the unique architecture of the mammalian ovary, the relatively small size and undifferentiated status of primordial follicles and the large numbers of primordial follicles as hundreds of thousands of these early staged follicles are present in young ovaries of all mammalian females. Furthermore, the primordial follicles are localised in the periphery of the ovarian cortex. This means that ovarian cortex can be harvested by the laparoscopic recovery of multiple small ovarian biopsies or after the surgical removal of a whole ovary from animals of high genetic merit or from rare species. An added advantage of this method is that tissue can also be harvested from post mortem samples. The primordial follicles can then be banked in slices of cortex at liquid nitrogen temperatures for as long as required and accessed when convenient. After long-term storage, tissue can be thawed and used to restore or extend natural fertility by autografting the tissue back into the body at either orthotopic or heterotopic sites (Picton *et al.*, 2001). Alternatively, where autografting is not possible the tissue can be xenografted into mice with Severe Combined Immune Deficiency

(SCID) as a means to growth the oocytes to maturity *in vivo* for later IVF and embryo production.

Confirmation that ovarian tissue cryopreservation together with grafting strategies are effective methods for preserving female fertility has been provided in both rodents and ruminants. Although one report suggests that whole ovarian freezing may be successful in sheep (Revel *et al.*, 2001), the majority of data suggest that in large animals and humans whole ovaries are too large to allow successful freezing and transplantation without producing widespread necrosis during re-vasularisation of the grafted tissue (Nugent *et al.*, 1997). An alternative approach for large animals is therefore to autograft or xenograft small pieces of ovarian cortex with a large surface area which will permit adequate gaseous diffusion and so prevent ischaemia until re-vascularisation of the graft has taken place. In support of this concept, encouraging results have been obtained in sheep with the restoration of natural fertility and the production of live offspring after orthotopic autografting (Gosden *et al.*, 1994; Salle *et al.*, 2002). Subsequent experiments, designed to evaluate the longevity and endocrine consequences of the autografting procedure, have confirmed that the grafts take with 100% efficiency and can remain patent for at least 22 months (Baird *et al.*, 1999). Endocrinologically, despite normal patterns of ovarian follicular development and reproductive cycles, cryopreservation followed by autografting resulted in a marked increase in circulating gonadotrophin concentrations, that could be attributed in part to a marked decrease in ovarian inhibin secretion (Baird *et al.*, 1999). Despite these encouraging results the successful application of this technology to the conservation of rare species may be hampered by the fact that the ovary is not an immunologically privileged site. This means that allografting is unlikely to be effective as the transplanted tissue will be rejected. Where auto- or allo-grafting of cryopreserved tissue is not an option, such as in the case of postmortem tissue banking, xenografting may prove valuable. Follicular development has, for example, been demonstrated to proceed to the early antral stages when humans ovarian cortex is xenografted under the kidney capsule of SCID mice (Oktay *et al.*, 1998). The strategy envisioned for rare species could therefore involve cryopreservation and xenografting followed by the collection of oocytes grown in antral follicles for subsequent IVM and IVF. Although controversial, this strategy could be justified as a means to conserve the female germ plasm of endangered species.

Follicle culture

Finally, follicle culture systems are now being developed which will utilise primordial follicles from cryopreserved ovarian tissue as a source of oocytes for *in vitro* growth and maturation, IVF and IVP. Significant progress has been made using pre-antral follicles harvested from fresh (Spears et al., 1994 and Cortvrindt et al., 1996) or cryopreserved (Newton et al., 2001) rodent ovaries and two studies have even produced live offspring after the complete *in vitro* growth and maturation of oocytes from primordial follicles (Eppig and O'Brien 1996, Obata et al., 2002). For large animals follicle culture systems have proved to be less efficient as oocytes require many months to complete their normal growth sequence (Picton et al., 1998; 2003a). Nonetheless, antral cavity formation and the induction of steroidogenesis has been demonstrated in long-term cultures of oocyte-granulosa cell complexes harvested from both fresh and cryopreserved sheep ovarian tissue (Newton et al., 1999) and antral cavity formation has been reported in cultures of bovine preantral follicles (Gutierrez et al., 2000).

To fully exploit the potential of primordial follicle banking as a means to preserve female germ plasm it is necessary to develop strategies which will support primordial follicle growth *in vitro*. In combination with ovarian cryopreservation, the development of technologies to grow and mature oocytes from primordial follicles holds many attractions for clinical practice and animal production technology as a small ovarian biopsy may contain many hundreds of primordial follicles. Unfortunately the methods which support the growth of isolated preantral follicles are ineffective when applied to the culture of primordial follicles. The most effective approach to date appears to be to leave the primordial follicles embedded within the ovarian stromal tissue and to culture small fragments of cortex rather than isolated follicles. Although cortical slice culture systems have been developed in a number of species (for example: Cushman et al., 2002; Wandji et al., 1996; Hovatta et al., 1997, Picton et al., 1999) the methodology must be tailored to the specific requirements of primordial follicles for each species. Consequently, progress in primordial follicle culture is hampered by our limited understanding of the mechanisms regulating early oocyte development. Along with improvements in the cryotechnology, more efficient culture methods must be developed to support the activation and long-term culture of primordial follicles and the *in vitro* maturation of the oocytes they contain.

Future challenges

Assisted reproduction and associated technologies now provide us with the means to preserve the germ plasm of farm animals of high genetic merit either directly through immature or mature oocyte cryopreservation or through embryo freezing. Furthermore, in conjunction with the freeze storage of oocytes and ovarian tissue harvested by surgical intervention or from postmortem samples, xenografting and/or follicle culture followed by IVM/IVF offers the possibility of preserving the oocytes of rare species. However, considerable effort is still required to improve the efficiency and safety of these techniques. While the frameworks are in place for the development of whole follicle culture systems, extensive basic research into the biology of oogenesis in different species is still required to improve the chances of success of this approach.

References

Anderiesz, C., Ferraretti, A-P., Magli, C., Fiorentino, A., Fortini, D., Gianaroli, L., Jones, G.M. and Trounson, A.O. 2000. Effect of recombinant human chorionic gonadotrophins on human, bovine and murine oocyte meiosis, fertilisation and embryonic development *in vitro. Human Reproduction* 15:1140-1148.

Anderson, E. and Albertini, D.F. 1976. Gap junctions between the oocyte and companion follicle cells in the mammalian ovary. *Journal of Cell Biology* 71: 680-686.

Baird, D.T., Webb, R., Campbell, B.K., Harkness, L.M. and Gosden, R.G. 1999. Long-term ovarian function in sheep after ovariectomy and transplantation of autografts stored at -196°C. *Endocrinology* 140: 462-71.

Boyazoglu, S.E., Landin-Avarenga, F.C., Verini-Supplizi, A. and Squires, E.L. 2000. Use of fetuin to mature equine oocytes for *in vitro* fertilisation. *Theriogenology* 53: 449.

Chen, L., Russel, P.T. and Larsen, W.J. 1993. Functional significant cumulus expansion in the mouse. Role for the preovulatory synthesis of hyaluronic acid within the cumulus mass. *Molecular Reproduction and Development* 34: 87-93.

Cortvrindt, R., Smitz, J. and Van Steirteghem, A.C. 1996. *In vitro* maturation, fertilization and embryo development of immature oocytes from early preantral follicles from prepubertal mice in a simplified culture system. *Human Reproduction* 11: 2656-66.

Cushman, R.A., Wahl, C.M. and Fortune, J.E. 2002. Bovine cortical pieces grafted to chick embryonic membranes: a model for studies on the activation of primordial follicles. *Human Reproduction* 17: 48-54.

Danfour, M.A. 2001. Influence of the environment on mammalian oocyte development. *Ph.D thesis, University of Leeds.*

Danfour, M.A., Picton, H.M. and Gosden, R.G. 1999b. Impact of culture and cellular environment on the maturation potential of bovine oocytes *in vitro. Journal of Reproduction and Fertility Abstract Series* 24: 44.

Danfour, M.A., Coulthard, H., and Picton, H.M. 2002. Impact of oocyte maturation environment on bovine embryo development *in vitro. Reproduction Abstract Series* 28: 63.

Eppig, J.J. and O'Brien, M.J. 1996. Development *in vitro* of mouse oocytes from primordial follicles *Biology of Reproduction* 54: 197-207.

Gandhi, A.P., Lane, M., Gardner, D.K., and Krisher, R.L. 2000. A single medium supports development of bovine embryos throughout maturation, fertilisation and culture. *Human Reproduction* 15: 395-401.

Gutierrez, C.G., Ralph, J.H., Telfer, E., Wilmut, I. and Webb, R. 2000. Growth and antrum formation of bovine preantral follicles in long-term culture *in vitro. Biology of Reproduction* 62: 1332-1328.

Hovatta, O., Wright, C., Krausz, T., Hardy, K. and Winston, R.M.L. 1999. Human primordial, primary and secondary ovarian follicles in long-term culture: effect of partial isolation. *Human Reproduction* 14:2 519-2524.

Hyttel, P., Callesen, H. and Greve, T. 1997. Oocyte growth, capacitation and final maturation in cattle. *Theriogenology* 47: 23-32.

Kito, S. and Bavister, B.D. 1997. Maturation of hamster oocytes under chemically defined conditions and sperm penetration through the zona pellucida. *Zygote* 4: 199-210.

Kotsuji, F., Kubo, M. and Tominaga, A. 1994. Effect of interactions between granulosa and theca cells on meiotic arrest in bovine oocytes. *Journal of Reproduction and Fertility* 100: 151-156.

Leese, H.J., Donnay, I. and Thompson, J.G. 1998. Human assisted conception: a cautionary tale. Lessons from domestic animals. *Human Reproduction Update* 13: 184-202.

Lonergan, P., Monaghan, D., Rizos, D., Boland, M.P. and Gordon, I. 1994. Effect of follicle size on bovine oocyte quality and developmental competence following maturation fertilisation and culture *in vitro. Molecular Reproduction and Development* 37: 48-53.

Meyling, A., Hove, H. and Jensen, A.M. 1990. Epidemiology of bovine virus diarrhoea virus. *Revue Scientifique et Technique* 9: 75-93.

Mhawi, A.J., Kanka, J. and Motlik, J. 1991. Follicle and oocyte growth in early postnatal calves: cytochemical, autoradiographical and electrion microscopical studies. *Reproduction Nutrition and Development* 31: 115-126.

Newton, H., Picton, H.M. and Gosden, R.G. 1999. *In vitro* growth of oocyte-granulosa cell complexes isolated from cryopreserved ovine tissue. *Journal of Reproduction and Fertility* 115: 141-150.

Newton, H., and Illingworth, P. 2001. *In-vitro* growth of murine pre-antral follicles after isolation from cryopreserved ovarian tissue. *Human Reproduction* 16: 423-9.

Nugent, D., Newton, H., Gallivan, L., Gosden, R.G. 1998. Protective effect of vitamin E on ischaemia-reperfusion injury in ovarian grafts. *Journal of Reproduction and Fertility* 114: 341-46.

Numabe, T., Oikawa, T., Kikuchi, T. and Horiuchi, T. 2000. Production efficiency of japanese black calves by transfer of bovine embryo produced *in vitro. Theriogenology* 54: 1409-1420.

Obata, Y., Kono, T. and Hatada, I. 2002. maturation of mouse fetal germ cells *in vitro*. *Nature* 418: 497-498.

Oktay, K., Newton, H., Mullen, J., and Gosden, R.G. 1998. Development of human primordial follicles to antral stages in SCID/ hpg mice stimulated with follicle stimulating hormone. *Human Reproduction* 13: 1133-38 .

Picton, H.M. 2001. Activation of follicle development: the primordial follicle. *Theriogenology* 55: 1193-1210.

Picton, H.M. 2002. Oocyte maturation *in vitro*. *Current Opinions in Obstetrics and Gynaecology* 14: 295-302.

Picton, H.M., Danfour, M.A., Harris, S.E., Chambers, E.L. and Huntriss, J. 2003a. Growth and maturation of oocytes *in vitro*. *Reproduction Supplement 61*. 445-462.

Picton, H.M., Gosden, R.G. and Leibo, S.P. 2003b. Cryopreservation of oocytes and ovarian tissue. *Medical, social and ethical aspects of assisted reproduction, WHO Technical Bulletin*.WHO, Geneva, pp 142-151.

Picton, H.M., Briggs, D. and Gosden, R.G. 1998. The molecular basis of oocyte growth and development. *Molecular and Cellular Endocrinology* 145: 27-37.

Picton, H.M., Mkandla, A., Salha, O., Wynn, P. and Gosden, R.G. 1999. Initiation of human primordial follicle growth *in vitro* in ultra-thin slices of ovarian cortex. *Human Reproduction* 13: 0-020.

Picton, H.M., Kim, S.S. and Gosden, R.G. 2000. Cryopreservation of Gonadal Tissue and Cells. *British Medical Bulletin* 56: 603-615.

Schramm, R.D. and Bavister, B.D. 1994. Follicle-stimulating hormone priming of rhesus monkeys enhances meiotic and developmental competence of oocytes matured *in vitro*. *Biology of Reproduction* 51: 904-12.

Spears, N., Boland, N.I., Murray, A.A. and Gosden, R.G. 1994. Mouse oocytes derived from *in vitro* grown primary ovarian follicles are fertile. *Human Reproduction* 9: 527-32.

Thompson, J.G., Allen, N.W., McGowan, L.T., Bell, A.C.S., Lambert, M.G. and Tervit, H.R. 1998. Effect of delayed supplementation of fetal calf serum to culture medium on bovine embryo development in vitro and following transfer. *Theriogenology* 49: 1239-1249.

Thompson, J.G., Gardner, D.K., Pugh, P.A., McMillan, W.H. and Tervit, H.R. 1995. Lamb birth weight is affected by culture system utilized during *in vitro* pre-elongation development of ovine embryos. *Biology of Reproduction* 53: 1385-1391.

Trounson, A.O. 2000. Effect of recombinant human chorionic gonadotrophins on human, bovine and murine oocyte meiosis, fertilisation and embryonic development *in vitro*. *Human Reproduction* 15: 140-1148.

rleadfulltextnow.

Trounson, A.O., Pushett, D.A., MacClellan, L.J., Lewis, I. and Garner, D.K. 1994. Current status in IVM/IVF and embryo culture in human and farm animals. *Theriogenology* 41: 57-66.

Vanderhyden, B.C. and Armstrong, D.T. 1989. Role of cumulus cells and serum on the *in vitro* maturation, fertilisation and subsequent development of rat oocytes. *Biology of Reproduction* 40: 720-728.

van Wagtendonk-de Leeuw, A.M., Mullaart, E., de Roos, A.P., Merton, J.S., den Daas, J.H., Kemp, B. and de Ruigh, L. 2000. Effects of different reproductive techniques: AI, MOET or IVP on health and welfare of bovine offspring. *Theriogenology* 53: 575-597.

Walker, S.K., Hartwich, K.M. and Seamark, R.F. 1996. The production of unusually large offspring following embryo manipulation: conception and challeges. *Theriogenology* 45: 111-120.

Wandji, S-A., Srsen, V., Voss, AK., Eppig, J.J. and Fortune, J.E. 1996. Initiation *in vitro* of growth of bovine primordial follicles. *Biology of Reproduction* 55: 942-948.

Wynn, P., Picton, H.M., Krapez, J.A., Rutherford, A.J., Balen, A.H. and Gosden, R.G. 1998. FSH pre-treatment promotes the numbers of human oocytes reaching metaphase II by *in vitro* maturation. *Human Reproduction* 13: 3132-3138.

Young, L.E. and Fairburn, H.R. 2000. Improving the safety of embryo technologies: possible role of genomic imprinting. *Theriogenology* 53: 627-648.

15

The integration of cloning by nuclear transfer in the conservation of animal genetic resources

D.N. Wells
AgResearch, PB 3123, Hamilton, New Zealand

Abstract

Cloning mammals from somatic cells by nuclear transfer has the potential to assist with the preservation of genetic diversity. An increasing number of species have been successfully cloned by this approach; however, present methods are inefficient with few cloned embryos resulting in healthy offspring. In those livestock species that have already been cloned, it is clearly feasible to use cloning to preserve endangered breeds (e.g. the last surviving Enderby Island cow). The opportunity exists to recover oocytes from these cloned heifers and use frozen Enderby Island sperm from deceased bulls for in vitro fertilisation and thus, expand the genetic diversity of this breed. Where there exists an adequate understanding of the reproductive biology and embryology of the species concerned and adequate sources of females to supply both recipient oocytes and surrogates to gestate the pregnancies, intra-specific nuclear transfer and embryo transfer can be utilised. However, when these requirements cannot be met, as is common for most endangered species, cloning technology invariably involves the use of inter-species nuclear transfer and embryo transfer. Even in intra-specific cloning the source of oocyte for nuclear transfer is an important consideration. Typically, cloned animals are only genomic copies of the founder if they possess mitochondrial DNA which differs from the original animal. Different maternal lineages of oocytes both within and between breeds significantly affect cloning efficiency and livestock production characteristics. Cloning should not distract conservation efforts from encouraging the use of indigenous livestock breeds with traits of adaptation to local environments, the preservation of wildlife habitats or the use of other forms of assisted reproduction. Whilst it is often difficult to justify cloning in animal conservation at present, the appropriate cryo-preservation of tissues and cells from a wide selection of biodiversity is of paramount importance. This provides an insurance against further losses of genetic variation from dwindling populations, disease epidemics or even

223

possible extinction. It would also complement the gene banking of gametes or embryos and can be performed more easily and cheaply. Future cloning from preserved somatic cells can reintroduce lost genes back into the breeding pool. With greater appreciation of the heritable attributes of traditional livestock breeds there is the desire to identify superior animals within these local populations and the genetic loci involved. Through clonal family performance testing, nuclear transfer can aid the selection of desirable genotypes and then the production of larger numbers of embryos or animals for natural breeding to more widely disseminate the desirable traits. With the identification of alleles conferring desirable attributes, transgenesis could be utilised to both improve traditional and industrial livestock breeds. This further emphasizes the importance of preserving global farm animal genetic resources.

Introduction

The term 'clone' is used in a variety of contexts in biology. Here it is defined as a set of genetically identical animals. However, the stringency of this definition depends upon the methods used to produce the clones.

The simplest methods of producing cloned mammals involve the manipulation of very early embryos just a few days after fertilisation, often before the cells have differentiated (or specialised). These embryo manipulations mimic those processes that sometimes occur in nature. Thus, it is possible to either separate individual blastomeres (or embryonic cells) at the 2- to 4-cell-stages using finely-controlled needles and pipettes. By allowing each blastomere to form an individual embryo and transferring these to suitably synchronised recipient females, genetically identical twins or quadruplets may result. Here each individual blastomere retains its inherent developmental totipotency; that is, the ability to generate an entire organism from a single cell, similar to the zygote. Another common method to produce genetically identical twins involves the bisection of more advanced embryos, at the morula or blastocyst stages, into two equal halves. The sets of animals so-produced by these methods from single founder embryos are true clones, since they share the same nuclear and mitochrondrial DNA.

A more relevant method applicable for the conservation of animals is the technique of nuclear transfer, or nuclear cloning (described below). The basic methodology was first established in the 1950s in amphibians (Briggs and King, 1952) as an experimental technique to investigate the nuclear equivalence of differentiated cell types (reviewed in Gurdon, 1986). Significantly, these early studies demonstrated that the nuclei of some differentiated cell types from tadpole stages retained their totipotency following nuclear transfer. Most notably, however, fully mature clones derived from the cells of adult frogs were never produced, with development arresting at tadpole stages. Advances in embryo culture and manipulation methods in the 1980s enabled the production of the first cloned mammals following nuclear transfer; but only from undifferentiated blastomeres (Willadsen, 1986) or from the pluripotent inner cell mass of blastocyst-stage embryos (Smith and Wilmut, 1989; Collas and Barnes, 1994). Thus, the numbers of cloned animals that could be produced was limited and they could only be produced from early embryonic stages. A revolution occurred in the nuclear transfer field in the mid-1990s with the pioneering work of Drs Keith Campbell and Ian Wilmut from the Roslin Institute in Scotland. They demonstrated the ability to clone from more differentiated cells than those used previously (Campbell *et al.*, 1996). The donor cells could come from a variety of sources including the somatic cells of adult animals, as most famously illustrated by 'Dolly' the sheep (Wilmut *et al.*, 1997). These somatic cells obtained from a biopsy can be cultured *in vitro* and maintained as primary cell lines. Thus, there is access to very large numbers of cells, they can be conveniently cryo-preserved and thawed, in addition to being genetically modified *in vitro* prior to nuclear transfer (Schnieke *et al.*, 1997).

Cloning by somatic cell nuclear transfer

Firstly, there is the choice of donor cell. Whilst a range of donor cell types have successfully resulted in cloned offspring following nuclear transfer, the ideal somatic donor cell type with the greatest cloning efficiency has yet to be identified (Oback and Wells, 2002). It may be expected that adult stem cell populations may result in higher cloning efficiencies than terminally differentiated cell types.

Whilst there are many variations on the theme, the conventional nuclear transfer process comprises a sequence of six main steps, as illustrated in Figure 1 and summarised below:

1) Mature unfertilised oocytes arrested at the metaphase II stage of meiosis are enucleated. These oocytes may be obtained either a few hours following ovulation in the female animal or after *in vitro* maturation of oocytes obtained by aspirating immature oocytes from the follicles of abattoir-derived ovaries. The enucleation process involves manipulation of oocytes in the microscope with finely controlled micro-surgical instruments to physically aspirate the metaphase II chromosomes (incorporating the oocytes' nuclear DNA) and the extruded first polar body. Thus, the genetic material of the oocyte is removed, resulting in what is termed a cytoplast (a cell containing only cytoplasmic material).

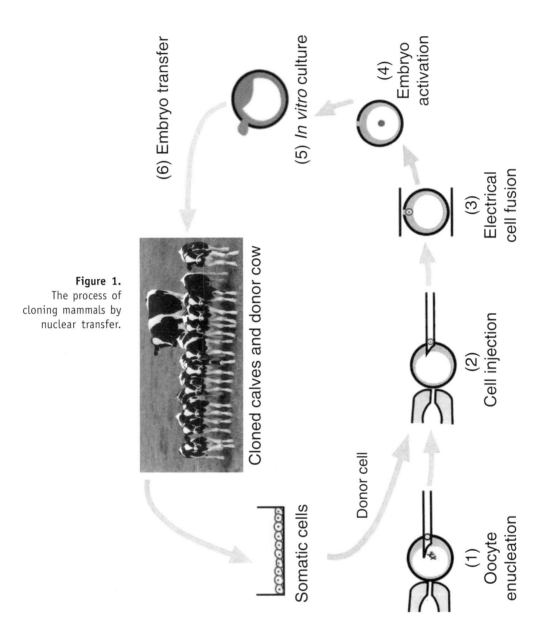

Figure 1.
The process of
cloning mammals by
nuclear transfer.

(1) Oocyte enucleation

(2) Cell injection

(3) Electrical cell fusion

(4) Embryo activation

(5) *In vitro* culture

(6) Embryo transfer

Donor cell

Somatic cells

Cloned calves and donor cow

2) Somatic donor cells are injected singularly underneath the outer *zona pellucida* (a protective membrane surrounding the oocyte and early embryo before implantation) and adjacent to the cytoplast membrane.

3) The cytoplast and donor cell are then fused together utilising pulses of direct current electricity. Following fusion, the genetic information contained within the nucleus of the donor cell therefore enters the cytoplast. This is the essence of the term 'nuclear transfer', whereby the genetic information from the oocyte is removed and is replaced with that from the donor cell. Interactions between the chromatin within the donor nucleus and factors present within the oocyte cytoplasm provide opportunity for epigenetic reprogramming and subsequent embryogenesis (see below).

4) The reconstructed 1-cell embryos are artificially activated using either specific chemical signals or direct current electrical pulses, in order to initiate embryonic development.

5) Following activation, the reconstructed embryos are typically cultured *in vitro* in a chemically-defined medium appropriate for the species.

6) In cattle, after seven days of culture, embryos that have developed into blastocysts of suitable quality (that is, embryos comprising around 150 cells) are transferred to the uteri of synchronised recipient females for development to term and the eventual birth of cloned offspring.

Current cloning efficiencies

An increasing number of species have been successfully cloned by somatic cell nuclear transfer (see Wilmut *et al.*, 2002). However, the present methods are inefficient with few cloned embryos resulting in healthy offspring. In the common farmed livestock species including sheep, cattle, goats and pigs, the cloning efficiency, as represented by that percentage of successfully reconstructed 1-cell embryos that develop in to viable cloned offspring, is currently in the range of <1-7% at best. To put these figures into context with methods of assisted sexual reproduction, the comparable efficiency of *in vitro* embryo production (IVP: comprising *in vitro* maturation, *in vitro* fertilisation and *in vitro* culture) in cattle, at least, is 18% viable calves from fertilised zygotes. That is, with present methods, cloning is at best one-third the efficiency of IVP in cattle. Both these reproductive methods compare rather unfavourably with artificial insemination which in the dairy industry is typically around 55%.

There are many factors which affect the efficiency of nuclear cloning. These include: the genotype of the donor animal; the type of donor cell with respect to its degree of differentiation; the length of time donor cells are cultured; the cell cycle stage of the donor cell and its coordination with that of the recipient cytoplast; the competency of the recipient oocyte for subsequent development; the detrimental effects of the various micro-manipulation methods used to reconstruct the embryos; the ability of the activation signals to faithfully induce the repetitive intracellular calcium oscillations to trigger normal embryogenesis; the species-specific culture media necessary to grow the embryo; the synchronicity with the recipient female at embryo transfer; and of course, the degree of reprogramming.

Reprogramming donor cells

One of the most critical aspects determining the success of nuclear cloning is epigenetic reprogramming. For normal development, the pattern of gene expression in the differentiated somatic cell must be completely reset to an embryonic state. Thus, the reconstructed embryo must proceed through the correct sequence of gene expression at the correct time, in the correct tissues and in the correct amounts for normal embryogenesis to occur. There is increasing evidence of errors in this intricate developmental process following nuclear transfer and these are considered to be the major cause of the low cloning efficiencies. Increased understanding of the molecular mechanisms involved in reprogramming and how this process may be further aided by treatments to remove the epigenetic constraints on the donor chromatin will lead to general improvements in cloning efficiency in the future. However, species- or breed-specific factors will still be important.

Whilst the majority of cloned embryos fail, there is evidence of complete reprogramming where up to 7% of reconstructed embryos can result in apparently normal, viable offspring. A range of studies have demonstrated that cloned animals can display normal behaviour, growth rates, reproduction, livestock production characteristics and lifespans (reviewed in Wells, 2003). However, the majority of clones do show inappropriate patterns of gene expression (Humpherys *et al.*, 1002; Rideout *et al.*, 2001; Wrenzycki *et al.*, 2001) and aberrant phenotypes. These epigenetic errors can manifest themselves at various stages of development from early embryos right through to adulthood. Furthermore, there is a wide spectrum of phenotypic outcomes ranging from those severe cases that are lethal to those that are more subtle and may not affect the phenotype of the animal.

Common clone-associated problems that are presumably the consequence of faulty reprogramming include abnormal placentation (Hill *et al.*, 2000), higher rates of pregnancy loss throughout gestation (Heyman *et al.*, 2002a), higher birthweights and higher rates of peri- and post-natal mortality (Heyman *et al.*, 2002b; Hill *et al.*, 1999). It appears that the incidence of these abnormalities is greater in sheep and cattle compared to pigs and goats (Wilmut, *et al.*, 2002). These issues raise animal welfare concerns that currently limit the acceptability and applicability of the technology.

Applications of cloning technology

There are a wide range of potential applications of nuclear transfer technology, including:

1) the multiplication of genetically superior animals in agriculture
2) production of transgenic animals for agriculture and medicine
3) animal research models
4) human cell based therapies
5) conservation of endangered livestock breeds and species

If the ethical costs associated with producing the few surviving healthy clones can be justified, some practical applications are feasible now. Others, however, will not be tolerated until the efficiencies improve markedly. The use of cloning for animal conservation is described below. Other applications of nuclear transfer technology are detailed elsewhere (see Wells 1999; 2003)

Conservation of endangered livestock breeds and species

There are a number of basic requirements that must be fulfilled to enable the cloning of any mammalian species. These include:

1) an understanding of the reproductive biology of the species concerned
2) adequate source of oocytes
3) suitable surrogate females to gestate the pregnancies
4) sufficient understanding of reprogramming
5) viable cells, or at least intact nuclei or chromosomes

The first three aspects are easily fulfilled in common farmed livestock species. This enables, for the cloning of endangered livestock breeds, intra-specific nuclear transfer and embryo transfer. This is in contrast to the situation with cloning endangered species which commonly will involve inter-species nuclear transfer and embryo transfer.

Cloning endangered breeds of livestock

AgResearch in association with the New Zealand Rare Breeds Conservation Society (NZRBCS) first exemplified this application of nuclear transfer technology by cloning the last surviving cow of the Enderby Island cattle breed (Wells *et al.*, 1998). The breed represented a feral herd adapted to the harsh sub-Antarctic conditions on Enderby Island which lies some 320km south of New Zealand. Cattle were first introduced over a century ago in a failed farming venture. Although the cow, affectionately named 'Lady', was the last living member of this breed, poor quality epididyimal sperm was cryo-preserved from 16 now-dead bulls. Before AgResearch became involved in the project in 1997, previous attempts of assisted reproduction (including artificial insemination and multiple ovulation and embryo transfer) in the three years after her capture from Enderby Island, had all failed to reproduce from Lady. AgResearch therefore initiated projects using more advanced reproductive technologies including: (1) ovum pick-up, *in vitro* embryo production and embryo transfer (OPU-IVP-ET) and (2) somatic cell cloning. From the follicular aspiration of Lady's ovaries, it was quickly appreciated that the quality of her oocytes was very poor. The reason was unknown and may have been due to either her age, genetic or environmental factors.

Priority was placed on preserving Lady's unique genetic resource by cyro-preserving samples of her somatic cells; specifically granulosa cells collected during the regular OPU sessions. After removal of the cumulus-oocyte complexes from the aspirated ovarian follicular fluid, the remaining granulosa cells were centrifuged, washed and seeded onto tissue culture plates to establish a primary cell line. Aliquots of these somatic cells were cryo-preserved and stored in liquid nitrogen. After thawing, these cells were used for nuclear transfer essentially as outlined in Fig. 1 (Wells, *et al.*, 1998). Recipient oocytes for nuclear transfer were obtained from the ovaries of commercially slaughtered cows from various local dairy and beef breeds. Likewise, recipient females used as surrogates were also sourced from alternate breeds.

This research culminated in the production of five cloned Enderby Island heifers derived from the nuclear DNA of Lady. Unfortunately two died at around two years of age. Nevertheless, this was the first step in maintaining the female genetics of this breed. Concurrent with the birth of the first clones, AgResearch was also successful with the birth of a purebred Enderby Island bull calf from OPU-IVP-ET using sperm from one of the dead bulls (W.H. Vivanco, unpublished; Figure 2). These Enderby Island cattle progeny have now reached sexual maturity and through the efforts of the NZRBCS, offspring have been produced from the natural mating of the cloned heifers to the Enderby bull. This represents an unusual genetic relationship as the bull is the 'son' of the Lady clones. Two of the three heifers have so far conceived with calves born in September and October 2002, respectively (www.rarebreeds.co.nz). Although this has been a significant milestone, it has not increased the genetic diversity of the breed. The opportunity exists however, to repeatedly harvest immature oocytes from the ovaries of the cloned heifers and to perform IVP with these oocytes and cryo-preserved sperm from the different Enderby bulls to more rapidly and extensively increase genetic diversity.

Figure 2.
(see text below figure).

Assisted reproductive technologies used to rescue the Enderby Island cattle breed. Granulosa cells recovered from Lady (centre), the last surviving animal of her breed, were used in nuclear transfer to produce cloned heifers (one of which is shown on the left). The bull calf (right) was produced by *in vitro* embryo production from an oocyte harvested from Lady following ovum-pick-up and *in vitro* fertilisation with sperm from a deceased Enderby Island bull.

In another example, the United States company Infigen Inc. in association with the Kelmscott Rare Breeds Foundation have successfully cloned the last sow of a particular lineage within the Gloucestershire Old Spot pig breed in North America (www.kelmscott.org).

As illustrated in the two examples above, it is relatively straightforward to use nuclear transfer in the conservation of breeds in livestock species that have already been cloned. This is primarily because of the adequate understanding of the reproductive biology of these species and the adequate supplies of females to provide recipient oocytes and surrogate females. However, it is important to note that in most situations the females will be sourced from alternative breeds within the species concerned. This point warrants reinvestigation of the cloning definition given in the introduction. Clones produced by nuclear transfer may not be completely genetically identical. In fact, nuclear clones are only genomic copies of the donor cells used for nuclear transfer if they possess mitochondrial DNA that differs from that of the original donor animal. Mitochondrial DNA is maternally inherited from the oocyte. Different maternal lineages both within and between breeds can have subtle differences in mitochondrial DNA that can influence livestock production traits (Schutz *et al.*, 1994; Mannen *et al.*, 1998). So, nuclear clones may not be true clones (unlike those from embryo splitting) and may not faithfully represent the full genotype (both nuclear and mitochrondrial) of the breed. Furthermore, somatic cell cloned animals may have mitochondrial DNA heteroplasmy whereby in each cell the majority of mitochondrial DNA is derived from the recipient oocyte with only a few percent from the donor cell in accordance with the ratio of mitochondrial DNA present in the two cells at the time of embryo reconstruction (Steinborn *et al.*, 2002). The developmental consequences of heteroplasmy and nuclear-mitochondrial DNA mis-matches following cloning are not fully appreciated at present.

Because of its maternal inheritance, the transmission of mitochondrial DNA from the recipient oocyte used in nuclear transfer, following future breeding, occurs only in female clones. In some cloned heteroplasmic females there may be rare instances where the original donor cell mitochondrial DNA genotype will segregate in the germline so that some oocytes and resulting offspring are in fact homoplasmic for the original donor cell mitochondrial genotype. Additional strategies such as injection of somatic mitochondrial DNA into the reconstructed embryo (Shitara *et al.*, 2000) may increase the contribution of mitochondrial DNA from the original donor animal and so increase the chance of recovering the original mitochondrial genotype upon subsequent breeding. Germline transmission from male clones should not corrupt the genotype of the breed following sexual reproduction, since paternal mitochondrial DNA is selectively destroyed within a few cleavage divisions after fertilisation (Sutovsky *et al.*, 2000).

Not all situations will involve inter-breed nuclear transfer and these confounding mitochondrial effects. In some applications for instance it may be possible to set aside a few surplus females from within the breed to recover oocytes via OPU, in order to perform intra-breed cloning to reintroduce animals back into the gene pool. These issues of mitochondrial DNA heteroplasmy and nuclear-oocyte cytoplasmic interactions become even more critical in efforts to clone endangered exotic species where inter-species cloning is typically required.

Cloning endangered mammalian species

Because the number of females is limiting once a species becomes endangered, a closely related species that is not itself threatened is typically required in order to supply sufficient qualities of compatible oocytes and surrogates for nuclear transfer. Moreover, there is often less knowledge pertaining to the reproductive biology of many exotic species that limits the applicability of cloning technology. There are two key issues for inter-species cloning: (1) the compatibility between the nucleus and the cytoplasm of the two species concerned following nuclear transfer and (2) the compatibility of placentation between the cloned conceptus and recipient uterus following embryo transfer.

The nuclear-cytoplasmic interactions necessary to ensure complete reprogramming and promote correct embryogenesis are initially regulated by the maternal factors present within the oocyte cytoplasm, but must ultimately come under the control of the donor nucleus from the species being cloned. Incompatibility between the nuclear and mitochondrial genomes can influence the development of the resulting organism and its general fitness (Nagao *et al.*, 1998). The more distantly related the two species are in evolutionary terms, the greater the likelihood of perturbed or failed development (Gurdon, 1986). Thus, it is important to also consider the cyro-preservation of oocytes (once the technology is more robust) and female embryos to supply the maternal cytoplasmic factors necessary for embryogenesis including the correct mitochondrial DNA for the species. Alternatively, ovarian tissue could be cryo-preserved offering at least two possibilities after thawing: (1) *in vitro* oocyte maturation from primordial ovarian follicles (Cortvrindt and Smitz, 2001; Obata *et al.*, 2002) or (2) xenografting in another species, for example in an immuno-compromised host animal (Liu *et al.*, 2001; Wolvekamp *et al.*, 2001). Having access to oocytes from the same species in the future enables intra-species nuclear transfer which is likely to be met with greater success.

There is a greater likelihood of success with inter-species cloning amongst those species that naturally hybridise or where inter-species embryo transfer has generated viable offspring. These situations indicate compatibility in nuclear-cytoplasmic interactions and in placentation between the two species concerned. There are a number of examples between members of the cervid, bovid and equid families, such as: hybrids produced between Sambar and Red deer (Muir *et al.*, 1997); *Bos gaurus* embryos transferred into domestic cattle recipients (Stover *et al* 1981; Hammer *et al.*, 2001); and Grant's zebra embryos transferred into domestic mares (Summers *et al.*, 1987). However, the placenta of the cloned fetus will likely be incompatible with the reproductive tract of the surrogate female in even closely-related species in terms of anatomy, physiology, immunology and endocrinology and simply fail to develop. Interestingly, the occurrence of placental anomalies is greater with some inter-species embryo transfers as seen with the incidence of hydrops following transfer of *Bos gaurus* embryos into *Bos taurus* recipients. The response of the recipient female to reject the xenografic conceptus can be ameliorated by reconstituting blastocysts so that the inner cell mass is of the foreign cloned species but the trophectoderm is of the same species as the recipient as this appears to provide the necessary immunological barrier (Anderson, 1988).

Experiments that have examined the viability of inter-species nuclear transfer embryos show that development to blastocyst stages proceeds relatively easily. Importantly, the time course for development is in accordance with the species of the donor nucleus indicating its control over early embryogenesis. Bovine oocyte cytoplasm has been shown to partially reprogram adult skin fibroblasts from a small range of mammalian species (sheep, pig, monkey and rat; Dominko et al., 1999). Although blastocysts resulted, normal pregnancies were not initiated. Similarly, only early pregnancies were initiated with Argali sheep fibroblasts (*Ovis ammon*) reconstructed with domestic sheep cytoplasts (White et al., 1999). Rather surprisingly, early implantation was reported following inter-species cloning in the giant panda (*Alluropoda melanoleuca*) using rabbit cytoplasts and domestic cat recipients (Chen et al., 2002); although these may not be the optimal combinations.

In the bovid family there is some indication that the more divergent the two species are, the more difficult inter-species cloning becomes. Using *Bos taurus* cytoplasts and recipients, development with donor nuclei from a small range of species has been compiled in Table 1. Predictably, viable cloned offspring were produced with donor nuclei from the *Bos indicus* sub-species (Meirelles et al., 2001) since hybridisation and embryo transfer occur readily. Likewise, a term calf was delivered using *Bos gaurus* nuclei (Lanza et al., 2000; Vogel, 2001) and a viable calf derived from *Bos javanicus* (Bali cattle). However, the more distantly related *Bubalus depressicornis* (a small water buffalo species) has apparently not so far resulted in pregnancies.

Table 1.
Successful inter-species cloning depends on the evolutionary relatedness of the species involved.

Donor species	Cytoplast & recipient species	Viable offspring	References
Bos indicus	Bos taurus	yes	Meirelles et al., 2001
Bos gaurus	Bos taurus	partial success[1]	Lanza et al., 2000 Vogel, 2001
Bos javanicus (Banteng)	Bos taurus	yes	Advanced Cell Technology (www.advancedcell.com)
Bubalus depressicornis (Anoa)	Bos taurus	no	Advanced Cell Technology (www.agbiotechnet.com)

[1] calf died two days after birth

The first report of long-term viable offspring following inter-species cloning was that of a mouflon (*Ovis orientalis musimon*) using *Ovis aries* cytoplasts and recipients (Loi *et al.*, 2001). This study was also significant because intact nuclei from non-viable cells were injected some 36 hours after death of the donor animal. Whilst this case proved successful and such emergency cases do arise, it highlights the prudence of preserving appropriate cells and tissue from living animals, because after death DNA will eventually degrade. It is for this reason that the resurrection of long extinct species by cloning as proposed for the woolly mammoth frozen in the Siberian permafrost (Stone, 1999), or the Tasmanian tiger from a century old pup fixed in alcohol in Australia (Nolch, 1999) using fragments of isolated DNA is the stuff of science fiction.

Cryo-banking genetic resources

For endangered livestock breeds and exotic species, somatic cells from a wide selection of males and females should be cryo-preserved as an 'insurance policy'. This would provide protection against possible further losses of genetic diversity in dwindling populations. It also provides insurance against catastrophes such as with disease epidemics, or even possible extinction with regionally adapted or localised breeds. It may also be appropriate to cryo-preserve samples of ovarian tissue from females to serve as potential sources of oocytes for nuclear transfer in the future. These strategies would complement the cryo-banking of sperm and embryos and could be performed more easily and cheaply; although the costs of cloning animals now and in the future will remain higher than artificial insemination and embryo transfer. Small tissue biopsies from living animals (or recently deceased animals in emergency situations) are straightforward to obtain and can be cryo-preserved directly or a primary cell line established and cells cryo-preserved. In practice, skin tissue and dermal fibroblasts will prove satisfactory and easily accessible. While the optimal cell types for cloning have yet to be identified, skin cells have been successful in a range of species. Cryo-preservation methods are fairly standard and routine with tissue and cells across the mammalian order, in contrast to the often specific freezing protocols required for gametes and embryos from different species and even individuals. As with the storage of other biological material, appropriate sanitary procedures need to be implemented to minimise potential biosecurity risks. To reduce maintenance costs associated with frozen storage in liquid nitrogen, it is feasible to consider freeze-drying cells to provide a source of intact nuclei for cloning that relies upon nuclear injection techniques rather than cellular fusion with live cells.

Recovering lost genetic diversity

Future advances in cloning and other associated technologies may make it possible to re-introduce animals (and their lost genes) from cryo-preserved cells back into the population. In this way it should be possible to at least maintain genetic diversity at present levels. This could be especially important for those animals that died before they passed on their genes to an adequate number of offspring. If an animal only produces one offspring in its lifetime, half of its contribution to the genetic diversity of the population is lost. Cloning could be used to effectively recover this by resurrecting animals from the cryo-preserved cells. There is confidence in using cloned animals as a conduit for genetic continuity in that the pregnancy and health complications that may have been present in some clones do not appear to be transmitted to subsequent offspring following sexual reproduction. This implies that any abnormal phenotypes present in surviving clones are epigenetic in nature and appear to be corrected during gametogenesis.

In situations where more conventional conservation efforts successfully increase the numbers of a once endangered livestock breed or exotic species in the future, some surplus females may be set aside to collect oocytes and be used as surrogates in cloning programmes with frozen somatic cells to reintroduce animals and increase the genetic diversity. Such intra-breed or intra-species nuclear transfer and embryo transfer would be more feasible and result in offspring with correct and compatible nuclear and mitochondrial genomes.

Animal selection and breeding programmes inevitably result in losses of genetic diversity. Cryo-preservation of tissue from animals provides a cheap way of conserving this variation, but suffers the common problem of identifying which (currently unfavourable) animals (based on the present selection criteria) to store material from. Intra-breed cloning could be easily used to reintroduce these animals back into the population in the future, if analyses of the tissue samples identifies alleles that are favoured for some new selection criterion.

An intriguing opportunity exists in the meat industry in conventional livestock production with the ability to clone animals from carcasses that have superior meat quality characteristics that represent heritable traits. At AgResearch we have exemplified this by cloning a set of bulls from a steer that following slaughter was assessed as having superior meat characteristics and genotype. Thus, these valuable genetics were recovered and the entire cloned males can now be used for breeding purposes.

Utilisation of genetic resources in transgenics

In addition to utilising intact genomes for cloning, particular alleles within the diversity of livestock breeds can be harnessed. Ultimately, future knowledge of the allelic variation within genes that are associated with favourable traits will allow direct introduction, or replacement by homologous recombination, of the desired alleles into the animal via transgenesis. Desirable alleles that confer favourable attributes such as pest and disease resistance or milk production traits for instance, can be isolated and introduced into the genome of another breed (or species) via genetic modification of somatic cells *in vitro*. Nuclear transfer with these modified cells can then generate cloned-transgenic offspring (Brophy *et al.*, 2003). Whilst it is acknowledged that safety and public acceptance aspects of this technology in food-producing animals need to be addressed, it is a direct way of improving a specific trait whilst retaining all the other beneficial attributes of the animal without the time and compromising effects of gene introgression through cross-breeding and subsequent back-crossing. However, it is important to appreciate that a particular allele may not have the same phenotypic effect in a different genetic background. This transgenic technology could be used to improve both traditional and industrial livestock breeds. An interesting test case for acceptability might be the introduction via transgenesis of alleles associated with scrapie resistance into a sheep breed where all individuals only posses the susceptible genotype.

Summary

Cloning should not distract from *in situ* conservation efforts. The sustainable use of traditional livestock breeds producing niche agricultural products, or with traits of hardiness adapted to particular local environmental conditions, along with the preservation of habitats and reduced hunting pressure are all of far greater importance in the conservation of animal genetic resources. Likewise, the use of other forms of assisted sexual reproduction (such as artificial insemination, multiple ovulation and embryo transfer, and *in vitro* embryo production) is more relevant than cloning. The *ex situ* cryo-preservation of genetic resources provides a critical insurance policy against possible future losses of diversity and somatic cells should be included along with gametes and embryos. However, cloning can be integrated into future breeding strategies to recover genetic diversity by performing nuclear transfer with preserved somatic cells to reintroduce lost genes back into the breeding population. This is probably the most effective use of cloning technology in animal conservation.

References

Anderson, G.B. 1988. Interspecific Pregnancy: barriers and prospects. *Biology of Reproduction* 38: 1-15.
Briggs, R. and King, T.J. 1952. Transplantation of living nuclei from

blastula cells into enucleated frogs' eggs. *Proceedings of the National Academy of Science USA.* 38: 455-463.

Brophy, B., Smolenski, G., Wheeler, T., Wells, D., L'Huillier, P. and Laible, G. 2003. Cloned transgenic cattle produce milk with higher levels of beta-casein and kappa-casein. *Nature Biotechnology* 21: 157-162.

Campbell, K.H., McWhir, J., Ritchie, W.A. and Wilmut, I. 1996. Sheep cloned by nuclear transfer from a cultured cell line. *Nature* 380: 64-66.

Chen, D.Y., Wen, D.C., Zhang, Y.P., Sun, Q.Y., Han, Z.M., Liu, Z.H., Shi, P., Li, J.S., Xiangyu, J.G., Lian, L., Kou, Z.H., Wu, Y.Q., Chen, Y.C., Wang, P.Y. and Zhang, H.M. 2002. Interspecies implantation and mitochondria fate of panda-rabbit cloned embryos. *Biology of Reproduction* 67: 637-642.

Collas, P. and Barnes, F.L. 1994. Nuclear transplantation by microinjection of inner cell mass and granulosa cell nuclei. *Molecular Reproduction and Development* 38: 264-267.

Cortvrindt, R. and Smitz, J. 2001. In vitro follicle growth: achievements in mammalian species. *Reproduction in Domestic Animals* 36: 3-9.

Dominko, T., Mitalipova, M., Haley, B., Beyhan, Z., Memili, E., McKusick, B. and First, N.L. 1999. Bovine oocyte cytoplasm supports development of embryos produced by nuclear transfer of somatic cell nuclei from various mammalian species. *Biology of Reproduction* 60: 1496-1502.

Gurdon, J.B. 1986. Nuclear transplantation in eggs and oocytes. *Journal of Cell Science,* Supplement 4: 287-318.

Hammer, C.J., Tyler, H.D., Loskutoff, N.M., Armstrong, D.L., Funk, D.J., Lindsey, B.R. and Simmons, L.G. 2001. Compromised development of calves (Bos gaurus) derived from in vitro-generated embryos and transferred interspecifically into domestic cattle (Bos taurus). *Theriogenology* 55: 1447-1455.

Heyman, Y., Chavatte-Palmer, P., LeBourhis, D., Camous, S., Vignon, X. and Renard, J.P. 2002a. Frequency and occurrence of late-gestation losses from cattle cloned embryos. *Biology of Reproduction* 66: 6-13.

Heyman, Y., Zhou, Q., Lebourhis, D., Chavatte-Palmer, P., Renard, J.P. and Vignon, X. 2002b. Novel approaches and hurdles to somatic cloning in cattle. *Cloning Stem Cells* 4: 47-55.

Hill, J.R., Burghardt, R.C., Jones, K., Long, C.R., Looney, C.R., Shin, T., Spencer, T.E., Thompson, J.A., Winger, Q.A. and Westhusin, M.E. 2000. Evidence for placental abnormality as the major cause of mortality in first-trimester somatic cell cloned bovine fetuses. *Biology of Reproduction* 63: 1787-1794.

Hill, J.R., Roussel, A.J., Cibelli, J.B., Edwards, J.F., Hooper, N.L., Miller, M.W., Thompson, J.A., Looney, C.R., Westhusin, M.E., Robl, J.M. and Stice, S.L. 1999. Clinical and pathologic features of cloned transgenic calves and fetuses (13 case studies). *Theriogenology* 51: 1451-1465.

Lanza, R.P., Cibelli, J., Diaz, F., Moraes, C., Farin, P.W., Farin C.E., Hammer, C.J., West, M.D. and Damiani, P. 2000. Cloning of an endangered species (*Bos gaurus*) using inter-species nuclear transfer. *Cloning* 2: 79-90.

Liu, J., Van der Elst, J., Van den Broecke, R. and Dhont, M. 2001. Live offspring by in vitro fertilization of oocytes from cryopreserved primordial mouse follicles after sequential in vivo transplantation and in vitro maturation. *Biology of Reproduction* 64: 171-178.

Loi, P., Ptak, G., Barboni, B., Fulka, J., Jr., Cappai, P. and Clinton, M. 2001. Genetic rescue of an endangered mammal by cross-species nuclear transfer using post-mortem somatic cells. *Nature Biotechnology* 19: 962-964.

Mannen, H., Kojima, T., Oyama, K., Mukai, F., Ishida, T. and Tsuji, S. 1998. Effect of mitochondrial DNA variation on carcass traits of Japanese Black cattle. *Journal of Animal Science* 76: 36-41.

Meirelles, F.V., Bordignon, V., Watanabe, Y., Watanabe, M., Dayan, A., Lobo, R.B., Garcia, J.M. and Smith, L.C. 2001. Complete replacement of the mitochondrial genotype in a Bos indicus calf reconstructed by nuclear transfer to a Bos taurus oocyte. *Genetics* 158: 351-356.

Muir, P.D., Semiadi, G., Asher, G.W., Broad, T.E., Tate, M.L. and Barry, T.N. 1997. Sambar deer (Cervus unicolor) x red deer (C. elaphus) interspecies hybrids. *Journal of Heredity* 88: 366-372.

Nagao, Y., Totsuka, Y., Atomi, Y., Kaneda, H., Lindahl, K.F., Imai, H. and Yonekawa, H. 1998. Decreased physical performance of congenic mice with mismatch between the nuclear and the mitochondrial genome. *Genes and Genetic Systems* 73: 21-27.

Nolch, G. 1999. Back from the dead. *Australasian Science* 20: 28-29.

Oback, B. and Wells, D. 2002. Donor cells for nuclear cloning: many are called, but few are chosen. *Cloning Stem Cells* 4: 147-168.

Obata, Y., Kono, T. and Hatada, I. 2002. Gene silencing: maturation of mouse fetal germ cells in vitro. *Nature* 418: 497.

Rideout, W.M., 3rd, Eggan, K. and Jaenisch, R. 2001. Nuclear cloning and epigenetic reprogramming of the genome. *Science* 293: 1093-1098.

Schnieke, A.E., Kind, A.J., Ritchie, W.A., Mycock, K., Scott, A.R., Ritchie, M., Wilmut, I., Colman, A. and Campbell, K.H. 1997. Human factor IX transgenic sheep produced by transfer of nuclei from transfected fetal fibroblasts. *Science* 278: 2130-2133.

Schutz, M.M., Freeman, A.E., Lindberg, G.L., Koehler, C.M. and Beitz, D.C. 1994. The effect of mitochondrial DNA on milk production and health of dairy cattle *Livestock Production Science* 37: 283-295.

Shitara, H., Kaneda, H., Sato, A., Inoue, K., Ogura, A., Yonekawa, H. and Hayashi, J.I. 2000. Selective and continuous elimination of mitochondria microinjected into mouse eggs from spermatids, but not from liver cells, occurs throughout embryogenesis. *Genetics* 156: 1277-1284.

Smith, L.C. and Wilmut, I. 1989. Influence of nuclear and cytoplasmic activity on the development in vivo of sheep embryos after nuclear transplantation. *Biology of Reproduction* 40: 1027-1035.

Steinborn, R., Schinogl, P., Wells, D.N., Bergthaler, A., Muller, M. and Brem, G. 2002. Coexistence of Bos taurus and B. indicus mitochondrial DNAs in nuclear transfer-derived somatic cattle clones. *Genetics* 162: 823-829.

Stone, R. 1999. Siberian mammoth find raises hopes, questions. *Science* 286: 876-877.

Stover, J., Evans, J. and Dolensek E.P. 1981. Inter-species embryo transfer from gaur to domestic Holstein. *Proceedings of the American Association of Zoo Veterinarians*, Seattle, Washington, USA. pp. 122-123.

Summers, P.M., Shephard, A.M., Hodges, J.K., Kydd, J., Boyle, M.S. and Allen, W.R. 1987. Successful transfer of the embryos of Przewalski's horses (*Equus przewalskii*) and Grant's zebra (*E. burchelli*) to domestic mares (*E. caballus*). *Journal of Reproduction and Fertility* 80: 13-20.

Sutovsky, P., Moreno, R.D., Ramalho-Santos, J., Dominko, T., Simerly, C. and Schatten, G. 2000. Ubiquitinated sperm mitochondria, selective proteolysis, and the regulation of mitochondrial inheritance in mammalian embryos. *Biology of Reproduction* 63: 582-590.

Vogel, G. 2001. Endangered species. Cloned gaur a short-lived success. *Science* 291: 409.

Wells, D. 1999. Animal cloning provides many new opportunities for livestock production and biomedicine in the future. *Agricultural Science* 12 (3): 22-27.

Wells, D.N. 2003. Cloning in livestock agriculture. In: *Reproduction in Domestic Ruminants V*, (*Reproduction* Supplement 61).Edited by B.K. Campbell, R. Webb, H. Dobson and C. Doberska. Society for Reproduction and Fertility, Cambridge, UK. pp. 131-150.

Wells, D.N., Misica, P.M., Tervit, H.R. and Vivanco, W.H. 1998. Adult somatic cell nuclear transfer is used to preserve the last surviving cow of the Enderby Island cattle breed. *Reproduction Fertility and Development* 10: 369-378.

White, K.L., Bunch, T.D., Mitalipov, S. and Reed, W.A. 1999. Establishment of pregnancy after the transfer of nuclear transfer embryos produced from the fusion of Argali (*Ovis ammon*) nuclei into domestic sheep (*Ovis aries*) enucleated oocytes. *Cloning* 1: 47-54.

Willadsen, S.M. 1986. Nuclear transplantation in sheep embryos. *Nature* 320: 63-65.

Wilmut, I., Beaujean, N., De Sousa, P.A., Dinnyes, A., King, T.J.,

Paterson, L.A., Wells, D.N. and Young, L.E. 2002. Somatic cell nuclear transfer. *Nature* 419: 583-587.

Wilmut, I., Schnieke, A.E., McWhir, J., Kind, A.J. and Campbell, K.H. 1997. Viable offspring derived from fetal and adult mammalian cells. *Nature* 385: 810-813.

Wolvekamp, M.C., Cleary, M.L., Cox, S.L., Shaw, J.M., Jenkin, G. and Trounson, A.O. 2001. Follicular development in cryopreserved Common Wombat ovarian tissue xenografted to Nude rats. *Animal Reproduction Science* 65: 135-147.

Wrenzycki, C., Wells, D., Herrmann, D., Miller, A., Oliver, J., Tervit, R. and Niemann, H. 2001. Nuclear transfer protocol affects messenger RNA expression patterns in cloned bovine blastocysts. *Biology of Reproduction* 65: 309-317.

16

Biosecurity strategies for conservation of farm animal genetic resources

A.E. Wrathall and H.A. Simmons
Veterinary Laboratories Agency, Weybridge, Addlestone, Surrey, KT15 3NB, UK

Abstract

The foot-and-mouth disease (FMD) epidemic in the U.K. in 2001 highlighted the threat of infectious diseases to rare and valuable livestock and stimulated a renewed interest in biosecurity. Not all diseases resemble FMD, however; transmission routes and pathological effects vary greatly, so biosecurity strategies must take this into account. Realism is also needed as to which diseases to exclude and which will have to be tolerated. The aim should be to minimise disease generally and to exclude those diseases that threaten existence of the livestock, or preclude their national or international movement. Achieving this requires a team effort, bearing in mind the livestock species involved, the farming system ('open' or 'closed') and the premises. Effective biosecurity demands that practically every aspect of farm life is controlled, including movements of people, vehicles, equipment, food, manure, animal carcasses and wildlife. Above all, biosecurity strategies must cover the disease risks associated with moving the livestock themselves, and this will require quarantine if adult or juvenile animals are imported into the herd or flock. Reproductive technologies such as artificial insemination and embryo transfer offer much safer ways for getting new genetic materials into herds/flocks for breeding than bringing in live animals. Embryo transfer is especially safe when the sanitary protocols promoted by the International Embryo Transfer Society (IETS) and advocated by the Office International des Epizooties (OIE: the 'World Organisation for Animal Health') are used. It can also allow the full genetic complement to be salvaged from infected animals. Cryobanking of genetic materials, especially embryos, is another valuable biosecurity strategy because it enables storage for contingencies such as epidemic disease or other catastrophes.

Introduction

A Cumbrian farmer, in a letter to his local newspaper, wrote 'My farm was taken out in April 2001 with foot-and-mouth disease (FMD). There had been no movements of livestock onto the holding since September 2000, yet seven months later FMD landed in one of my fields next to the main road. I commend a sensible approach to biosecurity on farm (such as disinfectant spraying vehicle tyres/wheel arches, washing out transport vehicles, human foot-dip baths, etc.) which any sane farmer would implement, coupled with similar biosecurity at auction marts, abattoirs, shows and other gatherings of livestock. Where did FMD appear from on my farm? Via the wind, wild birds or wild animals, all of which by-passed the biosecurity we had in place, or was it spread by humans as they drove along the road in their vehicles?' While comments and questions such as these are typical of many since the 2001 epidemic in U.K., it is not our intention here to discuss details of FMD transmission, or the pros and cons of biosecurity measures used during or since the epidemic.

'Open' and 'closed' herds and flocks

For the purpose of this paper 'farm animal genetic resources' include not only groups of rare breeds of livestock, but also other unique populations such as performance tested bull, boar and ram studs, commercially owned 'genetic nucleus' pig herds, and the DEFRA-owned flock of scrapie-free sheep imported from New Zealand. The latter contains sheep of most genotypes conferring susceptibility or resistance to scrapie, and its purpose is to supply healthy genotyped animals to outside establishments for research on scrapie and other transmissible spongiform encephalopathies (TSEs). Typically, rare livestock breeds are kept by hobby farmers in small herds/flocks on smallholdings, or as minority groups within herds/flocks of commercial livestock. From a biosecurity perspective most of these typical groups of rare breeds are 'open', i.e. no attempt is made to restrict entry of further animals, and only limited precautions are taken to avoid ingress of infectious diseases. The DEFRA flock of scrapie-free sheep, on the other hand, is closed to all but occasional inputs of sheep from New Zealand, and biosecurity measures to avoid entry of TSEs and other infectious diseases, particularly via animals or animal products, are strictly enforced. Obviously in some situations there is a 'semi-closed' policy and this, for example, applies to genetic nucleus pig herds for which bringing in 'new blood' for breeding is essential, although the type and disease status of the brought-in animals (or semen) has to be controlled. Just as biosecurity measures are important to prevent ingress of diseases into individual herds/flocks, the same is true for national livestock populations which are unique and valuable genetic resources. The importance of this was illustrated by the 2001 FMD epidemic when, as a probable consequence of illegal meat imports, the disease arrived in the U.K. and caused losses to agricultural and other industries amounting to approximately £12 billion (Anon. 2001). In some respects the national livestock population resembles a semi-closed herd/flock because not only does its disease status differ from that in other countries but importations of animals and animal products from other countries are restricted to a greater or lesser extent by biosecurity measures applied by the national veterinary authorities.

Disease status specification

All herds/flocks, whether owned by individual farmers, or nationally by countries, are free from at least some infectious diseases, and could be termed 'specific pathogen-free' (SPF). Most countries, and ideally owners too, should try to ascertain which specific pathogens are present and which are absent as they evolve strategies to remove or exclude those infections they consider important. For many years 'genetic nucleus' pig herds have made efforts to ensure absence of troublesome infections which undermine the rate of genetic selection and performance testing. Although initially expensive, creation of a primary SPF or 'minimal disease' (MD) status in such herds by hysterectomy and piglet removal, then artificially rearing the piglets to maturity on isolated premises is very effective for controlling conditions such as pneumonia and dysentery. Before pursuing a MD herd status, however, owners must ask themselves the question; 'can it be maintained?' Factors such as location and limited finance often preclude effective biosecurity strategies over the long term, so farmers and others should be realistic. Getting new blood into MD herds without introducing diseases from the general population is especially challenging. Pigs from other same-status MD herds ought to be safe, but their freedom from all infections, known and unknown, cannot be guaranteed. Neither can it be guaranteed that further batches of hysterectomy-derived offspring would not carry infections which had crossed the placenta prior to hysterectomy. Artificial insemination (AI), although relatively safe, does carry some risk, as was seen in the Netherlands in 1997-98 when classical swine fever virus was spread by AI with disastrous consequences (Hennecken et al., 2000). Another example occurred in our own Laboratory MD pig herd when, having inseminated some gilts with semen from an SPF boar stud, porcine parvovirus and Talfan virus (a pathogenic enterovirus) were inadvertently brought in and caused a major outbreak of foetal mortality (Parker et al., 1981). The MD pig herd illustrates in microcosm the biosecurity challenges facing livestock-producing countries as well as individual herd and flock owners with genetically valuable livestock.

Dealing with different kinds of infectious disease

As shown in Figure 1, characteristics of different infectious diseases vary greatly. Some, like FMD and classical swine fever, are highly contagious with rapid onset of clinical manifestations, whereas others progress slowly, taking weeks, months or even years before clinical signs are seen. With yet other diseases recrudescence of the pathogen occurs periodically, giving rise to recurrent bouts of illness. In a few instances some of the infected animals show no clinical illness at all, but nevertheless shed and transmit the pathogen to other animals. Obviously, the consequences that arise if genetically valuable animals are exposed to a disease depends very much on which disease is involved, its pathological effects and its perceived significance with regard to international trade. In its International Animal Health Code (the 'Code') (OIE, 2002) the Office International des Epizooties (also called World Organisation for Animal Health) lists diseases into three groups according to their significance for international trade, human health and animal welfare. List A includes the epidemic diseases like foot-and mouth-disease (FMD), classical swine fever and rinderpest which have massive impacts on trade and welfare, and which are the responsibility of national veterinary authorities to control; List B includes diseases such as brucellosis, tuberculosis (TB), scrapie and bovine spongiform encephalopathy (BSE), many of which have major but often insidious effects on productivity, and sometimes cause serious human health problems, so responsibility for their control falls on both the national authorities and the local farmer(s). Finally there are unlisted diseases such as bovine viral diarrhoea (BVD), streptococcal mastitis, and enzootic pneumonia of pigs with (as yet) little or no significant impact on human health or international trade, and which the farmers are usually expected to control themselves.

Aspects of disease transmission

To maintain freedom from infectious diseases requires knowledge of how infections are transmitted; it makes little sense to exclude or eradicate a specific infection if it may be brought back into the herd/flock very easily. Some pathogens, such as protozoa and parasitic worm eggs can stay alive in the environment for many years, and likewise bacteria such as *Clostridia* and the anthrax bacillus which may exist as spores. Certain viruses, including those causing FMD and Aujeszky's disease, can be airborne from herd to herd (Donaldson *et al.*, 2001; Gillespie *et al.*, 1996), whilst others such as bluetongue virus and porcine reproductive and respiratory syndrome virus may be carried by midges or mosquitoes (Mellor, 1994; Otake *et al.*, 2002). Mechanical transmission by flies and birds and is another possibility. Realistically,

Figure 1.
Illustration of some
types of infectious
disease.

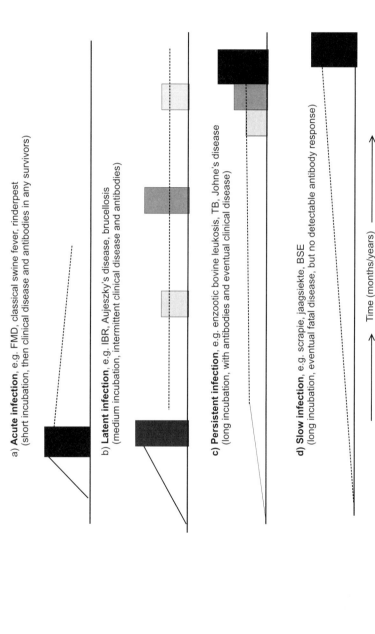

a) **Acute infection**, e.g. FMD, classical swine fever, rinderpest
(short incubation, then clinical disease and antibodies in any survivors)

b) **Latent infection**, e.g. IBR, Aujeszky's disease, brucellosis
(medium incubation, intermittent clinical disease and antibodies)

c) **Persistent infection**, e.g. enzootic bovine leukosis, TB, Johne's disease
(long incubation, with antibodies and eventual clinical disease)

d) **Slow infection**, e.g. scrapie, jaagsiekte, BSE
(long incubation, eventual fatal disease, but no detectable antibody response)

⟶ Time (months/years) ⟶

Key:
solid line = pathogen in blood (or other tissue); dotted line = pathogen not readily detectable; solid box = severe clinical disease; non-solid box = clinical/subclinical disease.

Antibody is usually detectable in a), b) and c).

apart from housing the animals continuously with filtered air systems, or locating them far away from other livestock, little can be done to protect against airborne infections and flying vectors. More achievable for housed but less so for extensively kept animals, are biosecurity measures to protect against wildlife such as deer, badgers, rabbits, hedgehogs and rodents, which can carry numerous diseases including BVD, tuberculosis, Johne's disease, FMD and leptospirosis, to name but a few. Wild animals (e.g. wild boar, and wild goats) of the same species as the livestock to be protected are a special problem. People and domestic pets can likewise actively or passively transmit diseases to livestock, so control of these may also be necessary. A number of serious diseases such as FMD, BSE, classical swine fever and salmonellosis can be food- or water-borne, and this is an especially important route of infection from a biosecurity standpoint. Inadequately cooked meats (as may occur in domestic swill) and, especially in the case of the TSEs, cooked or uncooked animal protein of any description must be strictly excluded from the diet of valuable herd/flocks. Pastures and crops, as well as silage, straw and cereal products made from these, are often contaminated by infected animal manures, so, particularly in the case of very resistant pathogens such as TSE agents, precautions may be needed. Likewise if manufactured from, or contaminated by, infected animal tissues, vaccines and other biological products such as pituitary hormones and catgut sutures may carry some risk. For the same reason effective disinfection of veterinary instruments should be a priority. Diseases can also, of course, be carried from farm-to-farm by a myriad of other contaminated items; vehicles, clothing and farm equipment, so, depending on the herd/flock and level of biosecurity required, it may be necessary either to exclude all but new items, or to ensure effective disinfection prior to entry. Realism and sense of proportion are needed here because not everything can be effectively disinfected, especially if TSEs are involved or suspected. The key point to remember is that by far the most common route whereby most conventional infectious diseases enter the herd/flock is via live adult or juvenile animals of the same species. Spread then occurs through their body excretions and secretions, and via aborted placentae, foetuses and offspring.

Health monitoring and management

Herds/flocks containing rare or high value genetic stock should undergo regular veterinary checks, using clinical examinations, productivity records and serological testing for specific diseases as necessary. Likewise all casualties plus some of the slaughter stock should undergo post-mortem examination to identify presence of diseases that are not clinically apparent. Critical to the success of these monitoring strategies is that serological, post-mortem and other samplings must conform to statistically acceptable criteria. Guidelines on the sample sizes necessary for confident interpretation of antibody tests and pathological lesions are available (Canon and Roe, 1982; Pointon et al., 1990). Results of monitoring should be evaluated by the farmer and veterinary surgeon working together, and then applied to manage and up-date health programmes, including use of vaccines and drugs (especially antibiotics). Finally they must be alert to national and local disease surveillance information and ensure they are prepared to up-grade biosecurity as and when it is necessary.

Vaccination

If specific disease agents are endemic in the herd/flock, surrounding region or country, and eradication or exclusion are impractical due to factors outside the owner's control, then alternative control strategies such as vaccination may have to be used. Some common bacterial infections in sheep such as those caused by *Clostridia* and *Pasteurella*, for example, can be routinely controlled by vaccination. Others, such as the often endemic infectious bovine rhinotracheitis (IBR) virus, may, as is typical of herpesvirus infections, remain latent in infected cattle but be intermittently excreted causing re-circulation and new outbreaks within a herd. If IBR eradication is impractical, vaccination with a suitable marker vaccine will allow use of blood tests to differentiate between animals with vaccinal and those with natural immunity; animals which have been infected by field virus (and may be latently infected) can thus be identified and if necessary culled (Mawhinney, 2002). This is not possible with traditional vaccines. The development of effective marker vaccines has been mooted as a key objective for control of FMD because they would allow vaccination whilst maintaining FMD-free status, thus enabling rapid disease control without compromising any future eradication policy. There are, of course, many other diseases for which vaccination is a good biosecurity strategy, and in open herds/flocks containing rare breeds it may be the only sensible option for protection against some specific pathogens.

Biosecurity strategies

Whatever biosecurity strategies are employed, it is important for both the farmer and his/her veterinary surgeon to take joint responsibility for them. Biosecurity protocols should preferably be written down as Standard Operating Procedures (SOPs) which are understood and agreed by all staff, then routinely followed. Independent verification that SOPs are being followed is also vital, so procedures and records should be regularly checked by a trusted person or organisation. In addition to the SOPs covering routine biosecurity matters, contingency plans should also be prepared for emergencies such as fire, flood, and national epidemics like FMD. This will ensure that sudden threats or disasters are dealt with promptly to avoid or minimise loss of valuable genetic resources. Another important aspect to be covered is animal identification.

Location, although seldom a matter of choice, has a major impact on biosecurity and ideally the genetically valuable livestock should be as far as possible from, and upwind of other livestock farms, especially large ones. Farm locations on a dead-end road are much more secure than those on busy highways that are likely to be contaminated by animals or animal products in transit. Numbers of entries and exits to the farm should be limited and visiting vehicles, including deliveries, must be strictly controlled. Public foot- or bridle-paths should if possible be diverted away or enclosed by high fences, but without alienating their legitimate users.

Buildings are significant because it is easier to protect small, intensively kept animals than large ones that are pastured over a wide area. Again, while seldom a matter of choice, rare breeding populations of pigs, sheep or goats could be kept in purpose-built, high security buildings with air filtration, similar to those used for laboratory animals. Whether purpose-built or not, where animals are housed, the buildings should provide best possible barriers against disease without prejudicing ventilation and other aspects of animal welfare and good management. In addition to the animal accommodation, the farm site should incorporate biosecure designs for delivery and storage of feed, manure handling, loading and unloading of animals, quarantine of new entrants, restraint and handling for procedures such as shearing, disease testing and embryo transfer, post-mortems and disposal of carcasses. Chemicals, drugs and equipment must have secure storage areas, and good facilities must also be provided for the staff, including changing rooms, showers, toilets, office and canteen. If possible, areas surrounding and between buildings should be concreted so as to allow rainwater and fluids to flow away. Construction materials should enable effective cleaning and disinfection, and preclude build-up of endemic disease agents. The entry of service personnel to repair buildings and large machinery should be minimised by use of new, good quality items. Likewise equipment and vehicles used to dispose of manure, carry feed or transport animals should belong to the farm or, if not, they must be properly washed and disinfected before coming into the premises.

Perimeter fences, if used, should be high enough and of a mesh size to effectively keep out unwanted people, animals and wildlife. Unless it is necessary to avoid publicity, signs 'prohibiting unauthorised access' and stating the presence of 'high-health status animals' should be put up at strategic positions, but with directions to where people can report, or telephone, to gain attention. Double fencing with at least three metres between fences, and closed-circuit television may be considered appropriate in high risk situations. Fences should be several metres away from the buildings to enable good observation and to keep vehicles, people and animals well away from the site. In the case of extensive farms that are spread over a large area, the perimeter should still be distinct and well identified by signs, plus fences or natural barriers to restrict vehicle access.

Disinfectants tend to be thought of as a first line of defence against pathogens, but they should not be. Disinfection mats and wheel dips, sprays, etc. require daily topping up and servicing, and have limited effectiveness unless they are preceded by thorough cleansing to remove all organic matter from the boots, wheels, vehicles, etc. to be disinfected. Such measures may have a place during emergencies, not least to create awareness of extreme biosecurity precautions, but their use should then be explained in writing and kept in place for a limited period. There are many different disinfectants, some (e.g. washing soda or formaldehyde solutions) being useful against viruses like those of FMD, and others such (e.g. 'chloros') being useful against the very resistant agents causing TSEs, but many of them are corrosive and most can cause pollution. Ideally, therefore, disinfection should be a targeted procedure that is used intermittently to reduce risks of disease transmission via items of non-new equipment that are brought onto the farm.

Entry of people should be restricted to those who work on-site and legitimate visitors. There should be a single main entrance located where incoming vehicles and visitors can be seen, ideally from the office. The outer door should be kept locked and a bell installed, plus a system to unlock the door without anyone having to go back through the shower. Although a shower cannot eliminate all infections carried by humans it is a physical barrier which raises awareness of biosecurity and can ensure that street clothes are removed and replaced by dedicated clothing supplied by the farm. Ideally the shower should be used by everyone, but, since this may be impractical, a sensible fall-back is to insist that those having had recent (within the past 48-72 hours) contact with livestock (especially livestock of the same species) are excluded or go through the shower, whilst others change their shoes and outer clothing. All such clothing should be supplied by and used only in the farm. Apart from veterinary surgeons, shearers and others with a legitimate reason, visitors should never handle the animals, and those who do touch them must shower-in or at least wash their hands thoroughly.

Livestock loading and transportation are high risk activities from a biosecurity viewpoint. Ideally there should be a purpose-built loading bay with facilities for washing and disinfection after each use, and effluent drainage which does not pass through the main buildings. Outgoing livestock should be moved into the loading bay by the farm staff who should leave them, re-enter the farm premises and lock the doors. The trucks used for transportation must arrive empty and have been thoroughly washed and disinfected beforehand by the haulier/driver who should also have sufficient manpower to load the waiting animals. If farm staff must help with loading they should exit after cleaning and disinfecting the bay, then re-enter the premises through the main entrance, changing their clothes and preferably showering too, before working with the resident animals.

Feed and bedding should be delivered in clean, disinfected vehicles and be piped through the perimeter fence into silos or bins, or hoisted over so that neither vehicle nor driver has to enter the premises. Where possible, and this applies particularly to extensively kept animals, foodstuffs should be home-grown using home-produced manures, or chemical fertilisers. Food waste ('swill') and animal by-products such as meat-and-bone meal which cannot be effectively sterilised against all pathogens (including those of TSEs) must never be used. Water should ideally come from the mains supply, or from farm springs or wells, and should undergo purification treatment to standards for human consumption.

Vehicles, unless new and for use on the herd/flock premises, should not come inside the perimeter. However, if entry of non-new vehicles is essential, all debris must be washed from their wheels, wheel arches and other potentially infected parts; these parts should then be soaked with a reliable disinfectant.

Equipment and other materials should ideally be bought new or, if the latter is not possible, should be properly cleaned and disinfected on entry to the premises. Many items come under this heading: veterinary drugs and instruments, tools for servicing and repairs, stationery and toiletries. Anything that has direct contact with the animals, or which cannot be disinfected without destroying its usefulness, needs particular care. High-cost items such as those used for shearing, artificial insemination, embryo transfer or pregnancy scanning, are often carried around by operators from farm to farm and can be relatively difficult to disinfect, so purchase of dedicated equipment should be considered. Surgical equipment poses particular risks as it may be used to penetrate tissues or even across the body wall. Injectable substances containing animal products, such as may be present in some vaccines and hormones, require special scrutiny to ensure absence of potential disease risks.

Manure and farm waste must be disposed of carefully so they do not provide avenues for disease entry. Slurry and manure should ideally be spread on land within the farm but, if not, arrangements should be made to pipe it through or lift it over the perimeter fence for disposal elsewhere. As mentioned above, it is preferable for dead and culled animals to be examined post-mortem, so they should be taken to a location outside the perimeter then transported in a dedicated vehicle to a local diagnostic laboratory. The vehicle must then be disinfected. Carcasses not examined in this way may have to be incinerated on the farm or taken to be incinerated or rendered. Since vehicles from knackeries and rendering companies should never come anywhere near to the farm, some other means of transport will have to be organised to remove carcasses to a pick-up venue.

Animals, other than the genetically valuable ones, should ideally be excluded, but this is rarely practical. On extensive farms, for example, dogs may be needed to manage sheep and cattle, and may contribute to biosecurity by deterring predators, trespassers and unwanted wildlife. Cats, too, may be kept to destroy vermin and, if people live on the farm, they may have pets or livestock other than those regarded as a rare genetic resource. Obviously these 'other animals', and the food they eat, can give rise to conflicts of interest which will have to be resolved pragmatically in conjunction with the veterinary surgeon. Likewise with regard to wild animals and birds, while it may be impossible to remove or control them on extensive farms, farmers and veterinarians should monitor their presence and be aware of the biosecurity risks they pose.

Entry of new livestock always poses the highest risks of bringing in diseases, but, even if herds/flocks aspire to a closed herd policy they usually require new 'blood' (genetic material) for breeding from time to time. In this paper we advocate using reproductive technologies such as embryo transfer for this purpose (see below), but we also recognise that some farmers prefer to import adult or juvenile animals. If such animals must come in it is important to bring them in through quarantine. They should come from trusted sources and be screened by testing for specific diseases prior to leaving their place of origin. Contamination can also occur during transportation, so vehicles must be thoroughly cleaned and disinfected before and after use, and routes between farms should be checked to avoid any obvious risks. Quarantine is complex and difficult to apply effectively in most herd/flocks, but its aim must be to detect specified diseases and to exclude them. Its application should be based on pathogenic principles, including incubation periods, immune reactions, and use of reliable tests, such as those described in the Manual of Standards of the OIE (OIE, 2000). It is important to note that serological and other tests for some diseases, such as those with long incubation periods (see Fig. 1), are not always effective. Sheep scrapie is a case in point since there are no serological tests for this, so it is preferable to test and accept only animals with appropriate scrapie genotypes (Wrathall, 1997). Other diseases are best detected at particular stages in the life cycle, for example enzootic abortion of ewes (ovine chlamydiosis) which is manifest most readily at lambing time. It is usual for animals to remain in quarantine for at least six weeks, but testing can begin after about 21 days when antibodies arising from exposure to certain infections at the time of entry will have had time to develop. The quarantine facility should ideally be located on another premises as far from the main herd/flock as possible, and have its own manure-handling facilities. If farm staff have to look after animals in quarantine they should attend them after the main herd/flock and not return the same day. Such staff should shower in and out on each occasion, or at least change into different sets of clothing. The entrance to the quarantine facility is one place where disinfectant foot baths might usefully be installed. Cleansing, disinfection and resting of the facility after batches of animals have been through is very important. A major dilemma with quarantine, particularly on-farm quarantine, is what to do if a serious disease is detected whilst the imported animals are held there? Farmers, in consultation with their veterinary surgeon, should prepare contingency plans specifying actions to be taken if such a disease is found. Actions will vary for different diseases. Use of reproductive technologies such as embryo transfer and, to a lesser degree AI, are much safer alternatives for introducing new 'blood'.

Reproductive technologies and biosecurity

Reproductive technologies such as artificial insemination, superovulation and embryo transfer, oocyte recovery and in vitro fertilisation have proved extremely valuable in livestock breeding and can also be used in the conservation of rare or otherwise valuable livestock. This is especially true when embryos from rare/valuable donors are cryobanked and later thawed for transfer into common breed recipients of the same species, thus enabling multiple individuals of the rare type to be born and reared. Unfortunately freezing and thawing of pig embryos is still ineffective, so cryobanking is not yet a viable option in this species, and embryos must be transferred within about 48 hours of collection. In species such as cattle, sheep, goats and deer where freezing is effective, however, cryobanking gametes and/or embryos has major benefits because:

a) it facilitates planned infusion of genetic diversity into inbred groups of animals;

b) it substantially reduces the risks of disease transmission (compared with introducing adult breeding animals) and eliminates the need for quarantine;

c) it provides long term insurance against social upheaval, natural disasters and disease epidemics (such as the recent FMD epidemic in the UK);

   d) it enables reduced animal numbers and thus the space
       requirements for a viable breeding population, and
   e) it eliminates much of the cost of long distance transport of post-
       natal animals.

Artificial insemination (AI) is the easiest and most available method to
bring new genes into a herd/flock but semen contains only half the
genetic component and, from a disease viewpoint, can carry many
pathogens including FMD virus, classical swine fever virus, and *Brucella
spp.* (Philpott, 1993). To reduce the risks, semen should be collected,
processed and banked by experienced personnel, and preferably it should
come from sires that are kept and monitored for diseases in an accredited
AI Centre. If inseminations are done by an outside person (inseminator)
he/she must enter the farm through the biosecurity cordon and all the
equipment used must be sterilised beforehand.

Embryo transfer is a particularly valuable technique because not only
does the embryo contain the individual's full genetic complement but
also it can usually be subjected to sanitary procedures to ensure virtual
freedom from infectious disease agents. These sanitary procedures, which
are detailed in the Manual of the International Embryo Transfer Society
(IETS) (IETS, 1998), include microscopic examination of the embryo to
ensure it's zona pellucida is intact and free from adherent debris, and
washing (at least 10 times) with media containing antibiotics, and
sometimes also containing the enzyme trypsin. The cryopreserved
embryo, packaged within its zona pellucida and suspended in a
metabolic resting state, is much better suited to long distance transport
than is the post-natal animal. Moreover, following transfer of the embryo,
it is gestated by its surrogate mother and provided at birth with antibodies
and other survival mechanisms to enhance its survival in the new
environment.

Until comparatively recently, for purposes of international trade, embryo
donors and recipients and even the resulting offspring, usually had to
undergo expensive test procedures, and were often kept in quarantine.
However, the IETS has campaigned since the early 1980s for embryos
to be considered as individual entities and certified with regard to any
disease risks largely on the basis of microscopical examination and
washing. Promoting the safety and welfare advantages of embryo transfer
has largely been done through the OIE which is the designated 'scientific
body of referral' (an arbitrator) to the World Trade Organisation on
safe trading of animals and animal products (including semen and
embryos). The health requirements and recommended rules for
international trade in embryos, which are set out in a series of Appendices
in the *Code* (OIE, 2002), are in fact based on inputs from the IETS.
These particular Appendices, which are periodically reviewed and
updated by the IETS Health and Safety Advisory Committee (formerly

the Import/Export Committee), cover *in vivo*-derived embryos of cattle, sheep/goats, pigs, horses, deer, South American camelids and laboratory rodents. Two other Appendices cover *in vitro*-produced and micromanipulated bovine embryos. Underpinning the Appendices in the *Code* is the IETS Manual (IETS, 1998) which describes procedures for embryo processing, and reviews their scientific basis in detail. A requirement central to all the Appendices for embryos is accountability of the 'Officially Approved Embryo Collection/Production Team' which should operate under the leadership of a 'Team Veterinarian' responsible for certifying that all official protocols for collection, sanitary processing and identification of the embryos have been followed. The OIE *Code* Appendices have been adopted by national veterinary authorities in many countries as a basis for international trade in embryos, and their principles have also been incorporated into legally binding Directives for trade within the European Union (EU), and for imports of embryos into the EU from non-EU countries.

Categorising diseases re. their transmission risk

The objective of embryo processing is to 'dis-infect', i.e. remove or inactivate, any potential pathogens that might be present, but without affecting embryo viability. The efficacy of such processing is well documented, and published data have enabled the Research Subcommittee of the IETS Health and Safety Advisory Committee to categorise many infectious diseases apropos their risk of being transmitted via *in vivo*-derived embryos. Categorisation is reviewed annually by the Sub-committee and, when appropriate, it is updated. The latest position is shown in Table 1.

With regard to the risk of disease transmission by embryo transfer, several of each infection type illustrated in Figure 1 have been studied to a greater or lesser degree and, as can be seen in Table 1, results are encouraging. Quantitative assessments of the risks of importing in vivo-derived embryos from zebu (*Bos indicus*) cattle in an area of South America with endemic FMD, bluetongue and vesicular stomatitis revealed that the disease risks are extremely small; less than 10^{-6} in most circumstances (Sutmoller and Wrathall, 1997). Other studies, particularly in the U.K., have also shown that despite major difficulties of diagnosing 'slow' (see Figure 1) infectious diseases like jaagsiekte (sheep pulmonary adenomatosis) and BSE, the risks of transmitting these by embryo transfer from infected animals are very low if recommended protocols, as published in the OIE *Code*, are applied (Parker *et al.*, 1998; Wrathall *et al.*, 2002). This emphasises the fact that if valuable herds/flocks are infected it may still be possible to salvage their genetic resources without risk of disease transmission.

Table 1. Categorisation of infectious diseases apropos the risk of their transmission through *in vivo*-derived embryos. Based on a review of available information by the Research Subcommittee of the IETS Health and Safety Advisory Committee in October 2002 (at St. Louis, Missouri, USA).

Category 1. Diseases or disease agents for which sufficient evidence has accrued to show that the risk of transmission is negligible provided that the embryos are properly handled* between collection and transfer:

> Aujeszky's disease (pseudorabies) (swine) - trypsin treatment required; Bovine spongiform encephalopathy (cattle); Bluetongue (cattle); *Brucella abortus* (cattle); Enzootic bovine leukosis; Foot-and-mouth disease (cattle); Infectious bovine rhinotracheitis - trypsin treatment required.

Category 2. Diseases or disease agents for which substantial evidence has accrued to show that the risk of transmission is negligible provided that the embryos are properly handled* between collection and transfer, but for which additional experimental data are required to verify existing data:

> Bluetongue (sheep); Classical swine fever (hog cholera); Scrapie (sheep).

Category 3. Diseases or disease agents for which preliminary evidence indicates that the risk of transmission is negligible provided that the embryos are properly handled* between collection and transfer, but for which additional *in vitro* and *in vivo* experimental data are required to substantiate the preliminary findings:

> Bovine immunodeficiency virus; Bovine spongiform encephalopathy (goats); Bovine viral diarrhoea; *Campylobacter fetus* (sheep); Caprine arthritis-encephalitis; Foot-and-mouth disease (swine, sheep, goats); *Haemophilus somnus* (cattle); *Neospora caninum (cattle)*; Ovine pulmonary adenomatosis; Porcine reproductive and respiratory disease syndrome; Rinderpest (cattle); Swine vesicular disease.

Category 4. Diseases or disease agents on which preliminary work has been conducted or is in progress:

> African swine fever; Akabane (cattle); Bluetongue (goats); Border disease (sheep); Bovine anaplasmosis; Bovine herpesvirus-4; *Chlamydophila psittaci* (cattle, sheep); Enterovirus (cattle, swine); *Escherichia coli* 09:K99 (cattle); *Leptospira spp.* (cattle, swine); Maedi/visna (sheep); *Mycobacterium bovis (bovine tuberculosis)* (cattle); *Mycobacterium paratuberculosis (Johne's disease)* (cattle); *Neospora caninum* (cattle); Ovine *epididymitis (Brucella ovis)*; Parainfluenza-3 virus (cattle); Porcine parvovirus; Scrapie (goats); *Ureaplasma/Mycoplasma spp.* (cattle, pigs, goats); Vesicular stomatitis (cattle, swine).

*Manual of the IETS (recommendations for sanitary handling) (1998)

For contingency purposes cryobanking of genetic materials, especially embryos, should also be considered as a biosecurity strategy since it enables the full genetic complement of valuable livestock to be recovered and stored indefinitely, or, alternatively, restored by transferring into common-breed recipients of the same species. Of crucial importance in this case is that such materials are collected, processed and cryobanked in such a way as to minimise any risks of their transmitting infectious diseases when they are thawed and used. While the disease control advantages of *in vivo*-derived embryos (those recovered from the dam's reproductive tract after *in vivo* fertilisation) are well researched, embryos fertilised and produced *in vitro*, or by genetic manipulation and nuclear transfer (cloning), may carry some additional disease risks which are currently under scrutiny in the IETS committees. However, the basic sanitary protocols set out in the IETS Manual (1998) for *in vivo*-derived embryos can still be applied to good advantage, and Appendices setting out recommended procedures for *in vitro*-produced and micromanipulated embryos have been published in the OIE *Code* (2002).

Conclusions

Biosecurity strategies for valuable livestock genetic resources should take account of the nature of infectious diseases and their many different modes of transmission. Farmers and others must be realistic about which diseases ought to be excluded from their herds/flocks and those they may have to tolerate and/or control; the aim should be to minimise disease, ensuring that diseases which threaten existence of the livestock, and those which preclude movements between herds/flocks, are excluded. To maintain health status, biosecurity strategies should be formulated by the farmer working with his/her veterinary surgeon and, when both parties are in agreement, the strategies should be translated into standard operating procedures (SOPs). The SOPs must take into account what livestock species are kept, whether the farming system is intensive or extensive, open or closed, and the nature of the premises, including construction of buildings and nature of the farm perimeter. Diseases already present should be identified and monitored, and those to be excluded specified, bearing in mind that some diseases will have to be tolerated and controlled. For closed and semi-closed herds/flocks in particular, SOPs should cover movements of people, vehicles, equipment, food, manure, animal carcasses, vermin, pets and wildlife. Contingencies for emergencies should also be prepared. Above all, the biosecurity SOPs should cover the risks associated with the movement of livestock themselves. Quarantine systems for any incoming animals should be formulated with great care because some subtle but nevertheless dangerous diseases are difficult to reliably detect and exclude if adult or juvenile animals are brought into the herd/flock.

Reproductive technologies such as AI and embryo transfer offer safer ways to move and introduce new genetic materials for breeding, and transferring embryos is especially safe when the sanitary protocols promoted by the IETS and advocated by the OIE are used. Research has demonstrated that with those provisos the risks of transmitting catastrophic diseases like FMD and BSE by embryo transfer are exceedingly small. Another reproductive technology with major biosecurity benefits is embryo cryopreservation which can enable long-term storage of the full genetic complement of valuable livestock with minimal risks of any diseases being present. This paper focuses mainly on biosecurity aspects of moving or storing *in vivo*-derived embryos for which the disease control advantages are well researched. There is less confidence with regard to embryos produced *in vitro*, or by genetic manipulation and cloning, and it is evident that higher disease risks may occur with these. Fortunately, sanitary protocols set out in the IETS Manual (IETS, 1998) for *in vivo*-derived embryos can also be applied to advantage for these other reproductive technologies, and codes of practice for *in vitro*-produced and micromanipulated embryos have been published in the OIE *Code* (OIE, 2002).

Acknowledgements

We are grateful to colleagues in the IETS Health and Safety Advisory Committee, and at VLA-Weybridge and ADAS Arthur Rickwood for their helpful discussions on a variety of aspects of biosecurity.

References

Anon. 2001. Editorial. *Nature Immunology* 2: 565.

Canon, R.M. and Roe, R.T. 1982. *Livestock Disease Surveys: A Field Manual for Veterinarians*. Australian Government Publishing Service, Canberra, Australia.

Donaldson, A.I., Alexandersen, S., Sorensen, J.H. and Mikkelsen, T. 2001. Relative risks of the uncontrollable (airborne) spread of FMD by different species. *Veterinary Record* 148: 602-604.

Gillespie, R.R., Hill, M..A. and Kanitz, C.L. 1996. Infection of pigs by aerosols of Aujeszky's disease virus and their shedding of the virus. *Research in Veterinary Science* 60: 228-233.

Hennecken, M., Stegeman, J.A., Elbers, A.R.W., Nes, A. van., Smak, J.A. and Verheijden, J.H.M. 2000. Transmission of classical swine fever virus by artificial insemination during the 1997-1998 epidemic in the Netherlands: a descriptive epidemiological study. *Veterinary Quarterly* 22 (4): 228-233.

International Embryo Transfer Society. 1998. *Manual of the International Embryo Transfer Society: A procedural guide and general* information for the use of embryo transfer technology, emphasising sanitary procedures. 3rd edition. Edited by D.A.Stringfellow and S.M.Seidel. IETS, Illinois , USA.

Mellor, P.S. 1994. Bluetongue. *State Veterinary Journal* 4: 7-10.

Mawhinney, I. 2002. Markers for the future of IBR control in the UK. *Cattle Practice* 10 (3): 213-217.

Office International des Epizooties (OIE). 2000. *Manual of Standards for Diagnostic Tests and Vaccines*. 4th edition. Office International des Epizooties, Paris, France.

Office International des Epizooties (OIE). 2002. *International Animal Health Code (Mammals, Birds and Bees)*. 11th edition. Office International des Epizooties, Paris, France.

Otake, S., Dee, S.A., Rossow, K.D., Moon, R.D. and Pijoan, C. 2002. Mechanical transmission of porcine reproductive and respiratory syndrome virus by mosquitoes, *Aedes vexans* (Meigen). *Canadian Journal of Veterinary Research* 66: 191-195.

Parker, B.N.J., Wrathall, A.E. and Cartwright, S.F. 1981. Accidental introduction of porcine parvovirus and Talfan virus into a group of minimal disease gilts and their effects on reproduction. *British Veterinary Journal* 137: 262-267.

Parker, B.N.J., Wrathall, A.E., Saunders, R.W., Dawson, M., Done, S.H., Francis, P.G., Dexter, I. and Bradley, R. 1998. Prevention of transmission of sheep pulmonary adenomatosis by embryo transfer. *Veterinary Record* 142: 687-689.

Philpott, M. 1993. The dangers of disease transmission by artificial insemination and embryo transfer. *British Veterinary Journal* 149: 339-369.

Pointon, A.M., Morrison, R.B., Hill, G. , Dargatz, D. and Dial, G. 1990. *Monitoring pathology in slaughtered stock: Guidelines for selecting sample size and interpreting results*. National Animal Health Monitoring System, United States Department of Agriculture, Washington DC, USA, pp. 257-264.

Sutmoller, P. and Wrathall, A.E. 1997. A quantitative assessment of the risk of transmission of foot-and-mouth disease, bluetongue and vesicular stomatitis by embryo transfer in cattle. *Preventive Veterinary Medicine* 32: 111-132.

Wrathall, A.E. 1997. Risks of transmitting scrapie and bovine spongiform encephalopathy by semen and embryos. *Revue Scientifique et Technique des Office International des Epizooties.* 16 (1): 240-264.

Wrathall, A.E. 2000. Risks of transmission of spongiform encephalopathies by reproductive technologies in domesticated ruminants. *Livestock Production Science* 62: 287-316.

Wrathall, A.E., Brown, K.F.D., Sayers, A.R., Wells, G.A.H., Simmons, M.M.,.Farrelly, S.S.J., Bellerby, P., Squirrell, J., Spencer, Y.I., Wells, M., Stack, M.J., Bastiman, B., Pullar, D., Scatcherd, J., Heasman, L., Parker, J., Hannam, D.A.R., Helliwell, D.W., Chree, A. and

Fraser, H. 2002. Studies on embryo transfer from cattle clinically affected with bovine spongiform encephalopathy (BSE). *Veterinary Record* 150: 365-378.

17

The role of rare and traditional breeds in conservation: the Grazing Animals Project

R.W. Small

School of Biological and Earth Sciences, Liverpool John Moores University, Byrom Street, Liverpool, L3 3AF, UK

Abstract

The landscape of the UK has been largely determined by past agricultural practices that have given rise to a range of anthropogenic habitats much valued by conservationists. Many of these have been created by, or for, grazing livestock. The suggestion that grazing and browsing animals were instrumental in 'cyclical succession' in the pre-agricultural period is also gaining ground. For these reasons the use of grazing animals in the management of conservation sites has become more common. Since its foundation in 1997 the Grazing Animals Project (GAP) has promoted and facilitated the use of grazing livestock in management of habitats for conservation.

In 2001 GAP produced, in consultation with animal welfare organizations, A Guide to Animal Welfare in Nature Conservation Grazing. The practical advice in, and approach of, this document is potentially invaluable not only to conservation managers and graziers but also to all keepers of livestock. Another GAP publication, the Breeds Profiles Handbook, gives brief descriptions of 55 breeds of livestock known, or anticipated, to be of value in conservation grazing. Many of these are rare or traditional breeds, as these have the characteristics that enable the stock to thrive on the nutritionally relatively poor forage afforded by many conservation sites. These characteristics are often identified as 'hardiness' and 'thriftiness', but are poorly defined except through the practical experience of conservation managers.

Conservation grazing is a relatively new niche, and one that cannot be filled by modern breeds or strains adapted to high-input, high-output systems. It is, therefore, a great opportunity for rare and traditional breeds, many of which developed in parallel with habitats now appreciated for their conservation value. This applies not only in the UK but also in other European countries. Moreover, recent developments, such as English Nature's Traditional Breeds Incentive for *Sites of Special Scientific Interest, several grazing projects funded by the Heritage Lottery Fund and the Limestone Country Life Project, suggest that this niche is no longer confined to nature reserves.*

Conservation grazing can contribute to genetic conservation by:

- *Enabling an increase in numbers and wider distribution of rare and traditional breeds.*
- *Allowing breeders to identify, and select, those individuals that fare best under relatively austere conditions.*
- *Providing an outlet, or providing additional grazing, for stock that could not otherwise be kept.*
- *Providing a market for good animals without reference to the show-ring.*
- *Providing a refuge for rare breeds from threats such as that posed by the National Scrapie Plan.*

Introduction

It is widely accepted that the landscape and habitats of much of western Europe have been determined by past, and present, agricultural use (Robinson and Sutherland, 2002). This is especially true of the UK where, with a few exceptions such as small, remote islands, mountain ledges and some maritime and wetland habitats, the entire land surface has been subject to some form of agriculture. For much of the north and west this has been, and still is, primarily grazing by domestic livestock. In the south and east, where arable farming has become predominant in recent decades, livestock was more widespread in past centuries as part of a mixed farming economy.

The inheritance from this long history of livestock farming is a range of habitats that have been created in whole or in part by, or for, grazing animals. At the extreme, habitats such as lowland heathland are the result of long periods of grazing on the vegetation growing on inherently nutrient deficient, often acid, soils (Bakker *et al.*, 1983; Tubbs, 1991; Bullock and Pakeman, 1997). The removal of the vegetation with its associated nutrients further impoverished these soils so that a distinct community of plants tolerant of nutrient deficiency developed. Maintenance of heathland communities relies on continued cropping of vegetation to offset nutrient inputs from the atmosphere or fixation of nitrogen by leguminous plants such as gorse *Ulex* spp. (Dolman and Sutherland, 1992).

Other habitats such as chalk downland and limestone grasslands also owe their species richness to low nutrient soils, combined with suppression of coarse grasses (e.g. tor grass *Brachypodium pinnatum*) by grazing.

In some farming systems the ability of livestock to collect widely dispersed nutrients from such grasslands was exploited by overnight folding of the animals, particularly sheep, on arable lands where their dung would fertilise the soils. Hay meadows too relied on inputs of farmyard manure to maintain their productivity; the meadows were needed to provide winter forage for livestock, but the aftermath was normally grazed and a dressing of farmyard manure applied in the spring. Cutting and grazing prevented dominance of coarse grasses and succession to woodland and favoured the spring and early summer flowering plants that are now so valued. With a few specific exceptions, such as some maritime grasslands and heaths, conservation of all UK grasslands and heaths requires grazing by large herbivores (Hearn, 1995).

For much of the early period of agriculture (from the Neolithic to the Mediaeval period) much grazing would have been in pasture-woodland habitats in which widely spaced trees, which could be pollarded to provide winter browse, were set in grazed grasslands. Remnants of such systems exist in sites such as Burnham Beeches, but also in the royal hunting forests such as the New Forest (Harding & Rose, 1986; Kirby et al., 1995). Grazing and browsing is an important element in the management of pasture woodlands (Read, 1993, 1994; Chatters and Sanderson, 1994); the New Forest is managed by essentially free-ranging herds of cattle and ponies (Pratt et. al., 1986; Putman et. al., 1987; Tubbs, 1997). For many years these pasture-woodlands were considered to have been carved out of the closed-canopy forest believed to be the climax community for western Europe. More recently these beliefs have been challenged by new hypotheses that suggest the role of wild grazing animals has been under-estimated (Vera, 2000).

In this view, rather than the rapid closure of glades created in woodland by fallen trees as saplings grow into the gaps, the grazing of wild herbivores would maintain the grassland community for much longer. Eventually grazing-resistant shrubs such as hawthorn *Crataegus monogyna* and blackthorn *Prunus spinosa* would provide enough protection from browsing to allow tree saplings to establish (Vera, 2000). Thus the dynamic processes of vegetation change would be a cyclical succession in which glades persisted for much longer and would represent a much greater proportion of the whole than previously envisaged. The climax community would not be a dark, primaeval forest but a much more open habitat, in which grazing and browsing animals were essential agents in the maintenance of the dynamic balance between woodland and grassland (Kampfe, 2000).

In western Europe the wild herbivores effecting this balance would have been red deer *Cervus elephas*, fallow deer *Dama dama* and roe deer *Capreolus capreolus*, elk *Alces alces*, European bison *Bison bonasus*, auroch *Bos primigenius* and wild horse *Equus caballus* (Yalden, 1982;

Kampfe, 2000), but not sheep *Ovis aries* or goats *Capra hircus* which were Neolithic introductions (Alderson, 1994; Kinsman, 2001). Wild boar *Sus scrofa* would also have been an important element in the fauna, as their rooting activities could uproot tree seedlings but also bury tree seeds leading to better germination and establishment (Kennedy, 1998). After many years in which grazing animals (particularly sheep) and woodlands were considered incompatible (e.g. Piggott, 1983; Putman *et al.* 1989; Putman, 1994) cattle grazing in woodlands is increasingly accepted as a management technique that can increase species diversity (Mitchell & Kirby, 1990; Mayle, 1999). However, to avoid damage to trees and allow some regeneration stocking rates must be as low as 0.05 livestock units per hectare (Hester *et al.*, 1999; Mayle, 1999).

Recent agricultural change

The 1947 Agriculture Act is often cited as the driver behind agricultural intensification in the UK (e.g. Robinson & Sutherland, 2002). The aim of the Act was to encourage self-sufficiency in food production, and farmers were paid subsidies to facilitate improvements in productivity. Later the EU's Common Agricultural Policy pursued a comparable aim, using similar mechanisms, on an EU-wide scale. As a result of these policies, and other pressures such as afforestation and urbanisation, large areas of semi-natural habitats have been lost, such as 97% of the species rich hay meadows and 80% of the lowland heath in England (English Nature, 2002). Bignal & McCracken (1996) estimate that more than 50% of Europe's most highly valued habitats occur on low intensity farmland. Although some of these are associated with arable systems (Robinson & Sutherland, 2002) many more are dependent on livestock. Ostermann (1998) reviewed the 198 habitat types listed in the 1992 EC Habitats and Species Directive: the favourable conservation status of 21 habitats was threatened by abandonment of grazing. In the UK 18 of the 39 key habitats under the UK Biodiversity Action Plan were dependent on livestock grazing (Bullock & Armstrong, 2000).

Perhaps the most significant factor in this intensification has been the widespread application of inorganic fertilisers. These were used not only in existing arable areas, but also to convert nutrient deficient soils such as lowland heath and chalk downland to arable production. Permanent pastures and hay meadows were frequently ploughed, fertilised and replaced with temporary leys usually dominated by perennial rye grass, *Lolium perenne*. These highly productive swards were used for silage production and for grazing by breeds that could make the most efficient use of the herbage. In turn, this led to breeding, or crossbreeding, for production characteristics amongst native breeds, but more often to introduction of breeds from continental Europe, especially in the beef and lowland sheep sectors.

This process of change contributed to the increasing rarity of some breeds that led to the formation of the Rare Breeds Survival Trust (RBST). Although few cattle and sheep breeds became extinct in the post-1947 period, several became critically endangered e.g. Gloucester cattle and Norfolk Horn sheep (Alderson, 1994). Breeders of other cattle, such as Lincoln Red, Aberdeen Angus and Hereford, attempted to respond to the challenge of the 'continentals' by infusions of genes from other breeds or strains e.g. Canadian Herefords. The remnant populations of purebred animals in these breeds are now recognized under the RBST's 'original population' category. Although crossbreeding increased the productivity of these breeds, it was often at the expense of their ability to finish on relatively poor pastures.

Thus at the same time as semi-natural grasslands and heaths were being lost to more productive leys so the traditional breeds were also becoming rare. Now when conservation managers attempt to maintain or restore those semi-natural habitats they need breeds that can thrive on the relatively poor forage. Where grazing is offered to local farmers or graziers it is often rejected as being inadequate for the needs of modern breeds. Alternatively, managers are compelled to use modern breeds or crosses through lack of alternatives, but most would prefer to use rare or traditional breeds (Small *et al.*, 1999). To circumvent these problems conservation agencies are increasingly acquiring their own flocks or herds e.g. the 'flying flock' in Norfolk (Tolhurst, 1994).

Conservation grazing

There is some debate amongst conservationists as to whether the aim of conservation management should be to maintain, enhance and perhaps extend the remnants of anthropogenic habitats such as lowland heaths, hay meadows and chalk downland, or to try to re-create the 'natural' habitats that pre-dated widespread agriculture. The former approach frequently relies on attempting to re-establish traditional farming practices, as these are seen as the primary force that created the habitats (Bignal & McCracken, 1996). Conservation managers may learn from farmers where traditional expertise is still available, or develop the techniques for themselves, or enter partnerships with farmers and graziers. This last approach is increasingly favoured as it allows both farmers and conservationists to do what they are best at and can contribute to the rural economy.

The alternative of re-creation of 'natural' habitats and their associated processes is embodied in the concept of 're-wilding' and is represented by re-introduction programmes such as that for the European beaver (*Castor fiber*) in Scotland. However, throughout much of the UK such schemes will be hampered by lack of extensive tracts of suitable habitat. Even where appropriate sites can be identified it will be impossible to re-create the full range of native herbivores because auroch and true wild horses are extinct (Gordon *et al.* 1990). This will also be a factor in the re-introduction of wild carnivores, notably wolf (*Canis lupus*), for which domestic sheep may be the only widely available prey. Nevertheless, the approach is being pioneered in the Oostvardersplassen reserve in the Netherlands where substantial herds of red deer, Heck cattle and Konik ponies roam over 5600ha (Kampfe, 2000). Minimal management of these herds is exercised and, in the absence of predators, populations are determined by winter mortality, as it is for Soay sheep on the island of Hirta (Clutton-Brock *et al.*, 1991). Human visitors to the Oostvardersplassen reserve are confined to the fringes, in part to prevent disturbance of the diverse and numerous bird species, but also to prevent close encounters with essentially wild cattle.

Thus whether the aim is to conserve habitats created by past agricultural activity or re-wilding, grazing and browsing animals are an important, often essential, component. In the absence of the wild progenitors of today's domestic breeds, or the inability to use feral herds where animal welfare and/or public safety are issues, conservation managers must turn to domestic breeds (Gordon *et al.*, 1990). As noted above, the post-war period has seen the emphasis on high input, high output production at the expense of breeds, or strains, with performance characteristics suited to the low intensity systems that conservation sites seek to employ or emulate. Conservation managers are therefore on the horns of a dilemma – whether to use the stock that is readily available locally, but which may not be suited to the conditions of the site, or to acquire stock for themselves (Small *et al.*, 1999). In areas with scant livestock farming, or where livestock farmers are unwilling to risk their stock on 'poor' grazing, acquisition may be the only option.

Grazing is used as a management technique on a wide variety of habitats (Small *et al.* 1999). Reviews of grazing on fens (Tolhurst, 1997), sea cliffs (Oates *et al.*, 1998), grasslands (Gibson, 1997) and heaths (Bacon, 1998; Lake *et al.*, 2001) have been undertaken. As well as the direct effects of grazing on vegetation, grazing animals can influence sward diversity through the transport of seeds (Fischer *et al.*, 1996), and have impacts on the fauna through dung deposition (Cox, 1999) and trampling (Key, 2000). Cattle dung in particular is an important source of invertebrate food for some bird species such as chough *Pyrrhocorax pyrrhocorax*, and grazing with cattle is encouraged in the UK breeding areas of this relatively rare bird. Grazing animals can have sometimes unexpected influences; for example, grazing of tidal salt marshes can reduce the growth of juvenile sea bass *Dicentrarchus labrax* (Laffaille *et al.*, 2000).

The Grazing Animals Project

The Grazing Animals Project (GAP) was established in 1997 as a result of two initiatives. Participants at a workshop at Liverpool John Moores University in June 1997 on *The Use of Rare Breeds in Conservation Management* recognised the need for a discussion group to share experience and develop best practice. Independently, the Forum for the Application of Conservation Techniques (FACT), a consortium of statutory and voluntary conservation agencies, identified the need to address the constraints that prevented effective implementation of conservation grazing schemes. The potential synergy from combining these two approaches was apparent and the two elements were merged to form GAP.

'Membership' of GAP is free and the mailing list, as at February 2004, had approaching 1000 names representing more than 270 organisations; many, but not all, members are practising conservation managers. A quarterly newsletter, *GAP News*, keeps members informed not only of GAP activities but wider issues relating to livestock and the countryside. Field visits to sites demonstrating aspects of management by grazing e.g. Elmley Royal Society for the Protection of Birds (RSPB) reserve, Ingleborough National Nature Reserve (NNR) and Martin Down NNR are arranged. Visits are also made to sites needing advice e.g. Portland Bill and Goss Moor.

The value of GAP has been recognised by English Nature, which has provided core funding since 1998. This financed a GAP Office to service membership, respond to enquiries, prepare materials and develop publications. In addition, specific projects have been funded by other organisations including Butterfly Conservation, Corporation of London, Countryside Council for Wales, Environment Agency, The National Trust and Royal Society for the Protection of Birds. In the first three-year contract period (1998-2001) three part-time contractors ran the GAP office and initiated a number of projects, such as the development of Local Grazing Schemes. Three conferences have been held (1999: Cambridge, 2001, 2003: Lancaster) and GAP members have contributed to other conferences, meetings and training courses. Practical workshops on cattle husbandry and organic grazing systems have been organised.

A free advertisement service in *GAP News* gave rise first to a separate publication, *Eco-Ads*, which then merged with a sister publication *Woodlots* to form *Eco-Lots*. This has now become mainly electronic with advertisers able to enter their own advertisements, although a paper version is still available to those without internet access. A number of other publications have been produced, such as *The Breed Profiles Handbook: a guide to the selection of livestock breeds for grazing wildlife sites* (Tolhurst & Oates, 2001) and *A Guide to Animal Welfare* in *Nature Conservation Grazing* (Tolhurst, 2001). The former contains general chapters on the use of cattle, sheep, equines, goats and pigs in conservation management, followed by profiles of 55 breeds. Each profile has a brief description of the breed plus sites where it has been used, with contact details for further information. The majority of profiles are of rare and traditional breeds as many of these, such as Hebridean sheep, Dexter cattle and Exmoor ponies, have established reputations in conservation grazing. However, there are also a few breeds which were thought to have potential and some 'commercial' types, as it is recognised that site managers frequently have to use whatever stock is locally available.

GAP recognised that in using livestock for conservation grazing animal welfare must be a priority, both for the animals and for good public relations. The *Guide to Animal Welfare in Nature Conservation Grazing* was developed in conjunction with a wide variety of animal protection/welfare organisations including the Royal Society for the Prevention of Cruelty to Animals, the Scottish Society for the Prevention of Cruelty to Animals, the Ulster Society for the Prevention of Cruelty to Animals, the Farm Animal Welfare Council (FAWC) etc. The guide first summarises the legal obligations and administrative requirements relating to animal welfare for each of the species likely to be used in conservation grazing. The main part of the guide is based on FAWC's *'five freedoms'*:

* freedom from hunger and thirst
* freedom from discomfort
* freedom from pain, injury and disease
* freedom to express normal behaviour
* freedom from fear and distress.

Again these are discussed for each of the livestock species with an indication of what is, and what is not, acceptable. There is a discussion of suitable types of animals for conservation grazing – not only the various species but also factors such as age, sex, background and husbandry (Tolhurst, 2001).

The guide adopts a novel 'risk assessment' approach to the use of animals on conservation sites. This links with each of the five freedoms and starts with an appraisal of the site to be grazed to identify potential causes of suffering. The guide includes a pro-forma for risk assessment with guidance on its completion and each identified hazard is given a score, depending on the probability of suffering (1-4) multiplied by the severity of suffering (1-5). Thus the level of risk ranges from 1-20. The descriptions for the probability and severity of suffering are shown in Tables 1 and 2. As with all risk assessments the initial scores are for the level of risk in the absence of any precautionary action(s). Appropriate action(s) to reduce the level of risk are identified before undertaking a re-assessment of the remaining risk when all the precautionary actions have been taken. The importance attached to the welfare guide was such that it was made available free of charge and copies were also included as a supplement to the *Breed Profiles Handbook*.

Table 1.
Description of the
Probability of
Suffering used in
the Grazing Animals
Project's *A Guide to
Animal Welfare in
Conservation Grazing*
(Tolhurst, 2001).

Probability of suffering	Description	Ranking
Improbable	Physically possible, but never known to happen, therefore very surprised	1
Possible	Occasional instances known or heard of, therefore little surprised	2
Likely	Known of with some frequency or might well happen	3
Very Likely	A common occurrence or surprised if didn't happen	4

Table 2.
Description of the
Severity of Suffering
used in the Grazing
Animals Project's *A
Guide to Animal
Welfare in
Conservation Grazing*
(Tolhurst, 2001).

Severity of suffering	Ranking
Minor suffering to one or more grazing animal	1
Major suffering to one grazing animal	2
Major suffering to several grazing animals	3
Death of one grazing animal	4
Death of several grazing animals	5

GAP also developed the concept of Local Grazing Schemes (LGS), which have the objective of integration of conservation grazing with local livestock farming and marketing, sometimes described as from 'grass to plate'. The idea is that conservation agencies work together, and with the local farming community, to establish the grazing needs in the local area and then consider ways in which those grazing needs can be best met. The pattern varies with each scheme to suit local needs and availability of stock etc. The first LGS to be established covered Hampshire, but there are now more than 30 in various stages of operation/development. Workshops and experience of early schemes enabled production of *Local Grazing Schemes: a best practice guide* (Grayson, 2001). Four proposed LGS were awarded £50,000 each from the Heritage Lottery Fund's (HLF) 'Your Heritage' FMD recovery fund and English Nature offered £5000 as the 10% contribution needed for further applications to the HLF under this scheme. One project seeks to encourage the use of traditional cattle breeds in the Cotswolds Area of Outstanding Natural Beauty (Small, 2002b) and another, the '21st Century Drovers Project' aims to establish cattle grazing on a range of conservation sites in the Northumberland National Park.

GAP is keen to learn from research into all aspects of conservation grazing. Particular areas of interest are grazing preferences, grazing behaviour, conservation grazing problems and their solutions. There is a great deal of anecdotal information on these topics, but little thoroughly researched data. Collation of the wealth of anecdotal evidence can be useful and GAP welcomes observations from practitioners or accounts of their experience of conservation grazing. GAP has conducted two questionnaire surveys. One was a general survey of conservation grazing schemes and was published as an English Nature Research Report (Small *et al.*, 1999). The second was into the impacts of the 2001 outbreak of Foot and Mouth Disease (FMD) on conservation grazing schemes. This was summarised as a brief report which GAP submitted to the post-FMD debate (Small, 2002a) and which complemented English Nature's own report on the impacts of FMD (Robertson *et al.*, 2001).

GAP let its second three-year contract to The Wildlife Trusts. Towards the end of the first contract GAP reviewed what had been achieved and what was still needed; from this a list of desirable projects for the period 2002-2005, with an estimated total cost of over £3.5m, was compiled. English Nature agreed to provide core funding for at least another 2 years. The GAP office re-located to the headquarters of The Wildlife Trusts at Newark and a GAP Project Co-ordinator, a GAP Services Co-ordinator and an assistant were appointed. The part-time LGS co-ordinator was re-appointed and was joined by five regional co-ordinators.

Although GAP focuses on the use of grazing animals on sites of wildlife interest, it is aware that it is not working in a vacuum and that there are many developments in the wider context of agriculture and the countryside. Indeed GAP is increasingly consulted on such matters and much time is taken up with deciding which new policies, consultations and discussions GAP should respond to, and then what should be said to ensure maximum beneficial effect. Agri-environment schemes such as Tir Gofal, Countryside Stewardship Scheme (CSS) and Environmentally Sensitive Areas (ESA) have been available for some time (since 1987 in the case of some ESAs). There is some evidence that these schemes are beginning to achieve measurable benefits (Carey *et al.*, 2002). Such schemes are an increasingly important part of government policy and under modulation provisions more money will be diverted from production subsidies to agri-environment schemes. CSS funds were expanded considerably in 2000, and the Arable Stewardship Scheme has been extended to all of England. The 'Curry' Report proposed a new, 'broad and shallow' tier available to all farmers for basic environmental practice; the pilot areas for this new scheme were announced in November 2002. GAP responded to the consultation on the Curry Report and expressed some reservations. For example, that the 'entry level' tier should not reward farmers who do little more than comply with existing legislation and the Code of Good Agricultural Practice. Rather it should 'send unequivocal messages to farmers about the value that society attaches to positive wildlife management'. GAP also contributed to the development of the Environmental Stewardship scheme which will replace ESAs and CSS; notably, this scheme will include conservation of farm animal genetic resources as a 'secondary objective'.

One way to attract more farmers to adopt wildlife friendly farming is to increase the premium they receive on their products. As well as organic options there are now a range of marketing initiatives from conservation agencies, for example:

- The Countryside Agency's *'Eat the View'* initiative
- The Wildlife Trusts' *'White and Wild'* milk brand
- The Rare Breeds Survival Trust's *'Traditional Breeds Meat Marketing Scheme'*
- The National Trust's marketing of Herdwick sheep products in the Lake District
- The RSPB's *'Farming for Birds'* and other projects.

All of these seek to encourage farmers, particularly the various agencies' tenant farmers, to adopt wildlife-friendly farming and to give them support in making that change. In 2002 English Nature launched the *'Traditional Breeds Incentive for Sites of Special Scientific Interest'*. This is a novel development, as agri-environment schemes do not normally try to influence the livestock kept by farmers. The Traditional Breeds Incentive pays farmers an additional premium to use one of 60 traditional breeds, which are identified as 'rare', 'distinctive' or 'adapted'. Payments are made for land within, or adjoining, a Site of Special Scientific Interest. Similar schemes have been funded from the EU's *Life* Fund e.g. the Limestone Country *Life* Project. This seeks to encourage a return to mixed farming in parts of the Yorkshire Dales National Park by funding the purchase of, and the infrastructure needed for keeping, traditional breeds of cattle (T. Thom, personal communication).

Thus although GAP is not directly concerned with the conservation of farm animal genetic resources, it has encouraged and facilitated the use of grazing animals. By the nature of many conservation sites, the breeds used need to be hardy and thrifty and many are rare or traditional.

Conservation grazing, rare breeds and genetic conservation

Since its formation in 1973, the Rare Breeds Survival Trust (RBST) has argued that the genetic resources of rare breeds should be conserved for uses that might not yet be apparent. Although rare breeds have been used in conservation grazing for almost 40 years (e.g. Soay at Aston Rowant NNR and Belted Galloway at Woodwalton Fen NNR (Small, 1994)), the scale of conservation grazing is now such that it could be considered one such unforeseen use. GAP estimates that over 600 conservation sites are grazed (although not all with rare or traditional breeds) and that as many as 1000 sites would benefit from grazing (Small *et al.*, 1999). In GAP's survey of conservation grazing schemes, breeds on the RBST Watchlist were used on 61% of responding sites, with Hebridean sheep being the most widely used sheep breed (on 18 of 108 sheep grazed sites) and Exmoor ponies the most widely used equine breed (on 9 of 30 sites using ponies). Rare breeds are also used for conservation grazing in the Netherlands (Cnossen, 2002) and elsewhere in mainland Europe.

Conservation grazing therefore offers significant potential as a use for rare breeds. This has already contributed to the growth in numbers of breeds such as the Hebridean. Larger numbers can help to ensure that the remaining genetic resource is maintained and more of the founder animals' contribution to the population is retained. Breeds such as the Exmoor pony may not as yet have experienced a significant increase in numbers (it is in category 2 (endangered) on the RBST Watchlist) but have benefited from a widening of their distribution. Purists may argue that the best Exmoors will be bred on their native moor, but to have a widespread population provides an insurance policy against localised disaster. This was evident in the 2001 FMD outbreak, where the rare breeds that suffered greatest losses were those with a considerable proportion of their total population in the hardest hit areas. Examples include Belted Galloway and Beef Shorthorn cattle in the Scottish Borders, Hill Radnor in the Welsh Borders and Whitefaced Woodland in the Pennines (Townsend *et al.*, 2002).

Many owners of rare breeds have relatively small land-holdings: in a survey of RBST members 25% had less than 4ha, and almost 50% less than 16ha (Townsend, 2002). Clearly this places a limit on the number of animals that can be kept, particularly of the larger species. Conservation grazing can provide additional land that allows more stock to be kept. Winter grazing is particularly valuable as it allows young stock, that would otherwise have been slaughtered, to be kept on and assessed on the basis of performance and, it must be admitted, appearance at maturity. Where an animal fails to 'make the grade' as breeding stock, the additional meat value of an older animal may make a significant contribution to the financial aspects of keeping rare breeds. From personal experience, the ability to overwinter Hebridean and North Ronaldsay ewe and ram lambs on a local nature reserve has allowed more informed selection of breeding stock, greater returns from both sale of breeding stock and meat and, not least, from Sheep Annual Premium paid on the ewe lambs.

Conservation grazing provides an outlet for stock, encouraging the retention of animals that may otherwise have gone for slaughter. In turn it increases demand for suitable stock, again leading to greater returns. To take just one example, the Nottinghamshire Wildlife Trust is aiming to amass a flock of 500 Hebridean sheep (P. Kemp, personal communication). Although some of these will be bred from within the existing flock many will be purchased over a period of several years. Such demand puts a 'bottom' in the market that increases returns to breeders, at least until the desired flock size has been reached.

Perhaps the most important aspect of the demand for suitable animals for conservation grazing is that it is generally without reference to the show ring. Many keepers of rare breeds see showing as part of the enjoyment of keeping rare breeds, and in itself showing is harmless. However, when combined with selling it can lead to over-emphasis on show animals, which attract high prices at the expense of less 'showy' individuals. It can also lead to breeding from fewer individuals in the pursuit of rosettes and/or higher prices, with potential narrowing of the genetic base of the breed. In general, conservation managers are not concerned with showing and often have a relatively tight budget for the purchase of stock. Consequently they will concentrate on features such as 'a good mouth' rather than on the colour of the rosettes.

As noted above, GAP places great emphasis on animal welfare. Nevertheless, it is clear that most conservation grazing sites are more challenging than agricultural leys, and hence provide an opportunity to select those animals that fare best under the challenge. This is not to imply that any animals should suffer or that the natural processes such as winter mortality be allowed to operate. These are ethical dilemmas that will need to be addressed should 're-wilding' be attempted as at Oostvardersplassen (Kampfe, 2000; Vera 2000). Until then, conservation grazing allows managers and breeders to select stock that thrive under the conditions imposed by the conservation site, thereby maintaining or enhancing the qualities of thrift and hardiness that are the main reasons for the selection of rare or traditional breeds. Arguably, this could lead to a reduction in the genetic diversity of the breed, but this seems unlikely as it merely reproduces either the anthropogenic selection pressures imposed by generations of farmers for those breeds with a commercial past, or the natural selection pressures that operated on those breeds that survived as feral or parkland flocks or herds.

Perhaps the greatest current threat to rare breeds of sheep comes from the National Scrapie Plan (NSP), the programme that aims to eliminate scrapie from the UK national sheep flock. In the main NSP, rams are tested for their scrapie susceptibility/resistance and on this basis are classified into one of five groups. Group 1 rams are homozygous ARR and considered genetically most resistant to scrapie; at the other end of the scale Group 5 rams are homozygous VRQ or heterozygous VRQ with AHQ, ARH or ARQ (DEFRA, 2001). Rams in Group 5 will not be allowed to breed from 2002 and the industry is encouraged to move towards Group 1 rams as quickly as possible. Some traditional hill breeds, such as Swaledale, Scottish Blackface and Welsh Mountain have a high proportion of Group 3 rams (combinations of ARQ, ARH and AHQ). The NSP represents a real threat to their genetic resources. In recognition of this threat the deadline for Group 3 rams was extended in November 2002 to allow their sale until 2006 and their use for breeding to 2008 (DEFRA, 2002).

The threat to some rare breeds is even greater. As with any breeding programme that focuses on a single trait there is very real danger that breeding for scrapie resistance will result in the loss of valuable genetic resources. The RBST prevailed upon DEFRA to set up a separate survey for rare breeds; under this scheme flock owners could have all their pedigree rare breed sheep (males and females) scrapie genotyped. The survey was conducted in spring and autumn 2002 and the results show that some primitive breeds, such as the North Ronaldsay and Castlemilk Moorit, are almost all (97.4% and 92.9% respectively) homozygous ARQ. However, the RBST has undertaken to review the results for each breed and, where possible, draw up breeding programmes for the national flock of those breeds. To do so relies on the co-operation of breeders, who must both provide the scrapie genotypes of individual animals and agree to follow the recommendations of the RBST in their use of breeding stock.

In this regard there seems to be a growing confusion between conservation grazing flocks and conservation flocks. When considering what might happen if a breed is shown to have a very low, or zero, frequency of scrapie resistant genotypes the concept of a conservation flock, i.e. a flock kept purely for breed preservation, has been mooted. This has been mis-interpreted in some quarters as conservation grazing flocks, but there is a difference. Many conservation grazing schemes derive some income from the stock, either directly from sales of animals or their products if the managing agency owns the stock, or from grazing rental if stock belong to farmers or graziers (Grayson, 1997). The latter will not keep sheep that cannot be sold as breeding stock or for food because of scrapie genotype status. It is true that conservation grazing objectives could often be achieved by sheep that live out their lives on the site, but few will be needed in this role of 'mouths' and breeding flocks would soon produce a surplus of worthless animals. In addition, few of the current breeders of these breeds would continue to keep them; these breeders may not make a living, or even a profit, from their sheep, but they do hope to have some return to offset their investment and some prospect of selling animals for breeding or food.

It must be clearly understood that confining scrapie susceptible breeds to conservation flocks relegates them to the status of zoo animals. They could be kept for their historical interest but, without any real prospect of selling surplus stock, only where the costs are offset either by visitors willing to pay for the privilege of seeing the animals (as in farm parks) or by government subsidy. It is possible that, if there are scrapie resistant individuals in the current flock, a breeding programme could increase the proportion of scrapie resistant genotypes over time, but only if breeders can be persuaded to participate in the programme while having little prospect of a financial return on their flocks. Again, government help would be needed.

It would be ironic if conservation grazing, a new niche for which rare breeds had been conserved since the formation of the RBST, was to become the sole repository for those genetic resources that many hundreds of small scale breeders had worked so hard to save. It is recognized that the NSP will probably only have serious consequences for a few sheep breeds, but their genetic resource is as potentially valuable as any other breed. However, in general conservation grazing provides real opportunities for many rare and traditional breeds and thus represents an unparalleled means to conserve both the live animals and their genetic resources.

Acknowledgements

I gratefully acknowledge the contribution of members of the GAP executive and the RBST to the development my understanding of conservation grazing, but in so doing I emphasise that the views expressed are my own and may not be those of GAP or RBST.

References

Bullock, D.J. and Armstrong, H.M. 2000. Grazing for environmental benefits. In: *Grazing management: the principles and practice of grazing, for profit and environmental gain, within temperate grassland systems,* Edited by A.J. Rook and P.D. Penning, British Grassland Society, Reading, UK. pp. 191-200.

Carey, P.D., Barnett, C.L., Greenslade, P.D., Hulmes, S., Garbutt, R.A., Warman, E.A., Myhill, D., Scott, R.J., Smart, S.M., Manchester, S.J., Robinson, J., Walker, K.J., Howard. D.C. and Firbank, L.G. 2002. A comparison of the ecological quality of land between an English agri-environment scheme and the countryside as a whole. *Biological Conservation* 108: 183-197.

Chatters, C. and Sanderson, N. 1994. Grazing lowland pasture woodland. *British Wildlife* 6 (2): 78-88.

Clutton-Brock, T.H., Price, O.F., Albon, S.D. and Jewell, P.A. 1991. Persistent instability and population regulation in Soay sheep. *Journal of Animal Ecology* 60: 593-608.

Cnossen, H.F. 2002. Rare livestock breeds and nature policy. *Vakblad natuurbeheer* Special Issue: Grazing and grazing animals: 48.

Cox, J. 1999. The nature conservation importance of dung. *British Wildlife* 11(1): 28-36.

DEFRA 2001. *National scrapie plan for Great Britain*. Department of the Environment, Food and Rural Affairs, London, UK.

DEFRA 2002. News release on www.defra.gov.uk/news/2002/021111c.htm

Dolman, P.M. and Sutherland, W.J. 1992. The ecological changes of Breckland grass heaths and the consequences of management. *Journal of Applied Ecology* 29: 402-413.

English Nature 2002. Information from www.english-nature.org.uk

Fischer, S.F., Poschlod, P. and Beinlich, B. 1996. Experimental studies on the dispersal of plants and animals on sheep in calcareous grasslands. *Journal of Applied Ecology* 33: 1206-1222.

Gibson, C.W.D. 1997. *The effects of horse and cattle grazing on English species-rich grasslands*. English Nature Research Reports No. 210. English Nature, Peterborough, UK.

Gordon, I.J., Duncan, P., Grillas, P. and Lecompte, T. 1990. The use of domestic herbivores in the conservation of the biological richness of European wetlands. *Bull. Ecol.* 31: 49-60.

Grayson, F.W. 1997. Does conservation farming work? *Enact* 5(4): 19-22.

Grayson, F. W. 2001. *Local Grazing Schemes: a best practice guide*. 2nd edn. Grazing Animals Project, Norwich, UK.

Harding, P.T. and Rose, F. 1986. *Pasture-woodlands in lowland Britain*. Institute of Terrestrial Ecology, Huntingdon, UK.

Hearn, K.A. 1995. Stock grazing of semi-natural habitats on National Trust land. *Biological Journal of the Linnean Society* 56(suppl.): 23-37.

Hester, A.J., Byrne, R. and Lund, J. (eds.) 1999. *Grazing management options for native woodlands*. Proceedings, Seminar Native Woodlands Group/Hill Land Use and Ecology Discussion Group, MLURI, Aberdeen 23rd April 1998.

Kampfe, H. 2000. The role of large grazing animals in nature conservation – a Dutch perspective. *British Wildlife* 12(1): 37-46.

Kennedy, D. 1998. Rooting for regeneration. *Enact* 6(4): 4-7.

Key, R. 2000. Bare ground and the conservation of invertebrates. *British Wildlife* 11(3): 183-191.

Kinsman, D.J.J. 2001. *Black sheep of Windermere: a history of the St. Kilda or Hebridean sheep*. Windy Hall Publications, Windermere, UK. pp. 6-19.

Kirby, K.J., Thomas, R.C., Key, R.S., McLean, I.F.G. and Hodgetts, N. 1995. Pasture-woodland and its conservation in Britain. *Biological Journal of the Linnean Society*. 56(suppl.): 135-153.

Laffaille, P., Lefeuvre, J-C. and Feunteun, E. 2000. Impact of sheep grazing on juvenile sea bass, *Dicentrarchus labrax* L., in tidal salt marshes. *Biological Conservation* 96: 271-277.

Lake, S., Bullock, J.M. and Hartley, S. 2001. Impacts of livestock grazing on lowland heathland. English Nature Research Report No. 422. English Nature, Peterborough, UK.

Mayle, B. 1999. *Domestic stock grazing to enhance woodland biodiversity*. Forestry Commission Information Note 28. Forestry Commission, Edinburgh, UK.

Mitchell, F.J.G. and Kirby, K.J. 1990. The impact of large herbivores on the conservation of semi-natural woods in the British uplands. *Forestry* 63: 333-353.

Oates, M., Harvey, H.J. and Glendell, M. 1998. *Grazing sea cliffs and dunes for nature conservation*. The National Trust, Cirencester, UK.

Ostermann, O.P. 1998. The need for management of nature conservation sites designated under Natura 2000. *Journal of Applied Ecology* 35: 968-973.

Piggott, C.D. 1983. Regeneration of oak-birch woodland following exclusion of sheep. *Journal of Ecology* 71: 629-646.

Pratt, R.M., Putnam, R.J., Ekins, J.R. and Edwards, P.J. 1986. Use of habitat by free-ranging cattle and ponies in the New Forest, Southern England. *Journal of Applied Ecology* 23: 539-557.

Putman, R.J. 1994. Effects of grazing and browsing by mammals on woodlands. *British Wildlife* 5(4): 205-210.

Putman, R.J., Edwards, P.J., Mann, J.C.E., How, R.C. and Hill, S.D. 1989. Vegetational and faunal changes in an area of heavily grazed woodland following relief of grazing. *Biological Conservation* 47: 13-32.

Putman, R. J., Pratt, R.M., Ekins, J.R. and Edwards, P.J. 1987. Food and feeding behaviour of cattle and ponies in the New Forest, Hampshire. *Journal of Applied Ecology* 24: 369-380.

Read, H. 1993. Rare breeds at Burnham Beeches. *the Ark* 20(7): 253-255.

Read, H. 1994. Native breeds in Burnham Beeches. *Enact* 2(4): 4-6.

Robertson, H. J., Crowle, A. and Hinton, G. 2001. *Interim assessment of the effects of the foot and mouth disease outbreak on England's biodiversity*. English Nature Research Report No. 430. English Nature, Peterborough, UK.

Robinson, R.A. and Sutherland, W.J. 2002. Post-war changes in arable farming and biodiversity in Great Britain. *Journal of Applied Ecology* 39: 157-176.

Small, R.W. 1994. Conservation and rare breeds of farm livestock. *British Wildlife* 6(1): 28-36.

Small, R.W. 2002a. *Impact of the 2001 outbreak of foot and mouth disease on conservation grazing schemes.* Grazing Animals Project. Norwich, UK.

Small, R.W. 2002b. Conservation grazing – a developing opportunity for rare breeds. *The Ark* 30(1): 27-28.

Small, R.W., Poulter, C. Jeffreys, D.A. and Bacon, J.C. 1999. *Towards sustainable grazing for biodiversity: an analysis of conservation grazing projects and their constraints.* English Nature Research Report No. 316. English Nature, Peterborough,UK.

Tolhurst, S.A. 1994. Flying flock on Norfolk's heaths. *Enact* 2(4): 18-20.

Tolhurst, S.A. 1997. Investigation into the use of domestic herbivores for fen grazing management; a document for discussion. Broads Authority, Ipswich, UK.

Tolhurst, S.A. (ed.) 2001. *A guide to animal welfare in nature conservation grazing.* Grazing Animals Project, Norwich, UK.

Tolhurst, S.A. and Oates, M. (eds.) 2001. *The breed profiles handbook: a guide to the selection of livestock breeds for grazing wildlife sites.* Grazing Animals Project. English Nature, Peterborough, UK.

Townsend, S. 2002. The RBST 2001 members' survey – an overview. *The Ark* 30(3): 98-99.

Townsend, S., Warren, S. and Wilson H. 2002. FMD – how the disease has impacted on different sectors of the rare breeds' world. *The Ark* 30(2): 65-68.

Tubbs, C.R. 1991. Grazing the lowland heaths. *British Wildlife* 2(5): 276-289.

Tubbs, C.R. 1997. The ecology of pastoralism in the New Forest. *British Wildlife* 9(1): 7-16.

Vera, F.W.M. 2000. *Grazing ecology and forest history.* CABI Publishing, Wallingford, UK.

Yalden, D. W. 1982. When did the mammal fauna of the British Isles arrive? *Mammal Review* 12(1): 1-57.

18

Use of molecular genetic techniques: a case study on the Iberian pig

M.A. Toro, E. Alves, C. Barragán, C. Castellanos, E. Fabuel, A. Fernández, C. Ovilo, M.C. Rodríguez and L. Silió
Departamento de Mejora Genética Animal, INIA, Carretera A Coruña km. 7, 28040 Madrid, Spain

Abstract

The usefulness of molecular genetic markers as a tool for the conservation, characterisation and differentiation of domestic animal populations are shown in the following text, that summarises diverse applications to Iberian pigs.

Pedigree and molecular coancestry

The concept of coancestry (or kinship) between two individuals plays a central role in animal breeding in order to estimate genetic parameters and to carry out genetic evaluations and in the field of conservation, to implement an optimal method of genetic management (Caballero and Toro, 2000). As a consequence of the great interest produced by the development of molecular markers, several estimators have been developed to measure pairwise coancestry using molecular information. Most of them are described in Lynch and Ritland (1999). Toro *et al.* (2002a) have compared their performance using pig and simulated data and some of the main results will be summarised below.

The coancestry coefficient (f) between individual X and Y was defined as the probability that two alleles taken at random, one from each individual, are identical by descent (Malecot, 1948). The molecular coancestry between two individuals X and Y (f_M) is obtained by applying this definition to the marker genes, that is, the probability that two alleles taken at random, one from each individual, are identical by state. There is a well-known relationship between the expected value of the coancestry coefficient and the molecular coancestry f_M: $\hat{f} = (f_M - \sum_i^n p_i^2) / (1 - \sum_i^n p_i^2)$, where n is the number of alleles and p_i is the frequency of the ith allele. The estimator calculated for each locus can be averaged over loci. This estimator performs better than the others proposed, at least for deep pedigrees.

Pig and marker data

The Iberian pigs considered belong to two strains (Guadyerbas and Torbiscal) of the early conservation programme established in 1945 at 'El Dehesón del Encinar' (Oropesa, Toledo, Spain). The herd was formed from four founder populations representative of the most important varieties of Iberian pigs, that were kept genetically isolated until 1963. Then, the four groups were slowly blended, resulting in the composite strain named Torbiscal (Rodrigañez *et al.*, 2000). One of the founder populations (Guadyerbas) was also maintained as a separate closed strain representative of the black hairless variety from the Guadiana Valley (Toro *et al.*, 2000a). The complete genealogy of all animals is available since 1945, being 18.9 (Guadyerbas) and 21.0 (Torbiscal) generations from the founders until the genotyped animals. PCR-amplified microsatellite markers were analyzed by capillary electrophoresis equipment with fluorescent detection. The number of genotyped microsatellites was 49, distributed among the 18 autosomal chromosomes. The number of alleles ranged from 2 to 9, with an average of 4.24 alleles and the heterozygosity values ranged from 0.19 to 0.79, with an average of 0.58.

Table 1.
Statistics of the pedigree and molecular coancestry coefficients and estimators of the pedigree coancestry in two lines of Iberian pigs (pig and simulated data): mean, mean square error (MSE) and correlation (ρ) with the pedigree coancestry value.

	All individuals			Guadyerbas			Torbiscal		
	Mean	MSE	ρ	Mean	MSE	ρ	Mean	MSE	ρ
Pig data									
Pedigree	0.16	0.00	1.00	0.39	0.00	1.00	0.16	0.00	1.00
f_M	0.41	0.07	0.93	0.55	0.03	0.63	0.43	0.07	0.39
$\hat{f}$	−0.01	0.04	0.90	0.23	0.03	0.57	0.03	0.02	0.37
$\hat{g}$	0.16	0.04	0.93	0.03	0.13	0.55	0.01	0.02	0.42
Simulated data									
f_M	0.30	0.02	0.96	0.49	0.04	0.58	0.30	0.02	0.58
$\hat{f}$	−0.01	0.03	0.96	0.28	0.01	0.57	−0.02	0.03	0.56
	0.16	<0.01	0.96	0.39	<0.01	0.58	0.16	<0.01	0.58

The value of the molecular coancestry (f_M) is obviously greater than the pedigree one and the estimator of the pedigree coefficient from markers ($\hat{f}$) underestimates the true value (Table 1). The correlation between the coefficients and the pedigree value was very high when all the animals were considered together and there was a wide range of pedigree coancestries (0.14-0.53). It was lower when the two strains were considered separately. The range of *f* values was only 0.34-0.53 in Guadyerbas and 0.14-0.36 in Torbiscal. One of the problems with all the estimators is that the matrix of coancestry coefficients obtained is

M.A. Toro et al.

not positive definite and therefore it cannot be used in many practical purposes such as genetic evaluations. To solve this problem we have proposed another estimator ($\hat{g}$) that consists of randomly generated coancestry matrices for all the individuals considered and from them selecting the one with the best correlation with the observed molecular coancestry matrix (Toro et al., 2002b). The random matrices of coancestries are generated assuming a three generation arbitrary pedigree.

Simulated data

In order to assess the dispersion of the coefficients and estimators, simulations were carried out using the gene dropping method and following the actual pedigree. A pig genome was simulated with 2125 loci distributed across 18 chromosomes of approximately the length indicated in the PigMap. The recombination fraction among loci on the same chromosome was 0.01 and the 49 markers were located at their map positions. At each marker locus there were 6 alleles at equal frequencies and in Hardy-Weinberg and linkage equilibrium. The number of simulation runs was always 100.

The simulation results also appear in Table 1. When the actual values of the gene frequencies are utilised, the results mimic those obtained with real pigs. They are very biased but their correlation with the pedigree is high for all animals and decreases for each strain. When the true values of the allelic frequencies in the base population are used, (italic figures) the estimator is unbiased, as expected, because it is based on the molecular coancestry coefficients.

Implications

One of the applications of coancestry coefficients in Conservation Genetics is to decide the gamete contributions (semen and oocytes doses) to a cryopreservation bank. Toro and Maki-Tanila (1999) showed that these contributions should minimise $w'Gw$, where w is the vector of gametic contributions of candidate animals ($\Sigma w_i = 1$) and G is the coancestry matrix for all individuals including reciprocal and self coancestries. For this reason it could be of interest to see how different the contributions are when different coefficients are used and their correlation with the contributions calculated when pedigree information is used. The minimum contribution of any individual was fixed to a value of 0.1% and the calculations were done using a simulated annealing algorithm.

The results of the gametic contributions are given in Table 2. The true f values attained are greater than the value when using the pedigree

283

value. The variance of the contributions also increases substantially. The most discouraging result is the low correlation between the contributions calculated with the estimators and with the pedigree information although it seems that imposing the restriction to the G matrix of being positive definite will substantially improve the results.

Table 2.
Differences among cryopreservation banks with gametic contributions calculated from diverse coancestry estimators.

	f	f_M	$\hat{f}$	$\hat{g}$
Estimated f value	0.140	0.385	−0.088	0.017
True f value	0.140	0.157	0.167	0.142
Variance of contributions	0.365	2.658	4.560	0.233
Correlation with best contributions	1.000	0.236	0.240	0.780

Conclusions

The coefficient of molecular coancestry f_M that follows the Malecot method seems to be the most natural to use as a measure of similarity. Furthermore, and due to this analogy, the estimators derived from it can be expected to give adequate estimates even if inbreeding is present (Eding and Meuwissen, 2001). However, as has been shown previously (Toro *et al.*, 2002a; Eding and Meuwissen, 2001), the lack of information on the allele frequencies in the base population induces a high bias of these estimators in populations with complex pedigrees, even with 49 markers. If the population we are dealing with has been maintained closed with a reasonable amount of mixing, predicting coancestries from markers will be poor as shown by the results of the much lower correlation within strains and it will be advisable to consider only first degree relatives. On the other hand, the high correlation between molecular and genealogical coancestry obtained when the two strains are considered together allows us to predict with accuracy which strain an animal comes from. In the application presented here the limitations of inferring genealogical coancestry from marker information when implementing cryopreservation banks need to be emphasized.

Marker-assisted conservation

Molecular markers are being advocated as a powerful tool for the genetic management of conservation programmes. There are two main decisions that have to be taken: the first is how to choose the breeding animals and how much they will contribute to the next generation. The second is how the matings will be organised. There is a consensus that to minimise the average coancestry of individuals of the next generation is the best criteria for the first decision. The problem to be solved is to minimise $c'Gc$ where c is the vector of contributions left by each candidate and G is the coancestry matrix. It can be solved using integer mathematical programming techniques, whose computational cost would be feasible in most practical situations but not for simulation work, where the algorithm should be used repeatedly. For this reason we have been using a simulated annealing algorithm that, although not assuring the optimal solution, generally behaved well.

The reasons for recommending this strategy have been summarised by Caballero and Toro (2000): a) it maximises the genetic diversity in terms of expected heterozygosity calculated using $1 - \sum_k \sum_i p_{ik}^2$, where p_{ik} is the average frequency in the selected population of the ith allele of kth locus; b) if the between-families coancestry is uniform the technique will equalise family size leading to optimal within-family selection, where among the dams mated with each sire, one is selected at random to contribute one son, another one to contribute two daughters and the remaining contribute one daughter each (Wang, 1997); and c) if the individuals are related the method will equalise not only contributions from founder individuals but also contributions from all previous generations.

When only pedigree information is available, global coancestry will be calculated from genealogies. On the other hand if only marker information is available we should minimise the global molecular coancestry, calculated applying the Malecot definition to markers, as explained before. When both pieces of information are available Toro *et al.* (1999) suggest calculating the coancestries conditional on marker information. The idea is to calculate the probability of identity by descent along all points in the genome given the pedigree and the marker information using a procedure based on Monte Carlo Markov chains. The same suggestion has been made by Wang (2002).

Simulation

The breeding population consisted of $N_s = 8$ sires and $N_d = 24$ dams. Each dam produced 3 progeny of each sex. These 72 offspring of each sex were candidates for selection to breed the next generation. This simulated nucleus mimics the conservation programme carried out in the Guadyerbas strain of Iberian pigs. A genome of 20 chromosomes, with 100 loci /chromosome, was simulated with all the alleles being different in the founder population and a variable number of markers and alleles were placed equally spaced. Besides the basic situations of random selection and classical (WFS) and optimal (OWFS) within family selection, several situations of minimisation of average group coancestry were considered depending on the information used to calculate coancestry: a) pedigree information only (PS); b) molecular information on all the genome (2000 loci, MS); c) marker information from 2 markers/chromosome with 10 alleles/marker (MS2); d) pedigree plus marker information from 2 markers/chromosome (PM2).

Results

The values of the average coancestry of evaluated individuals over 10 generations are given in Table 3. With no molecular information the true homozygosity values were almost identical to those calculated from pedigrees (results not shown). Optimal within-family selection (OWFS) was substantially more efficient (15% advantage) than classical within-family selection (WFS). The method of minimising the average group coancestry (PS) coincides basically with OWFS because the scheme is equilibrated. However it will be more robust against departures from ideal conditions: founder individuals differentially related, random offspring deaths or fluctuating populations sizes.

The method of minimum average group coancestry using all the molecular information (MS) produced the maximum advantage attainable. Although the algorithm utilised did not give the optimal solutions, the observed rate of true expected homozygosity is reduced almost by a half with respect to the situation where it is absent. When molecular information is used for selection, the inbreeding coefficient did not reflect the true homozygosity, and the discrepancy could be considerable. Furthermore, the rate of advance in homozygosity, unlike the rate of inbreeding, does not attain an asymptotic value after a short number of generations but it is continuously decreasing (Toro et al., 1999). The use of partial molecular information produces better results than OWFS but obviously there will be a substantial increase in the genotyping cost of the programme (144 individuals genotyped for 40 markers). The value of a marker is related to the number of alleles: 2 markers with 10 alleles are as valuable as 6 markers with 4 alleles.

Table 3.
Advance of the genome expected homozygosity over generations for different strategies of genetic management.

Generations	Random	WFS	OWFS	PS	MS	MS2	PM2
0	0.039	0.034	0.032	0.032	0.031	0.033	0.031
1	0.058	0.047	0.043	0.043	0.039	0.044	0.040
2	0.076	0.059	0.054	0.054	0.047	0.054	0.049
3	0.095	0.072	0.065	0.065	0.054	0.065	0.057
4	0.112	0.085	0.076	0.075	0.061	0.074	0.065
5	0.129	0.097	0.087	0.086	0.068	0.084	0.074
6	0.146	0.109	0.098	0.096	0.074	0.093	0.081
7	0.162	0.121	0.108	0.106	0.081	0.101	0.089
8	0.178	0.132	0.118	0.116	0.087	0.110	0.097
9	0.193	0.144	0.128	0.126	0.092	0.118	0.105
10	0.208	0.155	0.138	0.136	0.098	0.127	0.112
Ne	26	37	43	44	77	52	58

Conclusions

There are substantial benefits that could be gained by the use of molecular markers in the genetic management of a conservation programme. However, although the conclusions obtained through simulation probably have some generality, it should be recognised that some theoretical developments on marker assisted conservation are needed. For example, prediction of the rate of advance of true homozygosity by descent when the selected trait is heterozygosity itself, measured either by molecular and pedigree information, is lacking. On the other hand, the usefulness will depend critically on genotyping costs. Microsatellite DNA markers have been considered until now as the most useful markers, specially if multiplex genotyping is used, but in the near future other DNA polymorphism such as single nucleotide polymorphisms (SNP) could be the most adequate to be routinely scored. It is also interesting to emphasise that the adequate use of molecular tools requires the use of an optimal method of combining pedigree and molecular information that rely on sophisticated Monte Carlo Markov chains methods.

Analysis of genetic diversity in Iberian breed

In recent years, the old breed structure of Iberian pigs, with different local varieties, has been substituted by a pyramidal structure based on crossbreeding with Duroc pigs and strong dependence on a small number of elite herds supplying purebred Iberian animals to all the production area (Silió, 2000). In these circumstances, some ancestral varieties have disappeared, other ones are endangered or blended, leading to the need for a new design of the conservation programmes for these genetic resources.

Phylogenetic techniques based on genetic distances estimated from polymorphic microsatellite markers are the method of choice to assess the genetic diversity of livestock breeds (Barker, 1999). Thaon d'Arnoldi et al. (1998) emphasised the analysis of genetic distances by the Weitzman (1992) approach to measure the global diversity and the marginal loss of diversity attached to each breed. Conservation priorities can be determined, based on these results. Caballero and Toro (2002) consider these phylogenetic methods inappropriate for within-species breed conservation, because genetic variation within groups is ignored in this approach, although it may be of great importance for the management of livestock breeds. They have proposed other basic tools, measured from markers, to analyse genetic diversity in breeds or varieties of livestock species. Here, we assess the genetic diversity in populations of Iberian pigs and compare both approaches cited to establish conservation strategies for this breed (Fabuel et al., 2004).

Pig and marker data

Two of the groups of Iberian pigs considered in the present work, Guadyerbas and Torbiscal have been described previously. The remaining Iberian pig groups represent the main red (Retinto) and black hairless (Negro Lampiño) varieties and a black hairy variety (Entrepelado), whose piglets are chestnut colour at birth. Blood samples were collected from 173 pigs of both sexes in the Iberian breed herdbook. Due to their historical and present relations with the Iberian pigs, another 40 Duroc pigs were also sampled and analysed. Thirty six microsatellite markers, two on each autosome, were genotyped. They were chosen for their reproducibility, position along the chromosome, polymorphism and absence of null alleles. The number of alleles for each microsatellite ranged from 4 to 13.

Analysis of genetic diversity in a subdivided population

Here we follow closely the development of Caballero and Toro (2002) who expressed the genetic diversity as the complement to one of the average global coancestry $(\bar{f})$ as $GD_T = 1 - \bar{f} = 1 - \sum_{i,j=1}^{n} f_{ij} c_i c_j = 1 - \sum_{i=1}^{n} c_i \left[f_{ii} - \sum_{j=1}^{n} D_{ij} c_j \right]$,

where f_{ij} is the coancestry between populations i and j, D_{ij} the Nei's minimum distance between subpopulations i and j and c_i the weight given to each population that will be assumed to be equal. The right hand side of the previous equation shows how the average global coancestry depends on the within-subpopulation coancestry (first term in the brackets) and the average distance among subpopulations (second term in the brackets).

From the coancestry matrix between populations, the Nei's minimum distances and the Reynolds distances can be calculated and are included in Table 4. For all the populations of the Iberian breed the genetic distance to the Duroc population is greater than that to any of the other populations of the breed and for both distances the smallest value was obtained between Entrepelado and Retinto and the largest between Duroc and Guadyerbas. This indicates that the two breeds have a clear differentiation.

Table 4. Values of Reynolds genetic distance (above the diagonal), Nei minimum genetic distance (below the diagonal) and population self-coancestries (diagonal) between the analyzed Iberian and Duroc pig populations.

	Torbiscal	Guadyerbas	Retinto	Entrepelado	Lampiño	Duroc
Torbiscal	0.4147	0.2419	0.1386	0.1380	0.1331	0.2462
Guadyerbas	0.1641	0.5565	0.2492	0.2463	0.2141	0.3361
Retinto	0.1000	0.1827	0.3427	0.0391	0.0796	0.1912
Entrepelado	0.1004	0.1819	0.0270	0.3306	0.0760	0.1739
Lampiño	0.0970	0.1528	0.0577	0.0554	0.3219	0.1665
Duroc	0.2013	0.2762	0.1543	0.1386	0.1324	0.3524

When analysing the genetic diversity within a breed it is interesting to calculate the proportional contribution of each population to the global coancestry. The proportional contribution to the global coancestry of the Iberian breed of each analysed variety or strain is given in Table 5. The Guadyerbas strain is the one that would contribute most due to their own coancestry, but because it presents the highest genetic distance from the other strains, its total contribution is lower than the Retinto strain.

Table 5. Proportional contribution of each strain to the global coancestry of Iberian breed.

Strain	Within strain	Genetic distance from other strains	Total
Torbiscal	0.0743	0.0167	0.0576
Guadyerbas	0.1029	0.0259	0.0770
Retinto	0.0990	0.0192	0.0798
Entrepelado	0.0573	0.0120	0.0453
Lampiño	0.0558	0.0125	0.0433
	$\tilde{f} = 0.393$	$\bar{D} = 0.090$	$\tilde{f} = 0.304$

Another question is to ascertain the loss/gain of diversity if one or several populations are removed from the pool. This can be calculated by disregarding one (or more populations) and recalculating the global average coancestry from the remaining pool. An application of this technique to these Iberian pig populations is shown in Table 6. The removal of the Torbiscal and Guadyerbas increases the within-strain genetic diversity but decreases the average genetic distance. The global balance will be negative for Torbiscal and positive for Guadyerbas. That means that the removal of the Guadyerbas strain actually will improve the total genetic diversity of the breed. The removal of the other three strains will decrease the within-strain genetic diversity, but will increase the genetic distance, the balance being negative.

Table 6. Total genetic diversity and loss (-) or gain (+) of diversity when each one of the strains is removed.

Strain	Within strain genetic diversity	Average genetic distance	Total genetic diversity	Weitzman method
All strains	0.6067	0.0895	0.6962	0.5065
Torbiscal	+0.0054	−0.0073	−0.0019	−0.1386
Guadyerbas	+0.0408	−0.0348	+0.0060	−0.2492
Retinto	−0.0126	+0.0044	−0.0082	−0.0462
Entrepelado	−0.0157	+0.0048	−0.0109	−0.0391
Lampiño	−0.0178	+0.0050	−0.0128	−0.0796

With the Weitzman method, the removal of any strain will always decrease the total genetic diversity. The removal of the Gudayerbas strain will have the greatest impact on the decrease of variability and afterwards the loss of the Torbiscal strain will have the greatest impact.

It is clear that the results of assessing the conservation priorities either using the between-population diversity or the total diversity will produce different and sometimes opposite results. A practical proposal could be to consider an aggregate diversity, expressed as a linear combination of within-breed and between-breed diversity weighted in an appropriate manner, as suggested by Ollivier and Foulley (2002). The weights will depend on the scenario imagined for the long term use of the diversity. If the prospective is to use them in crossbreeding between populations, diversity should be prioritised. On the other hand, if we are thinking of the future creation of a synthetic population able to cope with a challenging environment, total diversity will be preferable.

Finally, if we consider the optimal contributions of each strain to produce a single synthetic or to build a germplasm bank, they can be calculated finding the appropriate values of c_i that would minimise the global genetic diversity or that maximise Nei's genetic distances or whatever combination of both we want to impose. An example appears in Table 7.

Table 7.
Optimal contributions that minimise global coancestry or maximise Nei's genetic distances.

	Contributions	
	Minimise global coancestry	Maximise Nei's genetic distance
Torbiscal	0.1284	0.2283
Guadyerbas	0.0438	0.4063
Retinto	0.1129	0.1730
Entrepelado	0.3023	0.1619
Lampiño	0.4127	0.0305

Conclusions

There are many aspects to conservation decisions, such as adaptive features, possession of specific traits of economic or scientific value or the historical or cultural value of the population or the degree of endangerment. Even in the context of genetic information there are also several aspects that could be considered: number of distinct alleles, genetic diversity in special regions of the nuclear or mitochondrial DNA. However, the genetic diversity or expected heterozygosity is considered a major criterion, but the priorities could differ depending on the emphasis in between- or within-population diversity.

Mitochondrial sequence variation in Iberian pigs

Animal mitochondrial DNA (mtDNA) is highly polymorphic and its inheritance is asexual, maternal (almost exclusively) and without intermolecular genetic recombination. The clonal transmission of mtDNA haplotypes without recombinational noise make it possible to discriminate maternal lineages within and among species (MacHugh and Bradley, 2001). The analysis of sequences of the most variable regions (Cytochrome B and Control region D-loop) allows a useful tool to study maternal relationships among the individuals and to characterise the genetic origin of domestic animal populations and breeds. In this sense, the analysis of pig mtDNA sequence variation has shown two independent domestication processes in Asia and Europe. The presence of Asian haplotypes in most of the European pig breeds has also confirmed the contribution of Asian pigs to the development of these breeds (Giuffra et al., 2000; Kim et al., 2002).

Figure 1.
Number of maternal lineages from the four founder strains (Ervideira, Caldeira, Campanario and Guadyerbas) surviving in the composite Torbiscal line over the period 1963-2000 inferred from pedigree analysis.

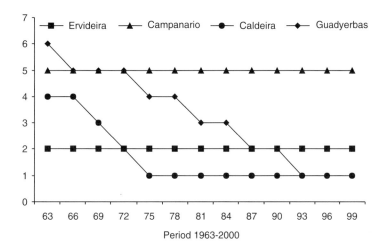

In order to begin these studies in Iberian pigs, mtDNA sequence variation was studied in Torbiscal pigs sampled to represent the nine surviving maternal lineages from four Portuguese and Spanish founder populations, according to a previous pedigree analysis (Figure 1). It should be noted that the complete pedigree goes back to the middle of the 20th century, when the diverse varieties of the Iberian breed were maintained with low genetic interchange among them. Five lineages correspond to founder sows of the Campanario strain, black hairless pigs from the variety named Lampiño de La Serena. Other two lineages correspond to the Erviderira strain, blond pigs from the now practically disappeared Dourado Alemtejano variety and the two remaining lineages to founder sows from the Guadyerbas and Caldeira, correspondent to varieties named Lampiño del Guadiana (black hairless) and Retinto (red), respectively.

The complete sequence of the D-loop region (1246 bp) and Cytochrome B (Cyt B) gene (1140 bp) were determined for the nine sampled maternal lineages by direct sequencing of PCR-products. We found the presence of nine SNP in the D-loop region and four SNP in Cyt B. The combination

Table 8. Variable positions in *D-loop* and *Cyt B* regions of mitochondrial DNA among founder maternal lineages surviving in the Torbiscal line of Iberian pigs. Sequence identities and deletions are indicated by dots and dashes. Nucleotide positions are numbered according to reference GenBank AJ002189 (Ursing and Arnason, 1998).

Haplotype	Genetic origin of maternal lineage	D-loop													Cytochrome B					
Nucleotide Positions		15544	15558	15578	15615	15694	15714	15715	15717	15741	15758	16127	16139	16141	14264	14309	14717	14740	15017	15264
	AJ002189	G	A	–	T	C	T	T	C	C	T	A	A	A	T	G	G	C	T	A
H1	Ervideira	A	.	.	.	.	.	C	.	.	C	G	G	G	.	A	A	.	.	.
H2	Guadyerbas	.	.	.	.	.	C	.	.	.	C	G	G	G	.	.	A	T	.	.
H3	Caldeira	.	.	.	.	.	C	.	.	.	C	G	G	G	.	.	A	.	.	.
H4	Campanario	.	.	A	.	.	C	.	.	.	C	G	G	G	.	.	A	.	.	.
H5	Campanario	.	T	.	.	.	C	.	.	T	C	.	.	.	.	.	.	.	.	.
H6	Campanario	.	.	.	.	.	C	.	.	.	C	.	G	G	.	.	A	.	.	G

of the 13 identified SNP allows to discern six haplotypes in the Torbiscal population (Table 8). Three haplotypes are specific for each one of the ancestral strains Ervideira (H1), Caldeira (H2) and Guadyerbas (H3). The other three were observed in lineages from the Campanario founder strain. The type H4 is common to the maternal lineages of three founder sows and the types H5 and H6 correspond to lineages of two different founder sows (Alves et al., 2002a). These results allow us to assess non-autosomal genetic variability in the composite Torbiscal line, related to differences among ancestral varieties. As a consequence, some changes to improve the preservation of this mitochondrial diversity should be introduced in the conservation programme.

These analysis were recently extended to another 17 Iberian pigs sampled from diverse farms, representative of the genetic types actually present in the breed. In the absence of complete genealogical data, the goal was to trace the migration between ancient and new populations. We found a total number of 19 SNP in these additional samples, including four new SNP in the D-loop region and three new SNP in the Cyt B gene. The combination of these SNP results in 12 mtDNA haplotypes. Four of them (H1, H2, H4 y H6) were previously identified in Torbiscal, showing common maternal ancestry. In the scarce farms breeding black hairless Iberian pigs, only types H2, H4 and H6 were found, assignable to ancient Lampiño varieties. However, haplotypes corresponding to the ancient varieties Lampiño (H4 and H6) or Dourado (H1) were also found in farms breeding red Iberian pigs. It confirm the increasing genetic interchange between herds and the consequent introgression among traditional varieties. Another interesting result is that two of the previously identified types in Torbiscal (H3 and H5) are exclusively present in this line, indicating its great interest as reservoir of the Iberian pig genetic resources. These results also allow us to discard the presence of Asian haplotypes and the maternal contribution of Asiatic pigs to the Iberian breed (Alves et al., 2003).

DNA tests for authentification of the raw material of Iberian pig products

Animal traceability from birth to market is increasingly being demanded by consumers as an element of food safety assurance systems. In this general context, the ability to discriminate the genetic origin of the raw material has a singular importance for high quality animal products. In many cases, the brand name of these products is directly associated to the species or breed from which they originated. Gandini and Oldenbroek (1999) have shown how a marketing link between product and breed can improve the economic profitability of local breeds. Part of the transforming Spanish industry is interested in distinguishing between cured products from purebred Iberian pigs and those from crossbred Iberian x Duroc pigs, within a market already initially differentiated. A DNA test of breed identity has clear advantages for the protection of the brand name of highly-prized products of purebred Iberian pigs.

Diagnostic markers and exclusion probability

Although the ideal diagnostic markers correspond to alleles that can be detected in only one population, this is not a frequent event. The term 'population-specific marker' usually defines the genetic markers that can be detected with large differences of allelic frequencies between the populations analysed. Microsatellites have been successfully used for distinguishing livestock species and breeds not very closely related (Blott *et al.*, 1999). More difficult problems arise when we try to differentiate purebred and crossbred animals from two related populations as is the case with Duroc and Iberian breeds. For example, only four out of the 49 microsatellites tested in our laboratory present alleles only detected in Duroc and never in Iberian pigs.

The use of Amplified Fragment Length Polymorphism (AFLP) have advantages in identifying diagnostic or specific markers. Although these markers are generally dominant, the AFLP technique does not require previous knowledge of the DNA sequence, generates reproducible fingerprinting profiles and allows the amplification of a high number of DNA fragments per reaction, enabling the detection of specific amplified fragments (Vos *et al.*, 1995). Thirteen out of the 15 primer combinations that were tested in DNA samples from Iberian and Duroc pigs showed good reproducible amplification patterns and were used to detect possible markers of the Duroc breed. These 13 primer combinations allowed the amplification of 588 fragments ranging between 35 and 400 bp, and out of them 139 were polymorphic. The aim of this study was to identify a set of AFLP markers present in the Duroc at variable frequencies and absent in the Iberian breed. Fourteen AFLP verified this condition and the nine most frequent could be used in a panel of markers to discriminate purebred and crossbred genotypes (Alves *et al.*, 2002b).

Furthermore, we have analyzed the DNA sequence of *MC1R* gene (*Extension* colour gene) in Iberian and Duroc pigs. The sequence analysis revealed a new allele/haplotype, *MC1R*7. We have genotyped *MC1R* alleles by PCR-RFLP test and fragment analysis of PCR products in a lot of animals of both breeds. The results confirm that the *MC1R*4 allele is fixed in Duroc and absent in Iberian pigs. The Retinto and Entrepelado Iberian varieties have present only *MC1R*6 and *MC1R*7 alleles, and black hairless pigs of the Lampiño variety are homozygous for the *MC1R*3 allele (Fernández *et al.*, 2002).

Table 9.
Exclusion probability (P_{EC}) for pure Iberian origin, for pigs with 1/4 Duroc genes, obtained using different number of markers and pooling from one to five individuals per sample.

No. of sampled pigs	1	2	3	5
DNA test based on				
MC1R + 4 microsatellites	0.761	0.943	0.986	0.998
MC1R + 9 AFLP	0.891	0.988	0.999	1.000
MC1R + 9 AFLP + 4 microsatellites	0.948	0.997	1.000	1.000

Being g_i the allelic frequencies for the ith marker and k the proportion of Duroc genes in the crossbred animals (usually, $k = 1/2$ or $1/4$), the exclusion probability of a purebred Iberian origin, conditional to the ith diagnostic marker, is $P_E^{(i)} = 2k\,g_i$. For example, the MC1R genotype allows a value of $P_E^{(i)} = 1$, for pigs with 1/2 Duroc genes ($k = 1/2$). But, a higher number of markers would be required to discriminate genotypes with a lower proportion of Duroc genes. With a total number of M markers, the whole exclusion probability of Iberian origin would be $P_{EC} = 1 - \prod_{i=1}^{M} (1 - P_{E(i)})$. The expected P_{EC} values for 1/4 Duroc crossbred pigs, using diverse DNA tests, based on different combinations of previously described markers are shown in Table 10.

Conclusions

These results highlight the potential of genetic markers as a useful tool for differentiating the genetic origin of products from purebred and crossbred Iberian pigs. Nevertheless the elevated costs and relative complexity of the AFLP technique make it somewhat cumbersome, especially if it has to be applied as a routine test to the quality control of thousands of samples, and more simple diagnostic markers must be investigated. Similar DNA tests could similarly be applied to protect the brand name of other animal food products.

Acknowledgements

We thank J. Fernández for computer software. Financial support has been provided by INIA grants RTA01-051 and RZ01-028-C2-1.

References

Alves, E., Ovilo, C., Rodríguez, M.C. and Silió, L. 2002a. Mitochondrial D-loop and Cytochrome B gene sequences as a tool to preserve the genetic diversity in an Iberian pig population. *International Conference on Animal Genetics*, August 2002, Göttingen.

Alves, E., Castellanos, C., Ovilo, C., Silió, L. and Rodríguez, M.C. 2002b. Differentiation of the raw material of the Iberian pig meat industry based on the use of amplified fragment length polymorphisms. *Meat Science* 61: 157-162.

Alves, E., Ovilo, C., Rodríguez, M.C. and Silió, L. 2003. Mitochondrial DNA sequence variation and phylogenetic relationships among Iberian pigs and other domestic and wild pig populations. *Animal Genetics* 34: 319-324.

Barker, J.S.F. 1999. Conservation of livestock breed diversity. *Animal Genetic Resources Information* 25: 33-43.

Blott, S.C., Williams, J.L. and Haley, C.S. 1999. Discriminating among cattle breeds using genetic markers. *Heredity* 82: 613-619.

Caballero, A. and Toro, M.A. 2000. Analysis of genetic diversity for the management of conserved subdivided populations. *Genetical Research* 75: 331-343.

Eding, H., Meuwissen, T.H.E. 2001. Marker based estimates of between and within population kinship for the conservation of genetic diversity. *Journal Animal Breeding Genetics* 118: 141-159.

Fabuel, E., Barragán, L., Silió, L., Rodríguez, M.C. and Toro, M.A. 2004. Analysis of genetic diversity and conservation priorities in Iberian pigs based on microsatellite markers. *Heredity* (in press).

Fernández, A., Óvilo, C., Castellanos, C., Rodríguez, M.C., Toro, M.A. and Silió, L. 2002. Identification of a new haplotype for the *MC1R* gene in Iberian pig. *International Conference on Animal Genetics*, August 2002, Göttingen.

Gandini, G. and Oldenbroek, J.K. 1999. Choosing the conservation strategy. In: *Genebanks and the conservation of farm animal genetic resource*. Edited by J.K. Oldenbroek, DLO Institute for Animal Sciences and Health. The Netherlands. pp. 75-90.

Giuffra, E., Kijas, J.M.H., Armager, V., Carlborg, O., Jeon, J.T. and Anderson, L. 2000. The origin of the domestic pig: Independent domestication and subsequent introgression. *Genetics* 154: 1785-1791.

Kim, K.I., Lee, J.H., Li, K., Zhang, Y.P., S.S., Gongora, J. and Moran, C. 2002. Phylogenetic relationships of Asian and European pig breeds determined by mitochondrial D-loop sequence polymorphism. *Animal Genetics* 33: 19-25.

Lynch, M. and Ritland, K. 1999. Estimation of the pairwise relatedness with molecular markers. *Genetics* 152: 1753-1766.

MacHugh, D.E. and Bradley, D.G. 2001. Livestock genetic origins: goats buck the trend. *Proceedings National Academy of Sciences USA.* 98: 5382-5384.

Malécot, G. 1948. *The Mathematiques de l'Heredité*. Masson et Cie, Paris, France.

Ollivier, L. and Foulley, J-L. 2002. Some suggestions on how to preserve both within and between genetic diversity. *Proceedings of the 53rd Annual Meeting of the European Association for Animal*

Production, Cairo, Egypt.

Rodrigañez, J., Toro, M.A. Rodriguez, C. and Silió, L. 2000. Alleles survival from Portuguese and Spanish strains in a population of Iberian pig. In: *Tradition and innovation in Mediterranean pig production*, Edited by J.A. Afonso and J.L. Tirapicos, CIHEAM-UE, Zaragoza, Spain. pp. 57-61.

Silió, L. 2000. Iberian pig breeding programme. In: *Developing breeding strategies for lower input animal production environments*. Edited by S. Galal, J. Boyazoglou and H. Hammond, International Committee for Animal Recording, Roma, Italy. pp. 511-520.

Toro, M.A. and Maki-Tanila, A. 1999. Establishing a conservation scheme. In: *Genebanks and the conservation of farm animal genetic resource*. Edited by J.K. Oldenbroek, DLO Institute for Animal Sciences and Health. The Netherlands. pp. 75-90.

Toro, M.A., Silió, L., Rodrigáñez, J., Rodríguez, M.C. and Fernández, J. 1999. Optimal use of genetic markers in conservation programes. *Genetics Selection Evolution* 31: 255-261.

Toro, M.A., Rodrigañez, J., Silió, L. and Rodriguez, C. 2000. Genealogical analysis of a closed herd of black hairless Iberian pigs. *Conservation Biology* 14: 1843-1851.

Toro, M.A., Barragán, C., Ovilo, C., Rodrigáñez, J., Rodríguez, M.C. and Silió, L. 2002a. Estimation of coancestry in Iberian pigs using molecular markers. *Conservation Genetics* 3: 309-320.

Toro, M.A., Barragán, C., Ovilo, C., Rodríguez, M.C. and Silió, L. 2002b. Estimation of coancestry from genetic markers in a population with complex pedigree. *Proceedings 7th Congress on Genetics Applied to Livestock Production* 33: 485-488.

Ursing, B.M. and Arnason, U. 1998. The complete mitochondrial DNA sequence of the pig. *Journal of Molecular Evolution* 47: 302-306.

Thaon d'Arnoldi, C., Foulley, J-L. and Ollivier, L. 1998. An overview of the Weitzman approach to diversity. *Genetics Selection Evolution* 30: 149-161.

Vos, P., Hogers, R., Bleeker, M., Reijans, M., van de Lee, T., Hornes, M., Frijters, A., Pot, J., Peleman, J., Kuiper, M., and Zabeau, M. 1995. AFLP: a new technique for DNA fingerprinting. *Nucleic Acids Research* 23: 4407-4414.

Wang, J. 1997. More efficient breeding systems for controlling inbreeding and effective size in animal populations. *Heredity* 79: 591-599.

Wang, J. 2002. Optimal marker-assisted selection to increase the effective size in small populations. *Genetics* 157: 867-874.

Weitzman, M.L. 1992. On diversity. *Quarterly Journal of Economics* 107: 363-405.

19

UK rare breeds: population genetic analyses and implications for applied conservation

S.J. Townsend

Rare Breeds Survival Trust, NAC, Stoneleigh Park, Warwickshire, CV8 2LG, UK

Abstract

Since the establishment of the Rare Breeds Survival Trust (RBST) in 1973, rare breed genetic conservation has only gradually emerged as a major key to the continued success and survival of populations. Thus, although basic population and demographic data have been recorded for all breeds listed by the Trust and most rare breeds have kept detailed pedigrees, there is still little information available about current population genetic structure and dynamics over time. Consequently, the majority of rare breed population meta-analyses based upon pedigree data are yet to be carried out. Transfer of rare breed records onto a computer database is therefore a current priority for the RBST, in addition to making available software to provide rare breed organisations with breed profiles to describe founder effects, effective population size (N_e), rate of inbreeding (ΔF), and kinship patterns. Results from rare breed pedigree analyses using this software can now be used to a) illustrate where and how loss of genetic diversity has taken place in rare breeds and b) enable decision processes concerning conservation strategy in future. Conservation strategy informed by such analyses can only be successfully implemented if due regard is given to the realities imposed by the need to maintain rare breed populations within an agricultural context, however. In light of these recent improvements to data recording and access to studies of population genetic structure, some of the traditional limitations to advances to rare breed conservation strategy may now need to be re-examined.

Introduction

Conservation genetic management of endangered populations today is assisted both by molecular and pedigree analyses (Paetku *et al.*, 1999; Cymbron *et al.*, 1999; Goldstein *et al.*, 1999; Caballero & Toro, 2000; Woodworth *et al.*, 2002). These are used to examine both inter- and intra-population relationships, and provide different levels of information for study of current population genetic variation, disease and character traits and historic population events. Here, pedigree analyses are considered as a tool for use in rare breed conservation.

Loss of genetic diversity

There are now 72 breeds (16 cattle, 12 horse/pony, 7 pig, 8 poultry and 29 sheep and goats), which meet the Rare Breeds Survival Trust (RBST) criteria for recognition as a rare breed, all of which record less than 3000 adult registered females, and fourteen of which have populations under 300 (RBST Watchlist 2002). It is well known that small breeding populations lose genetic diversity more quickly than large breeding populations, and sustained restriction in population size in particular can result in significant diversity loss (Taylor *et al.*, 1994; Frankham, 1996; Lande, 1998; Bouzat *et al.*, 1998). Loss of genetic diversity is closely associated with both a decrease in evolutionary capacity (e.g. the ability to harbour potential for trait selection) and an average increase in inbreeding from generation to generation, which can lead to reduced levels of fitness and infertility associated with inbreeding depression (Frankham & Weber, 2000; Bijlsma *et al.*, 2000; Hedrick *et al.*, 2000; Nieminen *et al.*, 2001). Human intervention to maintain and/or maximise genetic diversity in domestic populations can be achieved and subsequently sustained in theory (Caballero & Toro, 2002), however, since considerable pedigree information is usually available.

Use of pedigree analysis for conservation genetic management of UK rare breeds

Since the establishment of the RBST in 1973, rare breed genetic conservation has only gradually emerged as a major key to the continued success and survival of populations. In the first years of the rare breed movement, greatest emphasis was placed upon maintaining and/or increasing population number. While this approach can currently be vindicated as having been successful in ensuring survival of these breeds, it has also provided the principal mechanism in the live population through which genetic diversity has been maintained. Consequently, although basic population and demographic data has been recorded for all breeds listed by the Trust and most rare breeds have kept detailed pedigrees, there is still little information available about current population genetic structure and dynamics over time.

The absence of results from pedigree analyses that could have provided this information is partly due to unavailability of recording software in the past. Assessing population structure even in small populations is a difficult task using paper pedigree records alone, as it involves interpreting decades of annual entries by hand for many hundreds of animals. Thus, although pedigrees have been recorded for many UK breeds since the 19th century, and have been used since their foundation as a means of making individual breeding decisions, use of whole rare breed pedigrees to infer population genetic structure is not yet common practice.

Consequently, transfer of rare breed records from flock, herd and stud books onto computer database is a current priority for the RBST, in addition to providing software that can then use this information to create 'profiles' for rare Breed Societies that include estimates of basic population genetic parameters such as founder effect, effective population size (N_e), rate of inbreeding (ΔF), and kinship patterns. The software will be designed as an add-on to pedigree recording/breed management software already in use by many rare breed societies. As soon as the software becomes available, the main limiting factor on its use will be the lack of pedigree records still not transferred to electronic databases. Currently, the proportion of rare breeds with records in electronic format is 48%, 56% and 67% for sheep/goats, cattle and equines respectively (Townsend, 2002).

Results from preliminary rare breed pedigree analyses using RBST software are now beginning to illustrate some of the problems facing rare breeds. It can be shown that some of the smallest UK rare breed populations may have already lost between 10-20% of the diversity present at the time of RBST foundation, and are likely to lose diversity at least 40 times more rapidly than most large UK mainstream breeds in the next 30 years if current N_e sizes are maintained. Consequently, those rare breed populations that have been saved from extinction by the RBST during the last 30 years are now in need of careful conservation genetic management if their future survival and potential value as an indigenous genetic resource are to be ensured.

Future conservation goals and potential problems

Conservation strategy informed by pedigree analyses can only be successfully implemented if due regard is given to the realities imposed by the need to maintain rare breed populations within an agricultural context. Although pedigree analyses will allow conservation goals to be defined that keep loss of genetic diversity to a minimum, realistic management strategies then need to be created to achieve these goals before conservation strategy can be implemented.

There are a number of management strategies traditionally designed to minimise diversity loss that employ methods to minimise kinship, maximally avoid inbreeding or maximise N_e/N, for example. These methods were principally designed to be applied to whole captive populations, however, and may include mechanisms, such as equalising family size from generation to generation, which are unsuitable for fragmented livestock populations that are under differential selection or bred to accommodate market forces. There are ways of tailoring management strategies to suit application to livestock systems, such as using line breeding, selection systems that use conservation constraints or conservation breeding group networks, but all of these methods still require a high degree of breeder participation and commitment in order to be effective. Even without pedigree analyses available, there have been a number of these management schemes in place over the years, but most have experienced little long-term success. The Gloucestershire Old Spots pig breed is one of the few breeds to have maintained a longstanding programme, with a cyclic breeding system introduced to the breed in the 1960's that uses rotational mating from one generation to the next as a means of minimising the increase in inbreeding. When the system was implemented, there were only 120 sows in existence, however, mainly under the control of a single breeder, which meant that the system was relatively easy to implement and monitor. As the breed has become more successful, it is apparent that a decline in breeder participation has taken place, mainly due to increased diversification of breeder approach and population fragmentation.

Current position

It is clear that attitudes to genetic conservation among the rare breed community are central to the long-term success of any management strategy designed to maintain genetic diversity. These attitudes remain divided however – current indications (Townsend, 2002) show maintenance of genetic diversity coming second place to marketing, conformation to breed standard/type and increasing population number, for example. If these data are a true reflection of prevailing attitude, it is obvious that much work is required to put conservation of genetic diversity on an equal footing with these and other important factors before successful strategies can be implemented. It is therefore important to consider some of the reasons why it might have relatively low status on a rare breed owner's list of priorities – three common causes are considered here:

- Market forces: to ensure successful futures, rare breeds need to be maintained as part of the living countryside. This, in turn, means that rare breed owners need to earn income from their animals. Market forces are directly linked to this process, leading to different selection pressures within and between breeds. As a result, individual and group opinion often differs within and between Breed Societies as to what it is that should be conserved, and how it should be carried out.

- Legislation: there is currently enormous pressure and/or requirement to conform to current legislation – some examples include animal passport provision, animal movement restrictions and national breeding strategy imposed by the National Scrapie Plan.
- Breeder concern: there exists to a degree an underlying wariness of genetic research in any form among some rare breed owners. This can sometimes arise from fears that, once having been studied, inherent breed 'flaws' may be discovered or breeding plans may be removed from owner control, but also that any examination of genetic diversity automatically involves sampling of animals, which, after FMD, has many unpleasant associations.

Future developments

Efficient transfer of information obtained from forthcoming breed analyses is crucial to the future success of applied genetic conservation - until now, for example, whole breed analyses were usually not possible, meaning that the link between breed histories and genetic profile is not well known and rare breed owners are often unaware of the usefulness of such data. Once rare breed societies and owners have had time to assimilate knowledge from ongoing pedigree analyses, they will require help to formulate a conservation approach that combines the needs of rare breed owners with a strategy to minimise diversity loss. This process must be communicated effectively to individual breeders, and improvement to the rare breed society communication network is key to achieving this, in addition to the considerable resource and careful planning required for the success of emergent applied breeding programmes.

In conclusion, UK rare breed population genetic diversity and distribution is currently being examined for the first time on a large scale and considerable progress is being made in this area. Although the progression to an applied phase will not be automatic, gradual familiarisation with the status and history of current population genetic structure will mean that future strategy can now be well informed, and this information is vital for owners and rare breed organisations if effective conservation programmes are to be designed which can keep abreast of new challenges to long-term survival.

References

Bijlsma, R., Bundgaard, J., Boerema, A.C. 2000. Does inbreeding affect the extinction risk of small populations? Predictions from

Drosophila. *Journal of Evolutionary Biology* 13 (3): 502-514.

Bouzat, J.L., Lewin, H.A., Paige, K.N. 1998. The ghost of genetic diversity past: historical DNA analysis of the greater prairie chicken. *American Naturalist* 152: 1-6.

Caballero, A. and Toro, M.A. 2000. Interrelations between effective population size and other pedigree tools for the management of conserved populations. *Genetical Research* 75 (3): 331-343.

Caballero, A. and Toro, M.A. 2002. Analysis of genetic diversity for the management of conserved subdivided populations. *Conservation Genetics* 3 (3): 289-299.

Cymbron, T., Loftus, R.T., Malheiro, M.I., Bradley, D.G. 1999. Mitochondrial sequence variation suggests an African influence in Portuguese cattle *Proceedings of the Royal Society, London B* 266: 597-603.

Frankham, R. 1996. Relationship of genetic variation to population size in in wildlife. *Conservation Biology* 10: 1500-1508.

Frankham, R. and Weber, K.E. 2000. Nature of quantitative genetic variation. In: *Evolutionary Genetics: from molecules to morphology*. Edited by R.S. Singh & C.B. Krimbas. Cambridge University Press, Cambridge, UK. pp. 351-368.

Goldstein, D.B., Roemer, G.W., Smith, D.A., Reich, D.E., Bergman, A., Wayne, R.K. 1999. The use of microsatellite variation to infer population structure and demographic history in a natural model system. *Genetics* 151: 797-801.

Hedrick, P.W., Kalinowski, S.T. 2000. Inbreeding depression in conservation biology. *Annual Review of Ecology and Systematics* 31: 139-162.

Lande, R. 1998. Anthropogenic, ecological and genetic factors in extinction and conservation. *Researches on Population Ecology* 40 (3): 259-269.

Nieminen, M., Singer, M.C., Fortelius, W., Schops, K., Hanski, I. 2001. Experimental confirmation that inbreeding depression increases extinction risk in butterfly populations. *American Naturalist* 157 (2): 237-244.

Paetku, D., Amstrup, S.C., Born, E.W., Calvert, W., Derocher, A.E., Garner, G.W., Messier, F., Stirling, I., Taylor, M.K., Wiig, O., Strobeck, C. 1999. Genetic structure of the worlds polar bear populations. *Molecular Ecology* 8:1571-1584.

Taylor, A.C., Sherwin, W.B., Wayne, R.K. 1994. Genetic variation of microsatellite loci in a bottlenecked species: the northern hairy-nosed wombat. *Molecular Ecology* 3: 277-290.

Townsend, S. J. 2002. RBST Breed Society Survey, *RBST Publications*. Stoneleigh, UK

Woodworth, L.M., Montgomery, M.E., Briscoe, D.A., Frankham, R. 2002. Rapid genetic deterioration in captive populations: Causes and conservation implications. *Conservation Genetics* 3(3): 277-288.

20

A UK conservation success story: Longhorn cattle, a case study

E.L. Henson

British Longhorn Cattle Society/Grassroots Systems Ltd, PO Box 251, Exeter, EH2 8WX

Abstract

The Longhorn cattle breed has a long and prestigious history, dating back prior to the livestock pioneers of the 18th century. It was, for a period, the improved breed of choice in the Midland Counties. But the breed gradually fell from favour and, by the early 1970s, only 6 significant Longhorn herds remained in the UK. However, the Longhorn was one of many rare breeds to benefit from the growth of the rare breeds movement in the 1970s, led by the Rare Breeds Survival Trust. A number of factors have helped the breed to recover, including: an active breed society providing registrations and analyses based on these, promoting the breed, organising sales and shows and providing an important social framework for breeders and supporters; creation of a semen bank; niche marketing of meat and hides and the use of the breed in conservation grazing.

Introduction

I have worked with *in situ* conservation of rare breeds all my life, primarily in the UK, but also as the first director of the American Minor Breeds Conservancy and as adviser to the US National Academy of Science and the FAO. I believe that any successful conservation strategy involving farm animals is primarily dependant on both the animals and the people who keep them, and that the principal reasons for breed rarity, and therefore the effectiveness of strategies to ensure their survival, are based far less in scientific difference between breeds than in fashion, marketing, and the infrastructures which make their survival possible. Here, Longhorn Cattle are used as an example to illustrate the elements of what has been to date a successful rare breed conservation strategy in the UK.

The British Longhorn

The Longhorn breed (Figure 1) has a long and prestigious history dating back to a time before the great livestock pioneers of the 18th century made it, for a period, the improved breed of choice for the gentlemen farmers of the Midland Counties. Most well-known of these farmers was Robert Bakewell who, it is claimed, invented progeny testing, and hired out bulls to his neighbours and monitored their calves 'over the fence' before selecting those to bring back to Dishley Grange for use on his own herd. In time, the breed gradually fell from favour and was replaced by at first the Shorthorn and then a sequence of other improved or imported breeds. Finally, the development of intensive beef production methods in the late 1960's caused the breed to fall into terminal decline. Their large horns made them difficult to house and they did not compete well with Continental breeds under new intensive management conditions.

Figure 1.
The Longhorn breed of cattle.

The brink of extinction

By the early 1970's only 6 significant Longhorn herds remained, and during 1972 owners of three of these passed away and their herds were dispersed. Without the activities of the emerging rare breed movement this could well have signalled the imminent extinction of the breed, as had taken place with no less than twenty other unique and distinct UK breeds of British livestock during the first fifty years of the 20th century.

So what saved the Longhorn? How does a breed come back from such a precarious position to the one in which it finds itself today, with well over 1,000 registered females in numerous herds distributed the length and breadth of the country. There is little evidence that the recovery of the Longhorn breed is due simply to genetic superiority, although it is true that the breed does have some real attributes. It has an excellent record with respect to ease of calving, it has a reputation for being a good mother, and milks well to feed its calf. It is long lived and is an efficient converter of grass and other forage and browse into well-marbled quality beef. However, these qualities alone could not have saved it from extinction. Without the activities of the rare breed movement it would have joined the ranks of the 'lost breeds'.

Effective conservation strategy in the Longhorn breed

I believe that there are a number of key factors that combined to help not only save this breed for the purposes of conservation, but also enabled its breeders to rediscover a commercial niche. This market is in my view a critical part of any strategy for long-term survival.

Raising awareness

The rare breed movement, led by the Rare Breeds Survival Trust (RBST) in the 1970's increased the profile of rare breeds as a group, and made the case for their conservation for historical, aesthetic and biodiversity reasons. This created an interest and a market from conservation minded, often non-farming people who began to buy stock. Some of these people became active breeders, but even those who did not, supported the breed by providing a 'terminal' market.

Semen bank

The Breed Society established a semen bank that mirrored that proposed by the RBST, with between 10 and 25 bulls, unrelated for at least two generations, available at any time. Bulls were selected to represent as wide a pedigree spread as possible and in more recent years this has expanded to include inspection of both the bull and his daughters. A maximum of 50 straws could be sold from each bull in each year to prevent the excessive use of any one bull or family line. This strategy enabled small herd owners to keep and breed pure a small number of females without needing to own a bull, while avoiding the overuse of any one bull or family line.

Sales and shows

The RBST established the National Rare Breeds Show and Sale that was an opportunity for breeders to congregate, compare and exchange stock. For the first time the 'rarity value' of rare breeds began to have an effect, encouraging breeders to continue to breed pure and register stock. The Breed Society encouraged the establishment of show classes at major agricultural shows throughout the country and ran and continues to run workshops and member meetings to help newer breeders learn how to select, breed and present their stock. Showing is often seen as 'just a bit of fun' and may even be counterproductive in a rarer breed, but it does play a vital role in raising the profile of a breed among the agricultural public. This in turn has attracted commercial beef farmers to Longhorns.

Niche marketing

The RBST Traditional Breeds Meat Marketing Scheme and the Longhorn Cattle Societies' own Longhorn Meat Marketing Company have both been critical in providing a product market for the beef produced by its members. This is key to the breeds ultimate survival, which cannot be sustained long-term under the confines of a limited conservation breeding programme. Today almost all the Longhorn beef being produced by Society members is sold either to specialist butcher, or direct to customers mail order, and restaurants. This provides the breeders with a premium on their product that is the very best way to ensure they will continue to breed Longhorns. This has been further enhanced by a current market in Longhorn hide rugs, which adds considerable value to a Longhorn carcase over that of a Continental competitor.

Pedigree registration

The function of a Breed Society as a breed register is also essential to any pedigree breed, and the value of accurate, accessible records are increasingly important (Figure 2).

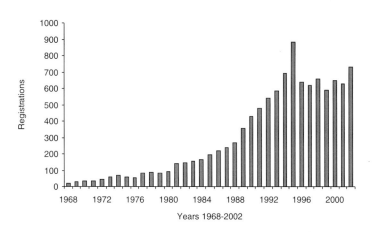

Figure 2.
The author uses the number of annual registrations as the measure of population trend in a breed. Live adult numbers can be estimates but may not accurately reflect the true effective population size when large numbers of adults are used only in commercial production.

In the case of the Longhorn there is a full pedigree electronic database, which tracks all living stock back to the 1950s. The Society uses Grassroots Systems pedigree software, which enables it to carry out all its administrative functions and track genetic characteristics, but with the addition of analysis software co-designed and produced by the RBST, it can also monitor the rate of change of inbreeding, effective population size, founder effects and kinship profiles. There are also some members recording and using best linear unbiased prediction (BLUP) estimated breeding values and the Society is able to carry out detailed analysis to select bulls for AI, track genetic changes within the population and monitor what is happening and where. This includes the annual update of births, deaths and transfers providing a vital tool during the recent foot and mouth crisis (Figure 3).

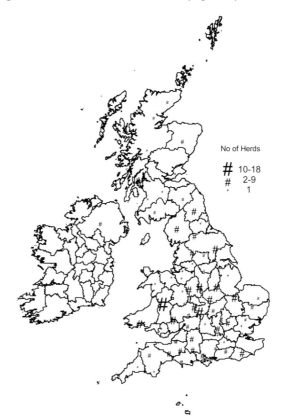

No of Herds

\# 10-18
\# 2-9
1

Figure 3.
Distribution of UK Longhorn Cattle Herds by postcode, November 2002.

Grazing Project

The Longhorn has also benefited from non-farming members who have had the vision to look outside the confines of basic farming structures for niche markets and roles for the breed. In particular, the Longhorn has proved itself extremely effective in conservation grazing of environmentally sensitive sites. The breed is now being used successfully in over 20 projects associated with the Grazing Animals Project (GAP), including a herd in Epping Forest, just north of London, of 100 Longhorns kept simply for their grazing style.

Social element

The Breed Society also plays a vital role in creating an enjoyable social structure within which pedigree members can operate. The Longhorn Breed Society has over 400 members, a quarter of whom regularly attend Society events. This same quarter of the membership is also the most active in showing, breeding bulls and in breed promotion. The pleasure members have in belonging to a Breed Society ultimately determines the success of the breed it represents. If members lose faith with their Society, they cease to be members and do not register their stock, with the result that the funding source dries up and breed promotion suffers.

Conclusion

I believe that successful breed conservation is dependant upon a combination of strategies:

- Pedigree data is essential to enable the planning of breeding strategies, to monitor and prevent introgression and raise the alarm if the breed is 'drifting' too rapidly.
- Semen banks are needed as a long-term conservation insurance tool and also to enable the small herd owner to breed pure. They may also be useful in breed improvement.
- Sales enable breeders to market their stock.
- Product markets create and maintain a realistic income for the breeders.
- Breed publicity and marketing maintains demand for breeding stock and breed products.
- The 'feel good factor' is required – breeds cannot thrive in the long-term without dedicated breeders to support and believe in them.

In my view therefore, the best way to save a breed, whether for its historical interest, aesthetic beauty, the 'genetic package' it represents, or its unique traits, is to create the environment in which the breeders can thrive, in a sustainable system.

21

Conserving animal genetic resources: making priority lists of British and Irish livestock breeds

S.J.G. Hall

Department of Biological Sciences, University of Lincoln, Riseholme Hall, Lincoln, LN2 2LG, UK

Abstract

Prioritisation of livestock breeds for conservation is agreed to depend upon the genetic distinctiveness of breeds, on census data and degree of endangerment, and on other factors relating to the present, future, or past function of the breeds in the livestock industry. How these factors can be combined to yield a prioritised list needs to be considered. An objective framework for prioritisation can be deduced if breeds are compared with each other by plotting genetic distinctiveness against distinctiveness of function. In this paper, the native British and Irish cattle breeds (n = 31 commercial, minority and rare breeds) have been prioritised in this way. Those with highest conservation priority are Chillingham, Gloucester, Guernsey, Jersey, Shetland and Irish Moiled. The 25 native British sheep breeds that are not on the Rare Breeds Survival Trust (RBST) Watchlist were also considered. The structure of the British sheep industry means that functional distinctiveness of breeds is not easily deduced. The only fully comparable characterisation data relate to wool fibre fineness class, so genetic distinctiveness was plotted against distinctiveness of this attribute. The non-rare breeds with highest conservation priority by this measure were Herdwick, Hampshire Down and Clun Forest.

Introduction

There are at least 5,000 breeds of the main domesticated mammal and bird species worldwide and it is estimated that 30% of them may be in danger of extinction (Barker, 2002). Debate has begun on how priority lists for conservation action might be drawn up. In Britain the Rare Breeds Survival Trust (RBST) has published a Watchlist, and inclusion of a breed depends on whether the breed has a distinctive history, whether registered adult females are numerically scarce, and on current population trends. Rare breeds are those with fewer than the following numbers of breeding females: cattle 1500, sheep 3000, pigs 1000, goats 1000, horses and ponies 3000, poultry 3000.

Clearly degree of endangerment is necessary for establishing priority lists and in many countries it may be sufficient, but the relative values of other forms of data need to be assessed critically (Ruane, 2000). With a view to maximising conserved evolutionary history, breeds might be selected for conservation on the basis of genetic distinctiveness, with microsatellites as the preferred method (Hall and Bradley, 1995). There has been considerable work on deduction of genetic distances among breeds and the consequences for species genetic diversity, of the loss of specific breeds, can be quantified (for example Barker *et al.*, 2001). At the same time, data on characterisation of breeds exist (Scherf, 2000, and www.fao.org/dadis). While the commercially important breeds are well characterised very few rare, local and threatened breeds are known in much detail and the data are of very variable extent and quality.

There is concern (Ruane, 1999; 2000) about whether genetic distancing work is fully adequate for prioritisation, and whether it is good value for money. The perception is that money has been made available for molecular work, but not for characterisation. It is generally agreed that genetic distinctiveness as measured by molecular methods is an important consideration. However, the weight to be placed on this aspect of breed identity, in comparison with aspects such as degree of endangerment, and traits of current economic value, of special landscape value, of cultural/historical value, needs further discussion (Ruane, 2000). Basic characterisation information in these traits may well be more cost-effective for conservation than molecular studies (Ruane 1999). In contrast, Barker (2002) argues that genetic distancing studies should be used to suggest which breeds should have priority for characterisation. This paper considers the native cattle and sheep breeds of Britain and Ireland (n = 32 and at least 59 breeds respectively (Hall and Moore, 1986, Hall and Clutton-Brock, 1988)). Of these, 16 and 27 respectively are on the RBST Watchlist.

Ruane (2000) has shown how Norwegian breeds can be classified according to seven criteria (degree of endangerment, traits of current economic value, special landscape value, traits of current scientific value, cultural/historical value, genetic uniqueness within species). The relative weighting of these criteria would be determined by a committee. The present study was undertaken because it may be valuable for such a committee to have access to a system which is relatively objective. To an extent the issue (of considering genetics and function together) has been foreshadowed in debates in conservation biology about evolutionarily significant units. For example Crandall *et al.* (2000) use the concept of exchangeability; populations can be compared for prioritisation in respect of their genetic distinctiveness and their ecological exchangeability. In this paper, I describe how this approach might be applied to livestock breeds.

Methods

For each species, each breed is compared with every other breed. One score is tabulated for genetic distinctiveness and another for functional distinctiveness. Genetic distinctiveness is inferred when breed history indicates that there has been little likelihood of substantial gene flow within the last 200 years (Hall and Clutton-Brock, 1988). Functional distinctiveness (the converse of exchangeability) is indicated if two breeds perform different economic, social and cultural functions.

Preliminary study suggested that while British and Irish cattle breeds are relatively easy to classify according to function the same is not true for sheep. In the latter, the prevalence and complexity of crossbreeding systems and the overwhelming economic importance of meat production has led to a pattern today of local maternal breeds and national sire breeds. Almost the only parameter that has been measured in a comparable way in many British sheep breeds relates to mean fibre fineness of their wool (NSA, 1982). This is assessed on a scale known as the 'Bradford count'. While wool fineness is of minor economic significance in British sheep farming today, it is highly heritable and repeatable (Simm, 1998) and it presumably reflects the joint action of natural and artificial selection in the past.

Here, all British and Irish native cattle breeds were considered because the numerical differences between rare, minority and most commercial breeds are not great and all require conservation monitoring because of the rapid population genetic changes that reproductive technology can produce. For example, traditional genotypes within the Aberdeen Angus, Lincoln Red and Hereford breeds are now included in the RBST Watchlist as original populations, while the number of British Friesians without north American Holstein introgression is probably declining. Only non-rare sheep breeds were considered, for two reasons: a) the lack of wool data for most rare breeds and b) the current practical need for coordinated policy in Britain on conservation of minority and local sheep breeds that are not on the RBST Watchlist.

Cattle

For each pairwise comparison, the questions asked were (1) are these two breeds genetically distinctive and (2) do they have distinct functions (for both questions, yes: score 1, no: score 0, intermediate: score 0.5). The maximum score for genetic distance would be 31, indicating that a breed is totally distinct from all others. For functional distinctiveness (non-exchangeability), a score of 31 would indicate a breed has a unique function. It was assessed by comparing economic function (there were 15 beef breeds, 9 dairy or dual purpose, 4 cottagers' or smallholders' breeds, and 4 breeds associated with a parkland or feral existence, while several are used as crossing sires for production of maternal crossbreds).

Sheep

For each pairwise comparison, genetic distinctiveness was noted in the same way as for cattle. When fibre fineness ranges overlapped, the score was 0, when they adjoined it was 0.5, and 1 indicated no similarity.

Results

Cattle

The breeds considered are listed in Table 1. Census population sizes (males and females combined) are given although they do not form part of the prioritisation procedure. Genetic distinctiveness is plotted against functional distinctiveness in Figure 1. Inclusion in a conservation priority list would be most justified for breeds that are both genetically and functionally distinctive, *i.e.* the listing would commence in the top right hand corner of Figure 1 and work downwards and to the left.

Sheep

The 25 non-rare sheep breeds are listed in Table 2. For each breed is given the fibre fineness range (Bradford count: the lower the figure, the coarser the wool), and census data. Genetic distinctiveness is plotted against uniqueness of fibre fineness range in Figure 2. As for cattle, greatest overall distinctiveness is exhibited by the breeds in the top right corner.

Discussion

Conservation of livestock biodiversity differs in very many ways from that of wild fauna and flora and one of these is the element of competition that exists between the societies and groups that represent different breeds. Thus prioritisation can be difficult and a transparent and objective procedure is necessary. The scheme presented here could be refined or extended, most notably regarding the deduction of functional distinctiveness. The general principle could also be applied to within-

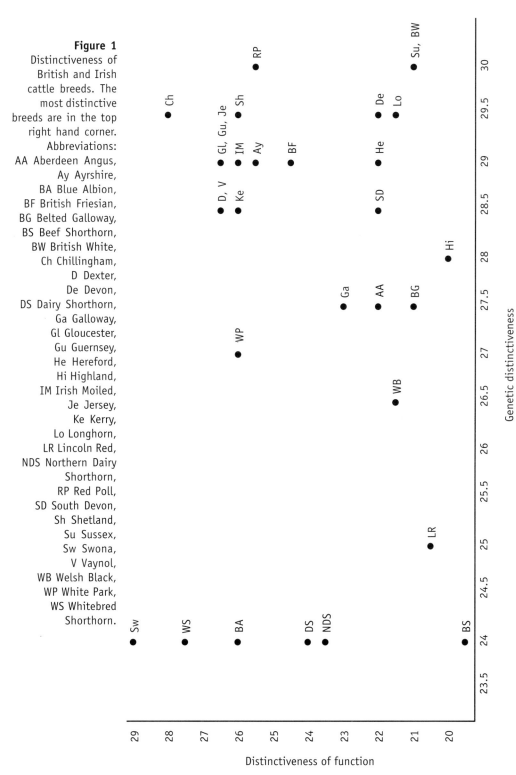

Figure 1
Distinctiveness of
British and Irish
cattle breeds. The
most distinctive
breeds are in the top
right hand corner.
Abbreviations:
AA Aberdeen Angus,
Ay Ayrshire,
BA Blue Albion,
BF British Friesian,
BG Belted Galloway,
BS Beef Shorthorn,
BW British White,
Ch Chillingham,
D Dexter,
De Devon,
DS Dairy Shorthorn,
Ga Galloway,
Gl Gloucester,
Gu Guernsey,
He Hereford,
Hi Highland,
IM Irish Moiled,
Je Jersey,
Ke Kerry,
Lo Longhorn,
LR Lincoln Red,
NDS Northern Dairy
Shorthorn,
RP Red Poll,
SD South Devon,
Sh Shetland,
Su Sussex,
Sw Swona,
V Vaynol,
WB Welsh Black,
WP White Park,
WS Whitebred
Shorthorn.

315

	Genetic distinctiveness	Functional distinctiveness	Census
Aberdeen Angus *beef and crossing*	28.5	22.5	8100
Ayrshire *high yield dairy*	30	26.5	60600
Beef Shorthorn *beef*	25	19.5	1740
Belted Galloway *parkland and beef*	28.5	21.5	1120
Blue Albion *medium yield dairy*	25	27	69
British Friesian *high yield dairy*	30	25.5	
British White *parkland and beef*	31	21.5	1284
Chillingham *feral parkland*	30.5	28.5	43
Dairy Shorthorn *high yield dairy*	25	25	4060
Devon *beef*	30.5	22.5	2100
Dexter *smallholder's cow*	29.5	27.5	2240
Galloway *beef*	28.5	23	5160
Gloucester *smallholder's cow*	30	27.5	763
Guernsey *high butterfat dairy*	30	27.5	3250
Hereford (Traditional) *beef and crossing*	30	22.5	434
Hereford (Commercial) *beef and crossing*	As above	As above	7000
Highland *parkland and beef*	29	20.5	4400
Irish Moiled *smallholder's cow*	30	27	249
Jersey *high butterfat dairy*	30	27.5	6500
Kerry *smallholder's cow*	29.5	27	94
Lincoln Red *beef*	26	21	800
Longhorn *parkland and beef*	30.5	22	1784
Luing *hardy beef*	27	23	1602
Northern Dairy Shorthorn *high yield dairy*	25	25	
Red Poll *medium yield dairy*	31	26.5	952
Shetland *smallholder's cow*	30.5	27	324
South Devon *beef*	29.5	22.5	12320
Sussex *beef and crossing*	31	21.5	2242
Swona *feral island*	25	30	15
Vaynol *parkland*	29.5	27.5	23
Welsh Black *beef*	29.5	22.5	9396
White Park *parkland*	28	27	467
Whitebred Shorthorn *crossing*	25	28.5	230

Table 1. British and Irish cattle breeds, with current agricultural function.

Note: spaces indicate data not available. Census figures are from the UK Genetic resource database (personal communication, M. Roper) at June 2000, except for Swona and Chillingham (personal communications from C. Annal and A. Widdows, respectively, August 2002).

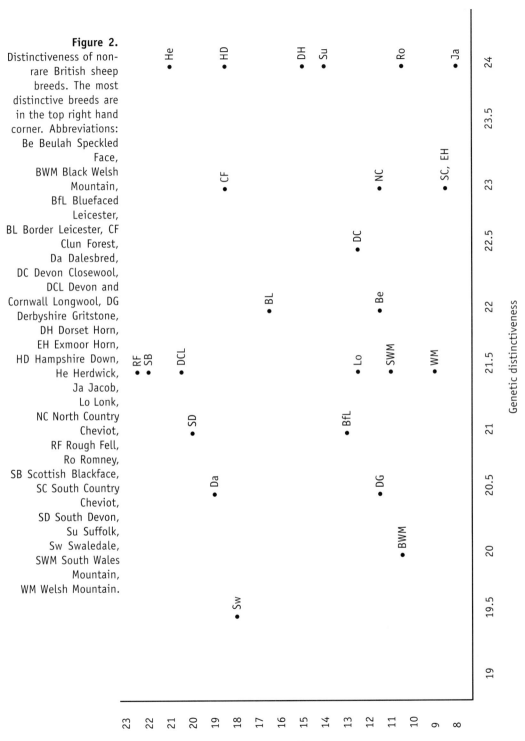

Figure 2.
Distinctiveness of non-rare British sheep breeds. The most distinctive breeds are in the top right hand corner. Abbreviations:
Be Beulah Speckled Face,
BWM Black Welsh Mountain,
BfL Bluefaced Leicester,
BL Border Leicester, CF Clun Forest,
Da Dalesbred,
DC Devon Closewool,
DCL Devon and Cornwall Longwool, DG Derbyshire Gritstone,
DH Dorset Horn,
EH Exmoor Horn,
HD Hampshire Down,
He Herdwick,
Ja Jacob,
Lo Lonk,
NC North Country Cheviot,
RF Rough Fell,
Ro Romney,
SB Scottish Blackface,
SC South Country Cheviot,
SD South Devon,
Su Suffolk,
Sw Swaledale,
SWM South Wales Mountain,
WM Welsh Mountain.

Genetic distinctiveness

Distinctiveness of fibre fineness range

	Bradford count	Uniqueness of Bradford count range	Genetic distinctive-ness	census
Beulah Speckled Face	50-56	11.5	22	
Black Welsh Mountain	48-54	10.5	20	6866
Bluefaced Leicester	48-50	13	21	31500
Border Leicester	44-48	16.5	22	5670
Clun Forest	56-58	18.5	23	4663
Dalesbred	32-40	19	20.5	
Derbyshire Gritstone	50-56	11.5	20.5	3950
Devon and Cornwall Longwool	32-36	20.5	21.5	2894
Devon Closewool	46-50	12.5	22.5	4512
Dorset Horn	54-56	15	24	24652
Exmoor Horn	48-56	8.5	23	25065
Hampshire Down	56-58	18.5	24	3683
Herdwick	28-32	21	24	76000
Jacob	44-56	8	24	2031
Lonk	44-50	12.5	21.5	3645
North Country Cheviot	50-56	11.5	23	110000
Romney	48-54	10.5	24	12500
Rough Fell	v coarse	22.5	21.5	
Scottish Blackface	coarse	22	21.5	
South Country Cheviot	48-56	8.5	23	153000
South Devon	36-40	20	21	
South Wales Mountain	40-50	11	21.5	
Suffolk	54-58	14	24	89500
Swaledale	28-40	18	19.5	
Welsh Mountain	36-50	9	21.5	

Table 2. British sheep breeds not on the RBST Watchlist. Bradford count for each breed is given, with the uniqueness score derived (see text) from the numbers of other breeds also having overlapping, adjoining or distinct range.

Note: spaces indicate data not available. Census figures are from the UK Genetic resource database (see Table 1). Scottish Blackface and Rough Fell are taken to overlap with Herdwick and Swaledale in relation to Bradford count.

breed prioritisation. Different herds or flocks could be assessed on the same two-axis system, if for example it was being decided which animals to collect semen or embryos from, or which breeding groups of particular conservation significance should be supported.

Specific results

The least distinctive cattle breeds are those of Shorthorn type, which are interrelated and which also have functional exchangeability with other beef and dairy cattle. The exception is the Swona cattle; though close genetically to Shorthorn and to Aberdeen Angus they have a distinctive ecology as a feral island herd (Hall and Moore, 1985). Of the ten non-

Shorthorn beef breeds, some are genetically distinctive, notably the Sussex, British White and Devon, but they are not functionally distinctive. The six breeds of highest overall distinctiveness are the following:

- Chillingham. Though not completely distinctive genetically by the criteria applied here (having contributed genes to the White Park) the cattle are only slightly exchangeable in functional terms with others, the Swona and to a lesser extent the White Park and Vaynol);
- Gloucester. Though probably with some genetic relationship to the Welsh Black and the Hereford, this is distinctive as a low-yielding specialised dairy breed;
- Guernsey and Jersey. No other breeds have the same high milk-fat attributes and the only genetic affinities they have are with each other, and probably with the South Devon;
- Irish Moiled, Shetland. Though probably linked genetically with other breeds of their native countries these breeds have distinctive functions as smallholders' cattle.

The most distinctive sheep breeds overall of the non-rare group are the Herdwick, Hampshire Down, Clun Forest and (perhaps) Border Leicester. The functional distinctiveness of the Scottish Blackface and Rough Fell from other breeds is noteworthy (Figure 2). The least distinctive group is probably the Welsh Mountain and its affiliates. This analysis has emphasised the importance of gathering contemporary characterisation data; wool characteristics may well repay further investigation as they have probably been shaped to a considerable extent by the environment as well as by commercial factors.

General

The general principle of this model is that genetic distinctiveness and functional distinctiveness are equally informative for the prioritisation process. Census data or degree of endangerment could be added on a third (z) axis but these parameters are very sensitive to breed structure, so it is suggested that they are used in a pragmatic way primarily to resolve conflicting claims of breeds which are close in genetic and functional distinctiveness. A breed can be classified according to the proportion of species genetic variation that would be lost if it dies out (for example Barker et al., 2001). This attribute of breeds could be included in the deduction of functional distinctiveness.

Ruane (2000) concluded that a committee of experts should be involved in prioritisation. The present study would facilitate the work of such a committee and make its deliberations and decisions more objective and transparent. This is very important if the private-sector breed enthusiasts who are critical to the survival of world livestock biodiversity, are to support its activities.

Acknowledgments

Mr. Mike Roper of the UK Department of the Environment, Food and Rural Affairs provided the national breed census data. Mr. Austen Widdows and Mr. Cyril Annal provided census information for Chillingham and Swona respectively. Dr. John Ruane made helpful comments on an earlier version of the manuscript. Financial support from the Kochan Trust and the Whitaker Trust is gratefully acknowledged.

References

Barker, J.S.F., Tan, S.G., Moore, S.S., Mukherjee, T.K., Matheson, J.-L. and Selvaraj, O.S. 2001. Genetic variation within and relationships among populations of Asian goats *(Capra hircus)*. *Journal of Animal Breeding and Genetics* 118: 213-233.

Barker, J.S.F. 2002. Relevance of animal genetic resources and differences to the plant sector. *Landbauforschung Völkenrode*, special issue 228: 15-21.

Crandall, K.A., Bininda-Emonds, O.R.P., Mace, G.M. and Wayne, R.K. 2000. Considering evolutionary processes in conservation biology. *Trends in Ecology and Evolution* 15: 290-295.

Hall, S. J. G. and Bradley, D. G. 1995. Conserving livestock breed biodiversity. *Trends in Ecology and Evolution* 10: 267-270.

Hall, S.J.G. and Clutton-Brock, J. 1988. *Two Hundred Years of British Farm Livestock*. British Museum (Natural History), London, UK.

Hall, S.J.G. and Moore, G.F. 1986. Feral cattle of Swona, Orkney Islands. *Mammal Review* 16: 89-96.

NSA. 1982. *British Sheep. Sixth edition*. National Sheep Association, Tring, Hertfordshire, UK.

Ruane, J. 1999. A critical review of the value of genetic distance studies in conservation of animal genetic resources. *Journal of Animal Breeding and Genetics* 116: 317-323.

Ruane, J. 2000. A framework for prioritising domestic animal breeds for conservation purposes at the national level: a Norwegian case study. *Conservation Biology* 14:1385-1393.

Scherf, B.D. 2000. *World Watch List for Domestic Animal Diversity. 3rd edition*. Food and Agriculture Organisation of the United Nations, Rome, Italy.

Simm, G. 1998. *Genetic Improvement of Cattle and Sheep*. CABI Publishing, Wallingford, UK.

Genetic diversity in European and Chinese pig breeds – the PigBioDiv project

S. Blott[3], M. SanCristobal[1], C. Chevalet[1], C.S. Haley[2], G. Russell[2], G. Plastow[3], K. Siggens[3], M.A.M. Groenen[4], M.-Y. Boscher[5], Y. Amigues[5], K. Hammond[6], G. Laval[1], D. Milan[1], A. Law[2], E. Fimland[12], R. Davoli[9], V. Russo[9], G. Gandini[10], A. Archibald[2], J.V. Delgado[11], M. Ramos[13], C. Désautés[7], L. Alderson[14], P. Glodek[8], J.-N. Meyer[8], J.-L. Foulley[15], L. Andersson[16], R. Cardellino[6], N. Li[17], L. Huang[18], K. Li[19] and L. Ollivier[15]

[1] INRA, Laboratoire de Génétique Cellulaire, 31326 Castanet Tolosan Cedex, France [2]Roslin Institute (Edinburgh), Midlothian, EH25 9PS, UK [3]Pig Improvement Group Limited (PIC), Fyfield Wick, Abingdon, UK [4]Wageningen Agricultural University, Costerweg 50, PO Box 9101, Wageningen, 6701 BH, The Netherlands [5]Labogena, Domaine de Vilvert, 78352 Jouy-en-Josas Cedex, France [6]Animal Genetic Resources Group, FAO, Viale delle Terme di Caracalla, Roma 00100, Italy [7]Agence de la Sélection Porcine, 149 rue de Bercy, 75595 Paris Cedex 12, France [8]Animal Genetics Institute, Albrecht-Thaer-Weg 3, 37075 Göttingen, Germany [9]DIPROVAL, Universita di Bologna, Via Rosselli 107, 42100 Coviolo- Reggio Emilia, Italy [10]Universita degli Studi di Milano, Via Celoria 10, 20133 Milano, Italy [11]Facultad de Veterinaria, Universidad de Cordoba, Avda Medina Azahara 9,14005 Cordoba, Spain [12]Nordic Gene Bank, Pb 5025, 1432 Aas, Norway [13]Universidade de Tras-os-Montes e Alto Douro, 5001 Vila Real Codex, Portugal [14]Rare Breeds Survival Trust, 6 Harnage, Shrewsbury, Shropshire, SY5 6EJ, UK [15]INRA, Station de Génétique Quantitative et Appliquée, 78352 Jouy en Josas Cedex, France [16]Swedish University of Agricultural Sciences, Husargatan 3, Uppsala, Sweden [17]China Agricultural University, Yuanmingyuan West Road 2, Haidian District, 100094 Beijing, P.R. China [18]Jiangxi Agricultural University, PO Box 85, MeiLing, 330045 NanChang, P.R. China [19]Huazhong Agricultural University, Shizishan Street 1, 430070 Wuhan, P.R. China

Introduction

Characterisation of genetic diversity in a large number of European pig populations has been undertaken with EC support. The populations sampled included local (rare) breeds, national varieties of the major international breeds, commercial lines and the Chinese Meishan breed. A second phase of the project will sample a further 50 Chinese breeds. Neutral genetic markers (AFLP and microsatellites), with individual or bulk typing, were used and compared.

Materials and methods

DNA from 59 European pig populations was extracted on samples of about 50 individuals per population. Individuals were typed for 50 microsatellites and for 148 AFLP bands. A subset of 25 populations was typed for 20 microsatellites on pools of DNA. Allele frequencies were estimated by direct allele counting for the co-dominant markers. Frequencies of AFLP negative alleles (absent bands) were obtained by taking the square root of absent band frequencies. Within-breed variability was summarised using standard statistics: expected and observed heterozygosity, mean observed and effective numbers of alleles, and F statistics. Between-breed diversity analysis was based on a bootstrapped Neighbor-Joining (NJ) tree derived from Reynolds distances (D_R). The standard distance of Nei (D_S) was also calculated.

Results

The mean observed (A_o) and effective numbers (A_e) of alleles per population, averaged over the 50 microsatellites, range from 2.9 to 7.2 and from 1.9 to 4 respectively, with means of 4.5 and 2.7. Significantly non-zero F_{IS} values were found in 22% of the populations, mainly for local breeds. One quarter of these non-zero values were negative, corresponding to synthetic lines created by crossing breeds or suggesting management to avoid inbreeding in local breeds. Between breed diversity analysis showed a star-shaped overall divergence pattern which suggests that genetic drift has had a strong influence on European breeds. Figure 1 shows that lines within breeds are grouped with high bootstrap values, apart from the Landrace group. The tree based on AFLP typings (not shown) gave a very similar topology.

Conclusions

An overview of the diversity structure of European pig populations was obtained using neutral genetic markers and a careful sampling scheme based on the general recommendations of the FAO. Insight into genetic diversity was derived from comparisons between two types of marker and two genetic distances. Bulked typing, while useful, produced biased estimates of genetic distance. During the second phase of the project 50 Chinese breeds will be added to the survey. Bulked genotyping of microsatellites will be investigated further. AFLPs will be replaced by SNPs, which have the advantage of co-dominance and also allow the comparison of diversity measured with functional polymorphisms compared to neutral markers. Further details of the project can be found at http://databases.roslin.ac.uk/pigbiodiv.

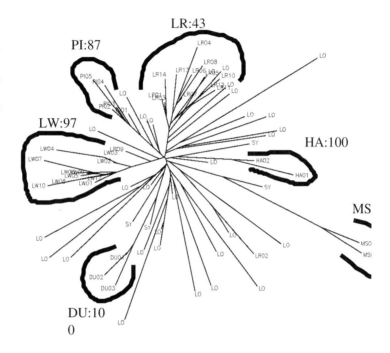

Figure 1. Neighbor joining tree based on microsatellite markers with bootstrap values for major breeds. LW: Large White, LR: Landrace, HA: Hampshire, DU: Duroc, PI: Piétrain, MS: Meishan.

Strategies to maximise the allelic diversity maintained in small conserved populations

J. Fernández[1], M.A. Toro[1] and A. Caballero[2]
[1]Departamento de Mejora Genética Animal, INIA, Carretera A Coruña km. 7, 28040 Madrid, Spain
[2]Facultad de Ciencias, Universidad de Vigo, 36200 Vigo, Spain

Introduction

One of the main objectives in conservation programmes is to maintain the highest levels of genetic variability, for the population to be able to face future environmental changes and to assure long-term response to selection, either natural or artificial (Oldenbroek, 1999, Barker, 2001). The classical measure of genetic diversity is the expected heterozygosity, or gene diversity (GD), but allelic diversity (AD), or the number of different alleles per locus, also has evolutionary importance. Most optimal strategies for conservation have aimed to maximise GD (e.g. Caballero and Toro 2000), but AD has received much less attention. The objective of the present study is to test the efficiency of the maintenance of allelic diversity of strategies based either on the allelic diversity itself or on the expected heterozygosity in a small population, using information from molecular markers.

Materials and methods

Computer simulations considered populations with $N = 8$ parents, half of each sex, per generation. Individuals' genomes consisted of 1 or 20 chromosomes with 100 multiallelic loci per chromosome. Base population individuals were assumed unrelated and not inbred, and all their alleles were different in each locus. Five unmanaged generations were simulated before the implementation of any strategy to mimic a more realistic scenario. Information from 2, 10 or 100 evenly spaced markers per chromosome were used for the management decisions. Fifteen non-overlapping generations were run under each of the different combinations of factors. Every generation the allelic diversity and the gene diversity of the population were calculated using all loci, and averaged over 100 replicates. Four strategies were compared differing in the criteria to choose the group of parents: i) minimisation of global pedigree coancestry weighted by the optimal contributions (PC); ii) the highest allelic diversity (AD); iii) maximisation of the expected gene diversity calculated from parents' frequencies and weighted by optimal contributions (GD); iv) minimisation of the expected number of alleles lost (EAL) in the offspring, expressed as   where p_{ij} is the probability of individual j not transmitting allele i (binomial probability), and A is the total number of different alleles. Optimisations were performed using a *simulated annealing* algorithm. Random mating was assumed throughout.

Results

Table 1 and 2 show the mean allelic and gene diversity of the population at generation 15. Values are percentages relative to the levels obtained by the PC method. This method is better than the use of limited molecular information (2 markers). In some situations, none of the methods outperform PC even with ten markers. Methods are less efficient in preserving genetic diversity (both allelic and gene diversity) for larger number of chromosomes. The allelic diversity (AD method) proves not to be a good tool on which to base management decisions. The GP method, in almost all situations, outperforms EAL method for the levels of gene diversity kept in the population and it is significantly worse for allelic diversity in only one case (20 chromosomes with two markers each).

Table 1. Mean allelic diversity at generation 15 (in percentage relative to PC method results).

	Markers	$\sum_{i=1}^{A}\prod_{j=1}^{N}p_{ij}$ AD	EAL	GD
1 chrom.	2	−20.9	−8.3	−7.8
	10	−20.7	17.0	16.2
	100	−19.3	26.7	24.0
20 chrom.	2	−18.8	−4.1	−7.2
	10	−17.5	−0.2	−0.2
	100	−17.7	0.8	0.1

Standard deviations range from 0.3 to 3.0

Table 2. Mean gene diversity at generation 15 (in percentage units relative to PC method results).

	Markers	AD	EAL	GD
1 chrom.	2	−30.0	−10.3	−8.8
	10	−29.3	20.6	28.9
	100	−27.6	31.9	41.1
20 chrom.	2	−26.8	−4.9	−6.9
	10	−25.1	0.2	2.1
	100	−24.9	1.3	2.4

Standard deviations range from 0.3 to 2.2

Conclusions

The results show that the most efficient method is GD, as it maintains similar levels of allelic diversity and higher levels of gene diversity than others methods based on AD itself. A further advantage of the use of gene diversity as the management criterion is that it also keeps lower levels of inbreeding.

References

Barker, J. S. F. 2001. Conservation and management of genetic diversity: a domestic animal perspective. *Canadian Journal of Forest Research* 31: 588-595.

Caballero, A., Toro, M.A. 2000. Interrelations between effective population size and other pedigree tools for the management of conserved populations. *Genetical Research* 75: 331-343.

Oldenbroek, J. K. 1999. *Genebanks and the Conservation of Farm Animal Genetic Resources*. DLO Institute for Animal Sciences and Health, Lelystad, The Netherlands.

Ovum recovery from ewes during the peak breeding season and transition to anoestrus

L.M. Mitchell[1], M.J.A. Mylne[2], J. Hunton[2], K. Matthews[2], T.G. McEvoy[1], J.J. Robinson[1] and W.S. Dingwall[1]
[1]Scottish Agricultural College, Craibstone Estate, Bucksburn, Aberdeen, AB21 9YA, UK
[2]Britbreed Ltd, Airfield Farm, Cousland, Dalkeith, Midlothian, EH22 2PE, UK

Introduction

Sheep are seasonally-polyoestrous short-day breeders. Although the domesticated breeds have longer breeding seasons than the feral breeds, their maximum ovulation rates are only achieved over a relatively short period. The effect of these seasonal shifts in ovarian response on the success of ovum recovery for genetic improvement or breed conservation is unknown. The aim of the present study was to assess the efficiency of ovum recovery procedures for genetic conservation outwith the normal breeding season.

Material and methods

Twenty mature Mule ewes underwent a standard oocyte recovery procedure during the peak breeding season (October) and transition to anoestrus (April). In each month, oestrus was synchronised (intravaginal progesterone, CIDR; Days 0-12) and all ewes received 7 i.m. injections of 1.125 mg oFSH at 12 h intervals beginning at 0800 h on Day 10. Oocyte-cumulus complexes were aspirated on Day 13, ~20 h after CIDR withdrawal, and were graded on the basis of their cytoplasmic integrity and cumulus cell investment (scale of 1-4; 1=excellent, 4=poor).

In a separate study, 50 mature Mule ewes underwent a standard embryo recovery procedure during October and April. In each month, oestrus was synchronised and all ewes received 8 i.m. injections of 1.125 mg oFSH as above. Artificial insemination was carried out on Day 14, ~46 h after CIDR withdrawal, using frozen-thawed semen collected on a single occasion prior to the start of the study (~125 x10^6 spermatozoa per ewe). Embryos were recovered on Day 20, graded according to their stage of development and morphology (scale of 1-4; 1=excellent, 4=poor) and cryopreserved in Ovum Culture Medium (OCM) containing 1.5 M ethylene glycol. During the following breeding season, grade 1 and 2 morulae and unexpanded blastocysts (October, n=147; April, n=138) were thawed and allowed to recover for 2-4 h in individual wells containing 500 ml of OCM + 5% heat-inactivated sheep serum in air. Their stage of development and morphology were then reassessed and 40 embryos from each month were transferred in singleton to synchronous recipients. Resulting pregnancies were monitored to term. Data were analysed by ANOVA and Chi square.

Results

For ewes undergoing oocyte recovery, the total number of ovarian follicles per ewe was not different in October compared to April (Table 1) but, in October, the proportion of follicles in the medium-size (3-5 mm) category was increased (P<0.05). Month did not affect the number of oocytes recovered per ewe or the proportion of recovered oocyte-cumulus complexes that were of grades 1 and 2 (October, 0.74; April, 0.62).

For ewes undergoing embryo recovery, month did not affect the numbers of corpora lutea per ewe ovulating or the numbers of ova recovered but, in October compared to April, a greater proportion (P<0.01) of recovered ova was fertilised and contained more than 16 cells (Table 2). The median stage of development of October and April embryos containing more than 16 cells was in each case early blastocyst and the proportions that were of grades 1 and 2 were not significantly different (0.89 vs. 0.87, respectively). Seven embryos from each month failed to re-expand following cryopreservation and thawing. For remaining October and April embryos, the median stage of development 2-4 hours post-thaw was in each case early blastocyst and mean (± s.e.) morphological grades were 2.4 ± 0.09 vs. 2.8 ± 0.11, respectively. For the 40 embryos from each month transferred in singleton to synchronous recipients, there was no significant difference in the proportions that established pregnancy (October, 0.78; April, 0.70).

Table 1. Oocyte recovery from ewes at contrasting stages of the breeding season.

Means ± s.e per ewe	Oct	Apr	s.e.	Sig.
No. ewes	20	20		
Total follicles	24.3	26.2	0.18	
Follicles 3-5 mm	17.6	12.8	0.21	*
Oocytes recovered	9.2	12.0	0.26	

Table 2. Embryo recovery from ewes at contrasting stages of the breeding season.

Means ± s.e. per ewe ovulating	Oct	Apr	s.e.	Sig.
No. ewes ovulating	48	45		
Ovulation rate (no. CL)	17.6	18.0	1.10	
Total ova recovered	15.3	15.1	0.03	
Embryos (>16 cell)	8.0	5.3	0.05	**

Conclusions

Results demonstrate that ewes undergoing oestrus synchronisation and ovarian stimulation during the late, compared to the peak breeding season yield fewer embryos as a possible consequence of seasonal changes in ovarian follicular development. Good quality embryos produced at contrasting stages of the breeding season do not differ in their ability to withstand cryopreservation and subsequently establish pregnancy in recipient ewes suggesting that ovum recovery procedures may be used for genetic conservation at all times of the year.

Acknowledgement

SAC receives funding from the Scottish Executive Environment and Rural Affairs Department.

Summaries of breakout group discussions

Implementing actions from the UK NCC Country Report on FAnGR

UK National Steering Committee on Farm Animal Genetic Resources

The group considered the creation, scope and remit of a UK National Steering Committee on Farm Animal Genetic Resources as recommended in the NCC report. It was felt that the committee should be independent of Government but be staffed by a Defra secretariat. The committee should report to the UK National Co-ordinator on FAnGR (currently within Defra) and similarly nominated individuals in the devolved Scottish Parliament, Wales and Northern Ireland Assemblies.

The scope of the committee's work should cover farm animal genetic resources, including cattle, sheep, pigs, poultry and equine but excluding wild animals, pets and other minor domestic species. The committee should remain distinct from other similar committees relating to plant and microbe genetic resources but interact with them to ensure good practice could be disseminated across the committee structure and areas of common activity such as public awareness could be co-ordinated.

The group considered that the FAnGR committee should comprise a core group of approximately 10 independent individuals with expertise across the subject areas of the committee's remit. Where specific areas of interest or specific topics were to be considered in depth then the committee could convene a series of sub-groups or specialist panels of additional individuals as required under the chairmanship of a core committee member. It was envisaged that these specialist panels could be formed in relation to either individual animal species (e.g. sheep) or specific topic areas (e.g. cryo-preservation).

The remit of the committee should include the following areas:

- provide advice on FAnGR issues to Defra, other Government departments, devolved institutions and non-governmental organisations (NGOs)

- act as a 'stakeholder forum' for all parties concerned with issues surrounding FAnGR

- set priorities for and advise on the implementation of the UK national action plan on FAnGR

- set R&D priorities in the area of FAnGR and feed those into the proposed Priorities Board recommended by the Policy Commission on the future of Farming and Food

- advise on a co-ordinated UK *in situ* and *ex situ* conservation programme for FAnGR and evaluate its implementation.

UK National Action Plan on Farm Animal Genetic Resources

The group also considered the implementation of a UK National Action Plan on Farm Animal Genetic Resources as recommended in the NCC report. It was envisaged that the plan would be co-ordinated by the UK National Co-ordinator on FAnGR, after receiving advice from the UK National Steering Committee on FAnGR.

The Nation Action Plan on FAnGR would include:

- providing a conduit for information gathering, central data collection and access to a library of knowledge on FAnGR;

- consider setting up a national website on FAnGR;

- enabling the *in situ* characterisation and utilisation of UK FAnGR with regard to product markets, environmental uses and regional adaptations;

- evaluating the relative merits of genetic diversity, phenotypic diversity and functional diversity in UK FAnGR and their impact on conservation action;

- establishing a basis for prioritising *ex situ* (e.g. cryo-preservation) and *in situ* conservation strategies;

- augmenting and developing the role of breed societies in UK FAnGR conservation and utilisation;

- assisting in the development of export opportunities for UK FAnGR;

- enabling and facilitating regional (European) collaboration in relation to FAnGR R&D, information databases and regulations;

- contributing to global collaboration and effort in FAnGR international treaty negotiations.

On reporting the groups' deliberations back to the full conference, concern was raised from the floor that serious consideration by Government and private sector funders would need to be given to the provision of meaningful long term funding if these worthy aspirations were to be fulfilled.

Mike Roper, Convenor

Prioritising breeds for conservation

1. It was agreed that prioritisation was likely to be necessary as even in a relatively wealthy country like the UK, private finance was probably insufficient to conserve all breeds. Economic factors are very important and many breeds simply cannot compete in the present-day economic climate. However, previously unanticipated uses of livestock were emerging, notably environmental conservation schemes. While a certain degree of breed conservation could be achieved as a side effect of such schemes, many breeds, including all breeds of pigs and poultry, cannot be envisaged as playing a part in them.

2. The primary criterion for prioritisation was seen as being the risk of extinction. It was emphasised that extinction could have demographic and genetic causes and monitoring of trends is very necessary. While an overall view is necessary, monitoring should be in the first instance by breed societies which should be enabled to do this by making available suitable computer software.

3. Utility should also be considered as a criterion for prioritisation. One reason for this is the public image of breed conservation; people do see livestock as performing certain functions and some reference to likely utility of conserved stocks would be expected to encourage the public to support the work. It may be desirable to classify utility as short-term, medium-term and speculative. Genetic studies on performance traits and disease resistance would provide valuable data here.

4. Local adaptation and distinctiveness, with their associated characteristics of cultural and social significance, should also be included in the criteria.

5. It was thought that today, genetic distance work is seen as being of less relevance to prioritisation, than it was even as recently as five years ago. Sheep breeds show relatively large between-breed microsatellite distances and these must be due to genetic drift due to breed structure and genetic bottlenecks, rather than to traits relevant to functional and historical characterisations.

6. Breed society activities were seen as being fundamental to conservation and it was thought that the degree of conservation support could be related somehow to the quality of the work that they do and to their degree of commitment though clearly this would have to be carefully managed.

7. The group considered briefly new ways of developing prioritisation protocols including the Weitzman approach and it was agreed that considerable further discussion is needed of the issues, both technical and policy.

Stephen Hall, Convenor

Future R&D needs

1.   Characterisation of breeds

- Characterise the traits
- Identify the genetic correlation
- Identify what we need to keep
- Climate Change – do we have breeds characteristics to deal with it, including potential new diseases
- Physiological growth, lamb growth, ability to deal with mineral excesses/deficiencies
- Identify characteristics that are marketable
- Environmental impact/value of animal
- Marketable value of animal

2.   Retention of genetic diversity

- Modelling & monitoring of genetic influence of ARR/ARR
- Other methods of separating susceptible flocks
- Alternative testing for scrapie
- Workshop between scrapie researchers and RBST & breed societies
- Crossbreeding – shouldn't be overlooked to maintain genetic diversity
- GIS database/national database of animal numbers

3.   Resolving technical issues

- Need a gene bank in case of crisis
- Technological issues need to be solved in terms of long term storage of genetic material.
- Need to bank sperm, eggs and embryos
- Saving genetic material would allow particular genes to be re-introduced if needed. However, need to know what the genes do.

4.   Sociological studies

- Sociological research into success of societies and how that influences the success of a breed
- What sustains/fuels a niche market?

Much of the group discussion focussed on scrapie and current scrapie policy. Many of the rare breeds have few animals of the desired genotype and breeders/societies are concerned that some native breeds could be lost.

Concerns were also raised as to whether breeding programmes would help to maintain genetic diversity.

Animal identification and traceability were also raised as important areas along with individual animal identification

Richard Small, Convenor

Index